lation during the past six centuries. The historical connection of population with the land, interregional migrations, the two long-range revolutions in land utilization and food production during the past millennium, institutional factors such as fiscal burden and land tenure, and major deterrents to population growth such as famines, floods, wars, and female infanticide — all these are discussed in the light of detailed local records. The author's redefinition of *mou* (Chinese acre) and criticism of the land statistics both of Western specialists and of the Chinese Communist government are of particular interest to China experts. "The laborious combing of materials concerning population and various correlated factors," says John K. Fairbank, "has put all future workers on this subject in his debt."

In his conclusion Dr. Ho compares population data with comments of contemporary observers and with economic and institutional changes and suggests ways for a reconstruction of China's historical data on population. Although dealing mainly with the past, the book has an important bearing on the understanding of one of the most explosive problems of the present world.

Dr. Ho is James Westfall Thompson Professor of History at the University of Chicago, a member of Academia Sinica, and author of *The Ladder of Success in Imperial China: Aspects of Social Mobility, 1368–1911* and *History of Landsmannschaften in China*. He has also contributed extensively to Western historical and social science journals as well as to Sinological journals.

HARVARD EAST ASIAN SERIES

1. *China's Early Industrialization: Sheng Hsuan-huai (1844–1916) and Mandarin Enterprise.* By Albert Feuerwerker.
2. *Intellectual Trends in the Ch'ing Period.* By Liang Ch'i-ch'ao. Translated by Immanuel C. Y. Hsü.
3. *Reform in Sung China: Wang An-shih (1021–1086) and His New Policies.* By James T. C. Liu.
4. *Studies on the Population of China, 1368–1953.* By Ping-ti Ho.
5. *China's Entrance into the Family of Nations: The Diplomatic Phase, 1858–1880.* By Immanuel C. Y. Hsü.
6. *The May Fourth Movement: Intellectual Revolution in Modern China.* By Chow Tse-tsung.
7. *Ch'ing Administrative Terms: A Translation of the Terminology of the Six Boards with Explanatory Notes.* Translated and edited by E-tu Zen Sun.
8. *Anglo-American Steamship Rivalry in China, 1862–1876.* By Kwang-Ching Liu.
9. *Local Government in China under the Ch'ing.* By T'ung-tsu Ch'ü.
10. *Communist China 1955–1959: Policy Documents with Analysis.* With a foreword by Robert R. Bowie and John K. Fairbank. (Prepared at Harvard University under the joint auspices of the Center for International Affairs and the East Asian Research Center.)
11. *China and Christianity: The Missionary Movement and the Growth of Chinese Antiforeignism, 1860–1870.* By Paul A. Cohen.
12. *China and the Helping Hand, 1937–1945.* By Arthur N. Young.
13. *Research Guide to the May Fourth Movement: Intellectual Revolution in Modern China, 1915–1924.* By Chow Tse-tsung.
14. *The United States and the Far Eastern Crises of 1933–1938: From the Manchurian Incident through the Initial Stage of the Undeclared Sino-Japanese War.* By Dorothy Borg.
15. *China and the West, 1858–1861: The Origins of the Tsungli Yamen.* By Masataka Banno.
16. *In Search of Wealth and Power: Yen Fu and the West.* By Benjamin Schwartz.
17. *The Origins of Entrepreneurship in Meiji Japan.* By Johannes Hirschmeier, S.V.D.
18. *Commissioner Lin and the Opium War.* By Hsin-pao Chang.
19. *Money and Monetary Policy in China, 1845–1895.* By Frank H. H. King.
20. *China's Wartime Finance and Inflation, 1937–1945.* By Arthur N. Young.
21. *Foreign Investment and Economic Development in China, 1840–1937.* By Chi-ming Hou.
22. *After Imperialism: The Search for a New Order in the Far East, 1921–1931.* By Akira Iriye.
23. *Foundations of Constitutional Government in Modern Japan, 1868–1900.* By George Akita.
24. *Political Thought in Early Meiji Japan, 1868–1889.* By Joseph Pittau, S.J.
25. *China's Struggle for Naval Development, 1839–1895.* By John L. Rawlinson.
26. *The Practice of Buddhism in China.* By Holmes H. Welch.
27. *Li Ta-chao and the Origins of Chinese Marxism.* By Maurice Meisner.

Harvard East Asian Studies 4

The Center for East Asian Studies at Harvard University administers postgraduate training programs and research projects designed to further scholarly understanding of China, Korea, Japan, and adjacent areas.

Studies on the Population of China, 1368-1953

Studies on the Population of China, 1368-1953

Ping-ti Ho

Harvard University Press, Cambridge, Massachusetts

Second Printing, 1967
Distributed in Great Britain by Oxford University Press, London

Publication of this volume was aided by a grant from the Ford Foundation. The Foundation is not, however, the author, owner, publisher, or proprietor of this publication and is not to be understood as approving by virtue of its grant any of the statements made or views expressed therein.

Library of Congress Catalog Card Number 59–12970
Printed in the United States of America

To the Memory of

JOHN BARTLET BREBNER

Eminent Historian, Revered Teacher,
and True Friend

Foreword

One of the great resources for the study of Chinese history in recent centuries is the enormous body of local histories (*fang chih*) or "gazetteers." More than 3,000 such multivolumed works are available in collections in this country, along with over 300 microfilm reproductions of rare early editions. They provide a ramified and wide-ranging record — geographic and economic treatises, historical narratives together with extensive documentation, collections of numerous biographies, and essays on customs and religious and other institutions — for all the major administrative areas of the Chinese empire, from the county on up to the province. Taken together with the large imperially sponsored encyclopedias and other compendia on major aspects of administration, these compilations present the modern social historian with a formidable challenge — how to winnow from such vast bodies of premodern source material the essential record concerning topics of modern interest.

The difficulties of this task for the past few centuries in China are comparable to those facing the medieval historian in the West. First of all, the gazetteers deal with institutions which have disappeared and use a technical terminology which is often not what it seems and can easily mislead the unwary. Furthermore these compilations, even when dated only a generation ago, represent the values of a previous and less-quantified age. Their "statistics" are not those of the modern-minded government statistician or economist, but rather the figures put down by literary scholars or tradition-bound clerks, meant to indicate merely an order of magnitude or to meet a ritualistic need for numerical data in the records. Instead of tables the modern researcher finds lists, sometimes with totals that do not tally. Instead of new efforts at periodic measurement, one finds the quotas accepted by officials or totals handed down through earlier records. Statistics, in this tradition, may reflect the chronicler's respect for the written word inherited from the past quite as much as they reflect the facts of the contemporary scene. Some compilers obviously have looked in their books rather than out the window. Part of the challenge is therefore to find the record of the rare observer, the truly inquiring mind, who saw his China with a fresh and skeptical eye.

The modern researcher in these materials must thus combine philological acumen with detailed institutional knowledge. In addition he must have the stamina of the modern historian who can wade through tons of documentation; for the available records of the Ch'ing period alone total hundreds of thousands of volumes and millions of pages.

Dr. Ping-ti Ho has brought to this task the requisite skills and a formidable assiduity. His studies of the Chinese local record concerning plant migrations, for example, have uncovered new evidence which reshapes the historical picture. The laborious combing out of materials concerning population and various correlated factors, which forms the basis for this volume, has put all future workers on this subject in his debt. Although concerned with population, Dr. Ho explicitly eschews quantitative analysis, for statistics of the modern or would-be-modern type — census data and government statistical reports designed for the purpose — are unavailable for China in the Ming and Ch'ing periods. As Dr. Ho's own preface indicates, he has reexamined the institutional structure and terminology which give meaning to the written evidence. This has led him to redefine key terms such as *ting* and *mou* which have often been accepted uncritically and to restudy the process by which official estimates of population were compiled in different periods.

Dr. Ho would be the last to claim that the Chinese record yields any firm basis for a modern effort at quantification of phenomena like births and deaths, migrations, crop production, malnutrition, infanticide, or any other of the many factors affecting the Chinese population over the last six centuries. On the contrary, he has excavated the record which is available, which may yield its indications to social historians and retain an illuminating, if also frustrating, interest for students of China's population.

April 1958 John K. Fairbank

Preface

It is recognized everywhere today that China's population is not only a national but a world problem. Interest in the facts and phases of its growth is therefore limited neither to demographers nor to Far Eastern experts.

In the history of population growth in China, no period is more important than the two centuries from 1650 to 1850. Under unusually favorable material conditions and the benevolent despotism of the early Manchu rulers, the population apparently trebled and reached perhaps 430,000,000 by 1850. In consequence, by the end of the eighteenth century China's resources had become so strained that her economy had begun to assume the pattern with which modern students are familiar.

In spite of various studies of the population of China by occidental and Chinese scholars, many aspects of the subject still remain obscure. The difficulty lies not in the absence of population data but, as Sir Alexander Carr-Saunders remarks, in knowing how to interpret them. Since 1952 I have been mainly concerned with interpreting the available data of the early Ch'ing period by tracing the changing institutional context of certain key population terms which led me as far back as 1368. In the course of my sampling through the vast body of Chinese local histories I was further led into a study of various factors that must have affected population growth. The first version of this study, on the period 1368–1850, was accepted for publication in the *Harvard-Yenching Monograph Series* in September 1954. With the aid and encouragement of the Chinese Economic and Political Studies program at Harvard the study has now been carried up to the period of the 1953 census.

This monograph aims to interpret the nature of different types of population data and to suggest tentative historical explanations as to how and why China's population has been able to grow in early modern and modern times. It therefore remains basically an essay in economic history and is not intended to be a demographic analysis, which must be undertaken by experts differently equipped than I.

In preparing this study I have benefited throughout the years from discussion and association with many teachers and friends. I am particularly indebted to Professors Franklin L. Ho and C. Martin Wilbur of Columbia University for their constructive criticism in the early stage; to my colleagues at the University of British Columbia, Professor Harry B. Hawthorn for constant consultation, Professor James O. St. Clair-Sobell for helping with an important Russian source on the 1953 Chinese census, and Mr. Neal Harlow for securing some vital research materials; to Mr. Howard Linton and Mr. Richard Howard of the East Asian Library of Columbia University, Dr. K'ai-ming Ch'iu and his staff members of the Harvard-Yenching Institute Library, and Dr. K. T. Wu and Mr. Joseph E. P. Wang of the Division of Orientalia of the Library of Congress, for their invaluable assistance. I must express my special gratitude to Dr. Ch'iu for having supplied me, for over five years, with numerous interlibrary loans.

Professor L. S. Yang of Harvard University and Mr. T'ung-tsu Ch'ü, formerly of Columbia University and now of Harvard University, have frequently been asked to check, at times even to copy, certain sources indispensable to the present study. The former has never failed to place at my disposal his encyclopedic knowledge of the economic history of China. My old classmate Professor Teh-chao Wang, while a visiting scholar from the Orient at Harvard, helped me to go through hundreds of nineteenth- and twentieth-century editions of Chinese local histories, thus enabling me to examine post-1850 materials as extensively as I had done for earlier periods. His contribution was particularly great in the preparation of Chapter VII on interregional migrations which, owing to the highly fragmentary nature of the sources, have so far been comparatively little known. To these old friends I owe a debt which can hardly be repaid.

I am grateful to the University of British Columbia, the Committee on Tsing Hua Fellowships for Teaching and Research, and the Chinese Economic and Political Studies program of Harvard University for successive grants-in-aid which enabled me to carry out my studies in the eastern United States during five consecutive summers.

My greatest debt, however, is to my wife Ching-lo, without whose prolonged self-denial this study could never have been undertaken.

Vancouver, B.C. P. T. Ho
August 1957

List of Tables

PART ONE

THE OFFICIAL POPULATION RECORD

CHAPTER I

The Nature of Ming Population Data

In the study of China's historical population problem and fiscal administration the reign of the Ming emperor T'ai-tsu (1368–1398) stands out as an important landmark. Coming from a poor family with memories of early wanderings and sufferings, the founder of the Ming dynasty understood the inherent economic injustice of the society better perhaps than any other ruler of the Ming and Ch'ing periods. "Mindful of the encroachment upon the meek by the powerful interests during the late Yüan period," comments the official history of the Ming dynasty, "he legislated on behalf of the poor and curbed the influence of the rich." [1] Whatever the limitations of his economic policy and the consequences of his unprecedentedly highhanded despotism, in the realm of fiscal administration he achieved two things that had not been successfully dealt with by the Sung and Yüan governments. The first was the compilation in 1381–1382, and the revision in 1391, of the files on labor services commonly called *Huang-ts'e* or Yellow Registers, which (with certain regional exceptions) were based on the enumeration of the entire population. For most parts of the country, therefore, the resulting population returns, with their brief information on age, sex, and occupation, bear a certain resemblance to modern census returns. It was not until after 1776 that China again had population enumerations comparable in nature and quality to those of the Ming T'ai-tsu period. T'ai-tsu's second achievement was the completion in some twenty years of the compilation of land-survey maps and land-tax handbooks called *Yü-lin t'u-ts'e*, literally the Fish-Scale Maps and Books. The former, though primarily a corvée and population register, also contained a summary of the land and other forms of property owned, and the amount of taxes and labor services borne by the entire household. The latter described each parcel of the land and listed the name of its owner. The Registers and the Books were thus the warp and weft of a complex fiscal structure which, with certain modifications in the course of time, lasted till the end of the Ch'ing period.

In most parts of China the population registrations of the period of Ming T'ai-tsu's reign are by definition and in practice very close to modern registrations, and therefore of value to a study of China's population in early modern times. A serious problem, however, lies in the fact that, although outwardly the system of population registration was continued after the death of the founding emperor in 1398, the emphases and methods of registration underwent important changes. The result was that later population returns became increasingly dissociated from true registrations and covered only portions of the entire population. The so-called population returns of certain regions in late Ming times, and of the whole country during the early-Ch'ing period, covered what in the last analysis may be called tax-paying units. The difficulty is increased by the lack of any clear explanations of such changes in the Ming and early-Ch'ing statutes and other government publications. For this reason, a major requisite to an understanding of China's historical population data is an uncovering of the changing institutional context of the population data throughout the period 1368–1850, particularly in the light of Chinese local histories.

The registrations of 1381 and 1391 had their antecedents. In the winter of 1370, when northwestern China and Szechwan had yet to be conquered, Emperor T'ai-tsu gave orders for a count to be taken in all territories already under his control. The method of registration, which had recently been successfully tried by a prefect of southern Anhwei,[2] was to issue to each household a *hu-t'ieh*, or household certificate, on which specific information was to be given as to members and property of the household. We are fortunate in having two reproductions of the pre-Yellow-Register household certificate, together with the text of the original proclamation in fourteenth-century vernacular. The proclamation[3] announced that on December 12, 1370 the Board of Revenue had respectfully received an imperial edict which stated:

The officials of the Board of Revenue will take notice that although the country is now at peace the government has not yet secured accurate information about the population. The *Chung-shu-sheng*[4] has been instructed to prepare population registers in duplicate for the whole empire. You, officials of the Board of Revenue, will send out proclamations ordering the provincial and local authorities to get all the population within their respective jurisdiction officially registered, and all the names of the people entered into the official registers. The number of persons of each household must all be written down without falsification. Each household is to be given a household certificate with a half seal on each side so that the two parts will tally. Since my powerful troops are no longer going out on campaigns, they are to be sent to every county, in order to make a household-to-household check of the returns. Those households whose tallies agree will be treated as subjects in good standing; if they do not agree, the family will be placed on the list of those liable for military service. If it is discovered in the course of checking that some local officials have falsified the returns, those officials are to be decapitated. Any common people who hide from the census will be punished according to law and will be drafted into the army. Let everyone respect and observe this. It is ordered that each household is to be given one-half of the tallied household certificate to be carefully kept.

"This," comments George Sarton, a modern historian of science, "is a good illustration of the thoroughness of the Chinese administration of those days . . . it increases our trust in those publications."

Through a study of the 1371 household certificate — that of the Lang family — which follows, certain facts become obvious. There can be little doubt that both the adults and juveniles of the household were registered on the certificate, for the genealogical tree of the Lang family, which is also extant, unmistakably shows that all the male offspring of the clan were descended from the bearer and his two sons.

The interesting thing to note is that this household of six consisted in fact of two families of three each. What was called a household in Ming and Ch'ing times was a customary Chinese compound family rather than a "natural" family in the Western sense. Throughout the Ming and Ch'ing periods the law provided that adult sons and grandsons be regarded as members of the same household unless by special permission of the head of the household they divided up the common property and lived under separate roofs.[5] This legal definition of the household is important in any study of the average size of the Chinese family in historical times.

The following is a sample of the 1371 household certificate:

The Lang Li-ch'ing Household

Place of Residence: Hsing-hua-ts'un (Apricot-Blossom-Village), Kuei-ch'ih County, Ch'ih-chou Prefecture.

Number of Males: 4
  Adult Males: 2
    The Bearer: 54 sui.
    His [adult] son, Kuei-ho: 28 sui.
  Male Children: 2
    The Bearer's second son, Kuei-mao, whose pet name is Kuan-yin-pao: 7 sui.
    The Bearer's grandson, Fo-pao: 7 sui.

Number of Females: 2
  The Bearer's wife, Ah Ts'ao: 42 sui.
  The Bearer's daughter-in-law, Ah Yin: 28 sui.

Property: A house of 5 rooms occupying 8/10 of a mou.

Hung-wu Fourth Year Month Day

Code Number and Signatures
Seals

The case of the Lang family and other data in local histories suggest the possibility that even before 1381, when the Yellow Registers were compiled on a national scale, the population reg-

Table 1. Population of Chi-hsi county.

Occupation	Census of 1371		Census of 1376	
	Households	Mouths	Households	Mouths
Military	386	2,925	547	3,845
Artisan	262	1,675	285	1,794
Commoner (*min*)	9,074	36,588	9,045	36,629
Scholar	36	146	36	146
Buddhist	25	76	25	63
Taoist	4	4	4	6
Total	9,787	41,414	9,942	42,483

Source: *Chi-hsi HC* (1581 ed.), 3.1b–2b.

istrations covered most of the population of localities under Ming T'ai-tsu's effective jurisdiction. On a local scale, the 1371 and 1376 censuses of Chi-hsi county in southern Anhwei, for instance, as shown in Table 1, offer an example. Following these records, it can be seen that in 1371 the average household had 4.22 persons, and five years later it averaged 4.27 persons. On a prefectural scale, Hui-chou, for instance, returned 117,110 households and 536,925 persons in 1371; and 120,762 households and 549,485 persons in 1376. The occupational breakdowns are also available.[6] In 1371 the average household had 4.58 persons, and in 1376, 4.55 persons. Likewise, Su-chou prefecture in 1371 had 473,862 households with a population of 1,947,871; in 1376 it had 506,543 households with a population of 2,160,463. The average household had 4.11 and 4.27 persons respectively.[7]

After the pacification of the whole Chinese world the founder of the Ming dynasty decided in 1380–1381 to set up permanent machinery with a view to undertaking future registrations, collecting taxes, and equalizing labor services. The machinery and methods of population enumeration are described in *HWHTK*.

The method is to organize 110 households into a *li*. Within each *li* ten persons who have the largest number of adult males [in their households] and who pay most taxes are to be chosen as the [ten *li*] headmen. The rest of the 100 households should be divided into ten *chia*. Each *chia* has ten persons [*i.e.*, heads of households]. The annual labor services are to be supervised by one *li* headman and ten *chia* headmen. Within the walled city [such a unit of 110 households] is called *fang*, in the outskirts of the city *hsiang*, and in the countryside *li*. The order [of performing such duties] should depend on the number of adult males [in the household] and the amount of tax. Each *li* should compile its own register. At the beginning of each register there should be a chart [summarizing the contents]. Within each *li* widowers, widows and orphans, not being liable to labor services, should be appended to the ten *chia* as *ch'i-ling* [*i.e.*, odds and ends]. The register should be revised once every ten years according to the changes in the number of adult males and tax payment. It should be made out in four copies: one to be submitted to the Board of Revenue, the other three retained by the provincial, prefectural and county authorities. The copy submitted to the Board of Revenue is to be bound with yellow covers and will therefore be called the Yellow Register. At the end of each year the registers should be sent to the eastern and western storehouses of Hou-hu ["Hinter Lake," modern Hsüan-wu Lake, Nanking] for safekeeping.[8]

In other words, the task of collecting taxes, registering population changes, and performing labor services fell annually on one of the ten *li* headmen and ten of the 100 households which formed the ten *chia*, thus forming a decennial rotation of services. Table 2 represents a summary of the registration of 1381–1382, with some interesting breakdowns, taken from the 1393 edition of the history of Yung-chou prefecture in southwestern Hunan.

Table 2. Population of Yung-chou in 1381–1382 (73,005 households).

Male		Female		Sex ratio[a]	Percentage of children	Mouths per household
Adults	Children	Adults	Children			
135,349	94,071	123,970	58,226	125.9	37.0	5.64

Source: *Yung-chou FC* (1393 ed.), 3, *passim.*
[a] Number of males per 100 females.

Whereas many local population enumerations of the later Ming period betray irregularities and even absurdities, these few available late-fourteenth-century enumerations, with their breakdowns, accord in general with our demographic knowledge of China.[9] Although by definition and in practice the population enumerations in the reign of Ming T'ai-tsu covered the entire population, cases of regional under-registration have been found. For instance, K'uai-chi county, part of modern Shao-hsing in Chekiang, returned 39,879 households but only 59,439 mouths during the period mentioned.[10] In 1381 Fu-chou prefecture in Fukien returned 94,514 households and 285,265 mouths — a number which, amounting to only one-half of the prefectural population during the late Yüan period — was interpreted by the editor of the 1613 edition of the prefectural history as the result of a lenient interpretation of the law whereby the fiscal burden of the area was reduced.[11] In the North China plain, where the effect of wars of the late Mongol period was keenly felt, the settlement and enumeration of the displaced proved to be a difficult task which took decades.[12] Although under-registration cannot definitely be established by the Ming local histories of the northern regions, which usually give only the totals without breakdowns, it is likely that the population of North China was somewhat under-registered during the reign of Ming T'ai-tsu. But it was in the peripheral areas that under-registration was most serious. Parts of Szechwan, Kwangsi, and Kwangtung had substantial aboriginal populations which constituted the majority in Yunnan and Kweichow. In Yunnan and Kweichow, which were not made into provinces until 1420, the law of 1381 provided that no Yellow Registers were to be compiled in areas ruled by tribal chieftains.[13] In fact, both the official historian of the Ming Yellow Registers and a leading modern authority believe that the few Yunnan and Kweichow Yellow Registers attributed to the reign of Ming T'ai-tsu were compiled after 1420.[14]

All in all, therefore, under-registration was by no means uncommon during the early Ming period. The actual population of China toward the end of the fourteenth century was probably over 65,000,000. Table 3 lists the official population returns.

It seems logical to agree with the comments of later Ming writers on population that China's population in the late fourteenth century was smaller than in the later Ming period, for the country had just gone through the decades of turmoil and wars that attended the downfall of the Mongol dynasty. The later Ming population returns, however, indicate a mildly falling population during the first half of the fifteenth century and then a stationary population fluctuating slightly around the 60,000,000 level. This will be explained presently.

Although for the greater part of the country the registrations in Ming T'ai-tsu's reign seem to have covered most of the population, soon after the compilation of the 1391 Yellow Registers the main interest of the government shifted to the fiscal side of the population problem. Prior to 1391, T'ai-tsu's insistence on successive enumerations of the entire population arose from his eagerness to make an equitable distribution of the labor services. But after the erection of the elaborate fiscal structure, based on the Yellow Registers and the Fish-Scale Maps and Books, he became engrossed in means of perpetuating it. The statutes of 1391 therefore attempted to avoid the reshuffling of the *li-chia* system as far as circumstances allowed,[15] because once the ten-year rotation of the

Table 3. Population of China in 1393.

Province	Households	Mouths	Mouths per household
Nan-Chihli [a]	1,912,833	10,755,938	5.62
Pei-Chihli [b]	334,792	1,926,595	5.75
Chekiang	2,138,225	10,487,567	4.90
Kiangsi	1,553,923	8,982,481	5.78
Hukuang [c]	775,851	4,702,660	6.06
Shantung	753,894	5,255,876	6.97
Shansi	595,444	4,072,127	6.84
Honan	315,617	1,912,542	6.07
Shensi [d]	294,526	2,316,569	7.87
Fukien	815,527	3,916,806	4.92
Kwangtung	675,599	3,007,932	4.57
Kwangsi	211,263	1,482,671	7.02
Szechwan	215,719	1,466,778	6.81
Yunnan	59,576	259,270	4.39
Total	10,652,789	60,545,812	5.68

Source: *Hou-hu chih*, 2; also Sun Ch'eng-tse, *Ch'un-ming meng-yü-lu* (1913 reprint), 35.6a–6b. *HWHTK*, 13.2892, gives 10,652,870 households.

[a] Before 1421, when the national capital was moved from Nanking to Peking, it was called Chihli, literally the Metropolitan province, which included Kiangsu and Anhwei.

[b] Before 1421 it was called Peiping, which covered the modern Hopei and a part of Chahar and Jehol.

[c] Hupei and Hunan.

[d] Shensi included Kansu.

li-chia service was fixed, any drastic change would upset the li-chia teamwork. The law of 1391 provided that when any household within the li fell into abject poverty or became extinct, a new household was to be recruited from those "odds and ends" of the same li, to avoid interfering with the functioning of the neighboring li. In the long run, however, human vicissitudes and natural changes in population frequently made the maintenance of the original li teamwork impossible. Furthermore, with the passing of the strong rulers of the early Ming, the gentry gradually reverted to various practices by which, in alliance with the underlings of the local government, they often succeeded in shifting a part or all of their burden of labor services and land tax to the poor. Their illegal methods of effecting such evasions became increasingly shrewd and varied.[16] The victims, unable to bear the extra burden, eventually resorted to desertion, which in turn adversely affected those who remained in the li. It was not too uncommon, from the middle of the fifteenth century on, for desertion of the li-chia personnel to reach the scale of village depopulation. Thus the original li-chia machinery became more and more unworkable. A further reason for the eventual break-up of the li-chia system was the corrosive influence of a continually rising money economy, made possible by the steady influx of silver following the coming of the Portuguese after 1514. The availability of a large amount of silver, together with the greater social and economic division of labor, made commutation of the li-chia services inevitable.[17] When labor services ceased to be performed by the villagers personally, both the li-chia machinery itself and its system of population registration became obsolete.

The laws of 1391 were another factor which brought about a most important change in the system of population registration. They provided that, in the future compilation of population returns, emphasis should be placed on males over ten years old. The list of male children over ten was to be arranged in order of age so as to ensure their punctual enlistment into the labor services when they reached sixteen. Although the statutes of 1391 did not clearly mention that females and children under ten could be underregistered, they nonetheless provided that "the aged, the physically unfit, children under ten, widows and immigrants may be included in the odds-and-ends households." [18] This regulation did not in itself contradict the original regulation of registering the entire population, as may be evidenced from the fact that many later Ming local histories, particularly those of the northern provinces, registered continual growth of the populations of their localities. Yet the statutes of 1391 certainly gave considerable leeway to local authorities as to the method of population registration so long as the significant portion of the population, *i.e.*, fiscal population, was duly covered by the decennial revision of the Yellow Registers. The population returns of Shanghai county in Ming times are illustrative of many localities in the southeast which tacitly turned from an enumeration of the entire population to that of *portions* of the population. The population data of Shanghai are summarized in Table 4. From the continual sharp drop in the number of females, in the average size of the household, and the sharp rise in the sex ratio, one can easily see the gradual change from registration aimed at the entire population to the registration of a portion of the entire population, although it is impossible to say that all the males given in the two sixteenth-century enumerations actually paid the land tax and performed the labor services.[19]

Table 4. Population of Shanghai county during Ming times.

Year	Households	Mouths	Males	Females	Persons per household	Number of males per 100 females
1391	114,326	532,803	278,874	253,929	4.66	109.8
1412	100,924	378,428	199,781	178,647	3.75	111.8
1520	93,023	260,821	179,524	81,297	2.80	220.8
1572	113,985	192,967	158,532	34,435	1.69	460.4

Source: *Shang-hai HC* (1588 ed.), 4, *passim.*

In any case, when the main emphasis of population registration was fiscal, the figures for mouths and females, and to a lesser extent also for households, were casual — so casual that Shao-hsing prefecture, which had a population of over 1,000,000 during the late fourteenth century, reported in 1586 a male population of 395,960 and a female population of 179,213.[20] Mien-yang county in south-central Hupei in 1391 returned 23,109 males and 24,201 females, but reported in 1522 a male population of 25,346 and a female population of only 13,876.[21] This change in the method of population registration was by no means confined to the Yangtze provinces. Hua-chou in Shensi, for instance, reported in 1572 a total population of 49,651, out of which only 14,166 were females. Out of the 35,515 males reported, 10,547 were liable to labor services.[22] Likewise, the magistrate of Ch'in-an county in Kansu thought it by no means unreasonable from the fiscal point of view to report in 1535 so disproportionate a population as one of 12,424 males and 6,099 females.[23] In the course of time the under-registration of females and male children under ten must have become prevalent, particularly in the southeast. This enabled the editor of the 1585 edition of the history of Shun-te county near Canton to say that "the omission of children in the official registers has been an old practice." [24] All this evidence, and much else from the southeastern provinces, not only reveals the tacit shift of emphasis in the system of population registration, but also accounts for the apparent decline or stagnation of population in certain areas — an appearance which, in the opinion of many editors of Ming local histories, is the exact opposite of the truth.

The third factor in the failure of later Ming registrations to cover the entire population was the evasion of population registration by people under the protection of the powerful rural interests. Although this practice may well have been fairly general in many parts of the country, it seems to have been particularly rampant in the lower Yangtze and southeastern coastal provinces.[25] In areas where the rates of land tax and quotas of labor services were high the peasants often found it advantageous to surrender a portion or the whole of their property to a powerful individual in exchange for his protection. In some areas where landlordism was a real economic and social force the influence of the rural scion was usually sufficient to secure for his protégés, chiefly through his connections with local government underlings, total exemptions from taxes and services. Since ownership of land was most concentrated in the lower Yangtze area, and certain parts of Fukien and Kwangtung,[26] omission and evasion of population registration due to gentry protection must have been serious. It was, in fact, so serious that an upright official of the lower Yangtze area testified in the early 1430's that this was one of the major causes of the apparent decline in population.[27]

The fourth factor that brought about a general under-registration of population in certain areas was official peculation. The deplorably low government pay for officials during the Ming period was nothing short of a scandal.[28] At first, owing to Ming T'ai-tsu's mass execution of corrupt officials and to the unprecedented generosity of promotion with which he rewarded the able and upright, the administration was reasonably honest and efficient; [29] but official venality was bound to increase as the successive rulers became less energetic. It was said that long before the middle of the sixteenth century the local government personnel of Liu-yang county in Hunan had privately appropriated a part of the labor-service payment, while working for a steady reduction of the corvée quota as well. Thus the number of various categories of households had been brought down from 12,680 in 1391 to 7,481 in 1551. The commoners had heartily approved such a practice, because the registered number of their households had been decreased from 7,460 in 1391 to a mere 152 in 1551.[30] Officialdom became even more money-conscious after the influx of European silver into China.[31] It was by no means purely coincidental that in Fukien, where trade with Europeans was most active, the appropriation by provincial and local officials of portions of public revenues from taxes and labor-service payments became almost conventionalized. The famous scholar Ho Ch'iao-yüan testified to this:

> Nowadays those of the people whose names are entered into the official registers are called the *kuan-ting* (officially-registered male adults liable to labor services). Those who are not entered into the registers are called the *ssu-ting* ("private," i.e., unregistered, ting). The official ting each pays about three-tenths of a tael to the local government. The private ting pay into the pockets of government underlings, who charge them roughly according to the size and economic status of their families. This is inevitable. When the authorities make the assessment, they base their estimate only on the amount of land tax [paid by the people]. Neither the numbers of the ting nor those of the mouths are authentic.[32]

It was said that in Shun-te county near Canton "the rich and powerful households ally themselves with the local-government personnel, so that sometimes only one out of several tens of actual ting is entered into the register." This was true also for other parts of Kwangtung.[33]

The fifth factor that helped to contribute to the under-registration of the population in a number of provinces was the merging of several clans into one in order to lighten, if not entirely escape, the fiscal burden. As has been pointed out above, the Ming laws required that grown sons and grandsons be regarded as members of the same household unless by special permission of their common sire they set up separate families. These laws undoubtedly encouraged in certain cases the merger of collaterals, a device which, though not strictly legal, was customarily acceptable. In Szechwan, which was remote from the two Ming capitals Nanking and Peking, this loophole in the laws was early exploited to the full by some large clans who, despite their different surnames, pretended to be members of the same household. In 1438 there was already a statute prohibiting this device in Szechwan.[34] Instead of being curbed by strong government action, the practice became more widespread with the continual deterioration in government efficiency. An upright early-sixteenth-century official testified:

It has been learned that the illegal practice of evading population enumeration in Szechwan is essentially the same as that in Hupei and Hunan, and is different from that of other places. Frequently over ten large clans merge under the guise of a single household. Actually they live as over one hundred separate families. All together they report but some ten ting.[35]

The unusually large sizes of the households in some northern provinces [36] gives reason to believe that this abuse was by no means confined to Szechwan, Hupei, and Hunan.

Still another factor responsible for under-registration was that people sometimes illegally changed, or secretly forsook, their given status. Ming T'ai-tsu's planned society was based on the traditional conception of social division of labor, with each social group performing its particular function. To ensure the maintenance of a stable society, the early Ming laws prescribed that professional status be hereditary. Thus the artisans were to be permanently registered under the *chiang-chi* or artisan status, members of the armed forces under the *chün-chi* or military status, the commoners under the *min-chi* or status of the common people. However, economic changes and the growth of population made the maintenance of such a rigid society impossible. As early as 1428, when the empire was in the midst of peace and prosperity, some "military" people had already managed to get away from their original status. They must have reached sufficiently large numbers to make a nation-wide registration of the chün-chi population necessary.[37] Statutes of 1438 and 1479 and of the Ching-t'ai period (1450–1456) repeatedly forbade chün-chi people to change their status.[38] The laws were, of course, of no avail, for "none of the social groups have suffered more from their services and been more eager to get rid of their status than the chün-chi people." [39]

Since the chün-chi people were registered with the Board of War and remained outside the Yellow Registers, they were not regularly registered decennially as were the commoners. Desertions from the chün-chi status might remain undetected over a considerable period. Yet in the long run the loss in the chün-chi population could not fail to be reflected in the official registers, for there were times when large-scale desertion called for a nationwide re-registration of the chün-chi people. It is unlikely, however, that the loss to the chün-chi registers was duly compensated by a proportionate gain in the ordinary Yellow Registers, for their main purpose in ridding themselves of their status was to escape services and fiscal burden altogther. While substantial numbers of chün-chi people had been living outside the meshes of the law as tax-free and service-free commoners, not a few of the commoners had been quitting their status and seeking the protection of the military garrisons, particularly in the frontier posts. By rendering services to the officers they escaped their fiscal burden as well as the li-chia population registration.[40] Similarly, many commoners surreptitiously changed their status into that of artisans, and vice versa. All this contributed to the loss in registered population.[41] So powerful were such economic and social forces that by the late Ming period the original occupational registration had become a dead letter.[42]

While these various factors accounting for population under-registration were due either to the loopholes in the law or to fraudulent practices on the part of the people, in certain areas under-registration was necessary from the standpoint of fiscal equity and sanctified by local custom. The 1585 edition of the history of Ch'ang-shan county in southwestern Chekiang explained:

It should be pointed out that some of the ting and mouth returns are a mere formality but some are true numbers on which the incidence of labor services falls. This is because although [the local authorities] do not dare to make returns far short of the government quotas, in allocating the burden of labor services they usually take into consideration people's ability to bear. Otherwise the people would flee from their homes and thereby bring about drastic reduction in numbers. In our locality, while there are nominally several tens of thousands of adult males in the official registers, those who are actually liable for labor services amount to only several thousands. This vast difference in number is by no means deliberate falsification. For the custom of sharing the burden of a ting by several adult males under the same roof can be compared to the custom of sharing one shih

(Chinese bushel) of land tax by several mou of land. Only by so doing can an equilibrium between fiscal burden and people's resources be established.[43]

It was but natural for the nominal total population of this locality to remain more or less constantly at the 73,000 level throughout the sixteenth century and for the significant fiscal population, or the labor-service performing ting, to be fixed at a low level of 7,300.

Small wonder, then, that under the combined circumstances described above the Ming population data drifted farther and farther away from the truth, and that the 1590 edition of the history of I-hsing county in southern Kiangsu contains the remark: "The changes in the numbers of households and mouths in the official registers are all unauthentic." [44] To comply with the early Ming regulations, central, provincial, and local authorities continued to compile household and mouth returns as of old, but the compilation became more and more a matter of formality. As long as a locality could bear more or less the same amount of taxes and services or manage to secure a reduction of the original fiscal quotas, the figures for households and mouths really meant very

Table 5. Population returns of Ch'ang-shu county.

Year	Households	Mouths
1371	62,285	247,104
1376	61,211	263,414
1391	67,077	284,671
1412	66,327	299,661
1432	77,665	315,959
1442	76,688	347,855
1452	82,055	358,836
1462	83,558	367,600
1472	87,474	381,577
1482	87,474	381,577
1492	88,044	381,577
1502 [a]	73,641	381,949
1522	73,641	381,949

Source: *Ch'ang-shu HC* (1538 ed.), 2.15b–17b.

[a] In 1497, 15,270 households and 58,000 mouths were transferred to T'ai-ts'ang county.

little. The formality of population registration can be demonstrated from the evidence of various local histories, as for example, Table 5. The fact that since the 1470's the numbers of households and mouths tended to repeat themselves reflects the fact that household and mouth had ceased to be of real fiscal significance. From the fiscal point of view the ting had become the base and primary concern of population registration although the 1538 edition of the history of Ch'ang-shu county, like most late Ming local histories, complied with the time-honored system by giving only household and mouth returns without yielding specific information on the ting.

The same tendency can be illustrated on a prefectural scale as Table 6 showing Ning-po, with breakdowns for its five constituent counties, shows. Except for the figures of Tz'u-hsi county, which show a steady decline, the population returns indicate either very slight changes or no change whatever in a full generation. By 1552 the registered population of the whole prefecture had dropped to only 45 per cent of that of 1391. In some counties of Shao-hsing prefecture the population returns began to be compiled arbitrarily during the Yung-lo period (1403–1424).[45]

Table 6. Population of Ning-po prefecture, 1522–1552.

County	1522		1532		1542		1552	
	Households	Mouths	Households	Mouths	Households	Mouths	Households	Mouths
Yin-hsien	58,345	193,380	58,350	193,385	58,355	193,395	58,361	193,412
Tz'u-hsi	21,000	37,525	19,300	32,501	18,732	27,455	16,000	23,365
Feng-hua	18,865	60,781	18,865	60,334	18,865	60,334	18,865	60,364
Ting-hai	12,517	37,450	13,026	38,808	14,017	38,701	14,017	38,722
Hsiang-shan	3,802	17,812	3,802	17,812	3,802	17,812	3,802	17,812

Source: *Ning-po FC* (1560 ed.), 11.33a–35b.

What is not mentioned in the cases of Ch'ang-shu and Ning-po is clearly explained in the history of Wu-chiang county in southern Kiangsu whose population may be seen in Table 7.

The editor carefully noted that all enumerations up to and including 1505 "contained both males and females." As to the sharp drop in the number of mouths in 1536, it is clearly explained that in that year the prefect during his recompilation of the registers omitted the households entirely and fixed the ting quota at 95,667, a figure that included only the *ch'eng-ting-nan-tzu*, or male adults

Table 7. Population returns of Wu-chiang county.

Year	Households	Mouths
1371	80,382	361,686
1376	81,572	368,288
1391	74,831	380,017
1412	74,831	259,101
1432	79,645	268,029
1442	72,708	268,029
1452	67,804	271,421
1462	68,365	272,691
1486	72,445	273,932
1505	81,916	267,100
1536	....	95,667
1557	86,860	259,657

Source: *Wu-chiang HC* (1561 ed.), 9.1a–2b.

actually liable to labor services. The mouth return of 1557 is further broken down as follows: (1) total number of males: 162,462; (2) ting: 160,044; (3) male children (*pu-ch'eng-ting*): 2,418; (4) total number of females: 97,195; (5) adult females: 96,205; and (6) female children: 990. From the table and from the 1557 breakdowns we can easily see that for decades, at least, adult females had been very much under-reported and children of both sexes almost neglected. What is most revealing is that the Chia-ching era (1522–1566), generally a period of fiscal reforms and tax readjustments, witnessed the fixing and revising of the ting quota, which had definitely become the hard core of the fiscal population. In reverence to the time-honored method of population registration, however, the 1557 enumeration reverted to households and mouths. For exactly the same reason the mouths of Hai-ning county in Chekiang, which had remained stationary at something over 200,000 since the beginning of the dynasty, suddenly dropped to 88,972 in 1522 and to 88,752 in 1532. From 1542 onward the number of mouths again was recorded as over 200,000.[46]

While Ch'ang-shu and Wu-chiang represented a minority of the localities in the southeast registering a mildly increasing or stationary population, most of the available late Ming southeastern local histories registered a steady decline. The population data of Lien-chiang county in Fukien, shown in Table 8, are indicative of what had been taking place in the system of population registration, in the southeast at least.

Table 8. Population returns of Lien-chiang county.

Year	Households	Mouths	Mouths per household
1381[a]	14,804	65,067	4.40
1482	6,028	16,928	2.81
1492	5,908	16,817	2.85
1512	6,902	16,913	2.45
1522	6,270	16,913	2.70
1578	6,378	14,802	2.32
1633	5,378	15,019.5	2.79

Source: *Lien-chiang HC* (1805 ed.), 2, *passim.*

[a] The returns for the period between 1381 and 1482, which show a slow and continual decrease, are omitted.

The editor commented that in 1578 the locality adopted the "single-whip" system, exempting all females from registration. From 1578, the land-tax and labor-service assessments were "based on ting alone and no longer on households." As is clearly shown by the data, ever since 1482 the ting had replaced the mouths as the significant population, although it was not until 1578 that the land tax and labor services were consolidated into a single payment.

In general, although the item ting was becoming more and more important in fiscal administration in the course of the sixteenth century, the ting figures are not given in the majority of late Ming local histories. This is because the compilation of the Yellow Registers, despite its ever-increasing confusion and obsolescence, remained outwardly a sacred institution to the very end of the Ming period. Since the Yellow Registers required, first of all, the figures for households and mouths, these figures must be periodically reported. This was sufficient in the early Ming period, because the mouths in theory and in practice already included the ting. For instance, Shun-t'ien prefecture, the Metropolitan area surrounding Peking, secured the permission of the emperor in 1571 to fix its ting quota at 147,300; yet in the last years of the sixteenth century it still reported the number of mouths, which exceeded the number of ting by half a million, without mentioning ting.[47] This also explains why Wu-chiang and Hai-ning counties, after briefly experimenting with the substitution of ting for mouths, reverted to the old established practice.

By far the most basic reason for the absence of ting figures in the majority of Ming local histories is that, with the growing anachronism of the Yellow Registers, there gradually arose a new kind of register, commonly called the *Fu-i ch'üan-shu*, literally The Complete Books of Taxes and Labor Services. The compilation of these more realistic handbooks by various local authorities generally synchronized with the important fiscal reform movement, known as the "single-whip" system of levy, which was in full swing from the second quarter of the sixteenth century. We are fortunate in having the microfilm reproductions of the 1611 edition of the *Fu-i ch'üan-shu* for Kiangsi province and the 1620 edition for Hui-chou prefecture.[48] The latter gives only the number of ting for the whole prefecture and for each of the counties, without bothering with the figures for households and mouths, which, according to the successive Ming editions of the prefectural histories, had remained invariably well above the 500,000 mark since the beginning of the dynasty. So far as population was concerned, the *Fu-i ch'üan-shu* reads: "*Hu-k'ou jen-ting*: 205,786 ting" for the whole prefecture. Thus in the fiscal sense what were literally the "households, mouths, population, and ting" simply boiled down to the ting.

The 1611 edition of the Kiangsi *Fu-i ch'üan-shu* is somewhat different. The total fiscal population for the province in that year was 2,485,931, of which 1,497,111 were ting and 988,820 were adult women. This was because in Kiangsi, Fukien, and Kwangtung during Ming and early Ch'ing times some of the adult women were liable to the salt gabelle, and therefore were decennially assessed. A most interesting thing is that what the *Fu-i ch'üan-shu* gives as the fiscal population for Jui-chou prefecture had been in fact the "entire" population of the prefecture since the late Chia-ching period (1522–1566).[49] The case of Jui-chou and the complete omission of the figures for households and mouths in the *Fu-i ch'üan-shu* indicate better than anything else the extent to which the original system of population registration had become outmoded.

Although during the late Ming period the ting theoretically meant a male adult between 16 and 60 liable to labor services or labor-service payment, already decimal digits were appearing in the late Ming ting returns of some counties. The case of Lien-chiang, already cited, is one. Likewise, Ch'ien-t'ang county in Hangchou prefecture recorded 51,900.5 ting.[50] As to the question of the exact nature of ting in late Ming times, there is no simple answer. The complexity of the nature of ting and its changing institutional context throughout late Ming and early Ch'ing times will be discussed in Chapter II.

Before concluding the study of the nature of Ming population data, the official figures for the national totals during the whole Ming period must be briefly discussed. According to data laboriously collected from the *Ming shih-lu* and other official sources and textually emended by a recent scholar, both the number of households and the number of mouths fluctuate, except in a few abnormal years, within a narrow range in the course of two and three-quarter centuries. The number of households fluctuates between a little over 10,000,000 and 9,000,000, and those of mouths between 60,000,000 and 50,000,000. Generally speaking, the figures are highest for the very beginning of the dynasty and lowest for the first sixty years of the fifteenth century. From his textual study and from his study of the numbers of households per li based on a few contemporary sources, this scholar has reached the conclusion that "Chinese statistics are characterized by a remarkably high level of accuracy."[51]

This is extremely doubtful in the light of our discussion, although the population data of the Ming T'ai-tsu period by definition covered the entire population. As has been pointed out, during the compilation of the first two Yellow Registers the population of the southwestern provinces was probably very much under-reported and regional under-registration has been discovered in parts of Chekiang and Fukien. Although the paucity of population data in the Ming T'ai-tsu period does not enable us to estimate accurately the total population of China during the late fourteenth century, it is safe to say that under-registration was by no means confined to the above-mentioned regions, and that the national population probably exceeded 65,000,000 to an unknown degree.

Defective as the data of the Ming T'ai-tsu period are, they are superior to the subsequent Ming population returns for the simple reason that they are by definition those of the entire population and respectable registrations were undoubtedly carried out in a large number of localities during the compilation of the first Yellow Registers. In the opinion of Liang Fang-chung, the leading authority on Ming fiscal history, only the first two or three Yellow Registers are of some use to population studies, while for many areas the subsequent registers were compiled with relatively little change. If any further evidence is needed to prove the unreliability of the later Ming Yellow Registers as sources of information on population it is provided by an early Ch'ing official who presented a memorial suggesting that the compilation of the useless Yellow Registers be discontinued. In the old Ming files he had discovered that in some of the 1642 Yellow Registers the population returns had been compiled in advance for the year 1651, seven years after the downfall of the Ming dynasty! [52]

From evidence presented in this chapter it is clear that population under-registration was becoming progressively more serious in the country in general and in the densely populated southeastern provinces in particular. The general pattern of population movements during the Ming period must be studied in the light of the expanding agricultural settlement, the increase in national food production, the rise of a more variegated economy, the comments of local histories, and the writings of qualified contemporary observers. After a detailed discussion of the material and political conditions affecting population growth and an examination of local and contemporary comments in Part II, it will become obvious that the population of China had been growing in a more or less linear fashion from 1368 to about 1600, despite the fact that the official Ming population data indicate a stationary population.

CHAPTER II

The Nature of Ting

In the course of time the ting became the most important, if not the sole, concern of Ming population registration. By the time the *Fu-i ch'üan-shu* were compiled by various local authorities, the ting had virtually superseded the households and mouths as the significant fiscal population. True, the household and mouth returns continued to be compiled up to the downfall of the Ming dynasty in 1644, but there can be little doubt that in most cases they were compiled arbitrarily. This inaccuracy in population registration was fully understood by the early Ch'ing government, which put an end to the compilation of household and mouth returns and based the first Ch'ing tax and labor-service assessments on the late Ming *Fu-i ch'üan-shu* of various prefectures and counties. Although the practice of compiling the Yellow Registers was continued after 1651, the scope of the Ch'ing Yellow Registers became much more broad than that of the Ming Registers. Whereas the Ming Yellow Registers were primarily fiscal handbooks based on population enumerations and property assessments, the Ch'ing Yellow Registers consisted of population and tax returns as well as routine expense accounts and reports on the purely administrative and disciplinary matters of practically all the major departments of the central government. Among the 6,602 volumes of the Ch'ing Yellow Registers preserved by the Peiping Palace Museum, only 193 deal with local population returns. For the period between 1651, when the first Yellow Registers were compiled by the new dynasty, and 1741, when important changes in the system of population registration were made, the only available population data are the annual ting returns.[1]

The importance of the official Ch'ing ting figures for the study of early Ch'ing population has long been accepted by occidental and Chinese writers on demography. They have been as a rule too ready to accept without detailed study the official definition of a ting as an adult male between sixteen and sixty years old who paid the ting tax. The charm of such an oversimplified definition has proved so irresistible that they have used the official Ch'ing ting figures to reconstruct the total population for late seventeenth- and early eighteenth-century China. Two early writers on Ch'ing population, E. H. Parker and W. W. Rockhill, entirely without explanation take ting and "families" and "taxed population" as almost identical terms. Their method of reconstructing China's historical population data is to multiply the ting figure for a particular year by four, five, six, or whatever appears to them to be most reasonable. Since the ting figure has usually been regarded as more or less tantamount to the "taxed population," Rockhill is convinced that the population data prior to 1712 (in which year the K'ang-hsi Emperor [1662–1722] fixed the national ting tax permanently on the basis of the 1711 ting returns) were too low and that the population data since 1712 were largely guesswork, "and where numbers are guessed they are always magnified." [2] W. F. Wilcox, whose influence over modern Chinese demographers has been considerable, accepts Rockhill's reasoning and analysis with certain revisions. His estimates that China's population was in the neighborhood of 65,000,000 in 1651 and below 350,000,000 by 1850 are regarded by a leading Chinese demographer [3] as the "most judicious." In order to appraise the value of these and other population estimates, it is imperative that the complexity of the technical term ting be fully investigated.

Before discovering the meaning of ting, we must know the nature and incidence of the ting tax during late Ming and early Ch'ing times. The ting tax is commonly believed to have been a poll tax or a poll tax on adult males. In reality, it consisted of a group of compulsory labor services which during the early Ming period were required of the people and which were gradually commuted into money payment.[4] As to the nature of the late Ming and early Ch'ing ting tax, no one has explained it more succinctly and with greater authority than Li Fu, governor general of Chihli, who stated in a memorial in 1726: "In my opinion, the present ting payment is the same as the labor services of ancient times." [5] In reviewing the historical process whereby the original ting services were commuted into cash payment we will have to go back to the early Ming corvée system.

According to the early Ming laws, the common people (except widowers, widows, the physically unfit, orphans, and men without property) were organized into units of 10 households, or chia, and units of 110 households, or li. Each year 11 of the 110 households performed the li-chia service, which during the early Ming period consisted mainly of the collection of taxes from the 110 households within the li. A rotation of the li-chia service was completed in ten years. The incidence of the li-chia service fell on the households. There soon came into being another category of labor services called the *chün-yao*, which consisted of various kinds of menial work for the local government. Unlike the li-chia service, the chün-yao service was assigned to the adult male. Strictly speaking, therefore, only the chün-yao service was the prototype of the later ting tax, as its incidence fell on the adult males.[6]

In the early Ming chün-yao assessment, however, consideration was given to the economic status of the household of which the adult male was a member. Not infrequently, therefore, households which had very few adult males but substantial landed property or mobile wealth were assigned a large number of ting. Since the ting were graded, the rich were usually assigned upper-grade ting who bore a heavier burden of service than the lower-grade ting. Conversely, poor households having a large number of adult males but meager property were assigned a very small number of lower-grade ting. Those adult males who had no property whatever or who were hired by others as agricultural laborers were exempted from the chün-yao assessment. It is important to note that even during the early Ming period the official ting figures did not represent the actual number of adult males in the country. The incidence of the chün-yao service fell partially on property, particularly landed property. Summing up the principle of ting assessment, an early sixteenth-century official was right in saying that it was based on "the number of adult males, the amount of landed property, other sources of income, domestic servants, horses, etc." [7] It was also common for a rich household to bear several kinds of service and for a number of poor households to share a single service.[8]

This clear-cut division of labor services between the household and the adult males did not exist for very long; gradually the importance of the household was overshadowed by that of the ting.

In the first place, the assessment of the household presented greater difficulties than that of the adult male. It was easier for a rich household to hide its wealth than to under-report its adult males. In the second place, although in theory it was the household that bore the incidence of the li-chia service, in practice the average small household was usually headed and supported by the same adult male who performed the chün-yao service. Only in cases where the household was unusually large and had several adult males might the incidence of the li-chia and chün-yao services be distinguished.[9] In the course of time, therefore, the distinction between the li-chia and chün-yao services became more indistinct and they tended to merge.

Moreover, the complexity of the early Ming fiscal structure, together with the practice by which the rich shifted their burden of labor services onto the poor, accounted for the impending collapse of the original fiscal system toward the end of the fifteenth century. Starting from the early sixteenth century, a long series of tax and corvée reforms took place in the country, mostly on a local scale. These reforms usually began with a partial merger of the labor services and then commuted them to money payment. The principle of commutation of labor services was indeed an old one, for the early chün-yao service was further divided into two kinds: the *li-ch'ai*, or services that had to be performed in person, and the *yin-ch'ai*, or services paid in cash. The former, which consisted mainly of such duties as checking the amount of tax money and measuring tax grain for the local government, frequently incurred heavy losses to the individual performing them, partly because of the differences in weights and measures and partly because of the exactions of local government underlings. The li-ch'ai was therefore usually assigned to men of substantial means, while the yin-ch'ai was generally assigned to the poor. Personal attendance to such duties being unpleasant, the rich naturally hired others to do them in their stead. This was already a kind of unofficial commutation which before too long became officially recognized.

By the sixteenth century the process of commutation was greatly facilitated by the influx of silver from European traders. The degree to which labor services were merged and commuted differed from place to place, but as the sixteenth century unfolded, the burden of labor services gradually shifted from the household to the adult male. The date of the merger of the li-chia and chün-yao services also varied from place to place; generally speaking the process took place earlier in the southeastern provinces, where the influence of a rising money economy was most noticeable. The labor services must have been merged sufficiently early to enable the editor of the 1585 edition of a Kwangtung local history to say "our reigning dynasty has always assessed the ting rather than the household." [10] Yangchou, the great trading center at the junction of the Grand Canal and the Yangtze, had practically discontinued household and mouth registration and had assessed all labor services to the ting since the third quarter of the sixteenth century.[11]

Had the fiscal reforms stopped at the merger of the two groups of labor services, the task of investigating the incidence of the ting tax would be much simpler. But, simultaneously with the amalgamation and commutation of labor services, in many localities the commuted labor services were being partially or substantially merged with the land tax. In a part of modern Shao-hsing county in Chekiang, for instance, 13 mou of first-grade land, 15 mou of second-grade land, and 100 mou of hilly land were converted outright into one ting.[12] On the other hand, in Wu-chin county in Kiangsu one officially registered ting was converted to 2 mou of land. The land had so overshadowed the ting as a contributor to the combined local tax payment that from 1535 a certain amount of land, rather than 110 households, was reckoned for fiscal purpose as a li.[13] In Ch'ang-shan county in southwestern Chekiang the common practice was to convert two officially registered ting into one *shih* (Chinese bushel) of land tax.[14] Whatever the methods of computing, one thing is clear: the incidence of labor services was at least partially shifted to the land or the land tax.

Specific information on the apportioning of the labor-service payment between the land and the ting is available for a number of localities. For general illustration the exact portions of the labor-service payment borne by the land and the ting in seven lower-Yangtze counties during the early seventeenth century are computed in Table 9. As the respective portions of the labor-service payment borne by the ting and the land became conventionalized, the ting quota also tended to remain more or less fixed, whatever the actual changes in the adult-male population. The danger of using the ting figure for the reconstruction of historical population will

Table 9. Percentages of labor-service payment borne by land and ting.

County	Percentage of corvée payment borne by the ting	Percentage of corvée payment borne by the land
Wu-hsien	34.5	65.5
Ch'ang-chou	27.6	72.4
Wu-chiang	23.2	76.8
Ch'ang-shu	11.6	88.4
K'un-shan	12.0	88.0
Chia-ting	14.8	85.2
T'ai-ts'ang	7.4	92.6

Source: Detailed figures and provisions are given in Ku Yen-wu, *THCKLPS*, 7.58a–61b. Because of numerous decimal digits in the figures my rough computations are only approximately accurate.

become most evident after an examination of the following discussion by an early seventeenth-century Chekiang local history:

> Evasions and omissions in population registration are common throughout the country and by no means confined to the southeast. Yet there are reasons why such fraudulent practices are most serious in the southeast. During the early years of the dynasty . . . labor services were apportioned according to the ting, the assessment of which had to be very careful. Later on [the ting assessment] was no longer based exclusively on the ting but partially on the land tax. As the southeast has been further developed, the apportioning of the labor services has come to depend more and more on the land, with the result that in the compilation of the Yellow Registers a certain amount of land, rather than an actual number of households, constitutes a li. In the north there is still actual population registration, because labor services are assessed to the ting. South of the Yangtze there is no longer actual population registration, as labor services are largely borne by the land.[15]

In some exceptional areas in the south the official ting figures had nothing whatever to do with the adult-male population, because the incidence of the ting tax had been completely shifted to the land before the turn of the sixteenth century.[16]

The movement for merging the labor services into the land tax, commonly called the *i-t'iao-pien*, or the "single-whip" system, took place in north China somewhat later than in the south. Since the land of north China was less fruitful than the soil of south China, the ting in northern provinces loomed large in the fiscal system. Indeed, during the second half of the sixteenth century, when the single-whip system was in full swing in south China, northern gentry offered considerable resistance to this fiscal reform, for their property was mainly land.[17] The regional difference is well explained by a late Ming edition of the history of Ssu-chou in the Huai river valley:

> The population of our locality is entered into the Yellow Registers, yet still we have to make periodic ting assessments. This is because the land south of the Yangtze is so fruitful that almost all labor services are charged to the land tax, while the ting tax plays a very minor role. There, the ting quota is based on those who are actually listed in the Yellow Registers. Probably only one out of every ten adult males is officially registered as ting. On the contrary, in regions north of the Yangtze the land is less fruitful. The common practice has been to apportion the labor-service taxes to the ting, only occasionally to the land tax. Ting as a rule bear two-thirds of the service taxes. Owing to the usually heavy service taxes, the number of adult males being registered as ting has to be large and the households have to be assessed carefully.[18]

However, neither the relative unfruitfulness of the land nor the resistance of the gentry could long delay the adoption of the single-whip system in north China. In the first place, the consolidation of taxes and services into a single payment was administratively such a great convenience that many local officials were in favor of it. In the second place, in many northern localities the scandal-ridden old system of household and ting assessments desperately needed thorough overhauling. Thirdly, the logic of such fiscal innovation, which would free the poor and landless from the usually heavy burden of ting tax, was irresistible. For these reasons, some southerners who served as provincial and local officials in north China paved the road for the single-whip system, which began to make headway in the northern provinces in the late sixteenth century.[19] The portion of labor-service tax shifted to the land was generally much smaller than in the south,[20] although by the beginning of the seventeenth century some northern counties had brought about a complete merger of the ting and land taxes.[21] According to a well-informed conservative member of the scholar-gentry of Honan, during the early seventeenth century the partial merger of ting and land had become fairly common throughout north China, with the exception of Chihli, which alone observed the time-honored separate assessments of ting and land.[22] After the single-whip system had struck root in north China, the discrepancy between the official ting quota and the actual adult-male population was bound to increase. In fact, in a twelve-article proclamation issued in 1606 to explain the new tax system, the magistrate of a Shantung county emphatically stated that prevention of any change in the ting quota was a requisite to the success of the single-whip system.[23]

All this, however, does not indicate that ting registration in north China was in process of becoming a mere formality, as it had definitely become in the south. Comments on the regional differences in the incidence of the ting tax, such as those cited, cannot easily be brushed aside. Also, various qualified early Ch'ing writers testified to the importance of household and ting assessments in north China, as the ting still bore the major portion of the labor-service tax. For example, Chang Yü-shu, concurrently grand secretary and president of the Board of Revenue, 1691–1711, stated that the ting and household assessments had been fairly strictly carried out in the northwest and "cases of evasion of the ting tax have been few." [24] Moreover, there must have been partial truth in the general principle, at least in north China, that the ting were recruited from the households.[25] No doubt this was true only in broad principle. It was probably truer in Chihli than in any other northern provinces. In Chihli, where the soil is relatively unfruitful and where the ting assessment had been more rigidly performed, there was a tangible, if not always accurate, relation between the number of households and the number of ting. For instance, Chihli reported 3,248,711 ting in 1724 and 3,071,975 households in 1753.[26] But elsewhere in northern and northwestern China the number of ting bore but faint relation to the number of households. The very fact that the rates of the ting tax were as high as four or five taels for each ting in some parts of Shansi and over eight or nine taels in some counties in Kansu, as compared with a fraction of a tael in most provinces, indicates that not every household in those districts was assigned a ting.[27] Instead of the ting payment being shared more or less equally among the households, the bulk of the tax was borne by rich merchant families, for which Shansi and parts of Kansu were famous during the late Ming and early Ch'ing periods. Even in Shansi, where the ting payment was far more important than the land tax, there were counties in which both had become completely merged since the middle of the seventeenth century.[28] On the other hand, in the Huai River area where, despite the lack of rich merchants and a fruitful soil, the local governments were in urgent fiscal need, the number of ting far exceeded that of the households. Cases had been known in which a county of some 10,000 households was assigned as many as 30,000 ting.[29] Even in north China as a whole, therefore, there is danger in using the official ting returns as an index to the actual numbers of households.

The method commonly used by modern scholars to reconstruct China's population data for the period 1651–1740 will be further discredited by a study of the ways in which the original Ch'ing ting quotas for various localities were fixed and subsequent assessments made. Like the early Ch'ing local quotas for the land tax, the "original" Ch'ing quotas for the ting tax were based on late Ming ting returns. Wherever possible, provincial and local authorities abided by the late Ming quotas. In localities where the effect of war and emigration was tangible, they made reductions. The following are a few of the rather numerous examples found in local histories. Ma-ch'eng county in northeastern Hupei had 10,605 registered households in 1619. At the opening of the Ch'ing period the number of households was registered at 10,704 and the ting quota was fixed at 10,605.[30] Tang-yang county in western Hupei had 2,030 registered ting in 1642. Six years later, after the change of dynasty, the Ch'ing "original" quota was fixed at 2,030, although it allowed a temporary exemption of "1,884 ting and 4/10 of a bushel of ting tax." [31] This indicates that in this locality the ting service had been converted to payment in kind. Lo-ch'üan county in north-central Shensi, after deducting those who had emigrated and those legally exempted from the late Ming quota of 2,010, fixed its "original" quota for the new dynasty at 1,642.[32] Ch'ang-sha prefecture of Hunan likewise deducted 34,062 from the quota of 1578 and fixed its "original" quota at 114,960.[33] In a large number of localities unaffected by the wars of the mid-seventeenth century, such as Shanghai,[34] the "original" Ch'ing quotas were exactly the same as those of the late Ming period.

If these fixed original Ch'ing quotas failed to represent the actual adult-male population, the subsequent quinquennial ting assessments by no means reflected the growth of population during early Ch'ing times. It was during the quinquennial ting assessment that difficulties sometimes arose. According to government regulations, a ting who had reached the age of 60 was to be deleted from the official register and a member of his family, or his next of kin, or some one from a large and substantial household in the neighborhood registered to replace him. Some local officials doubtless attempted to increase substantially the ting quotas of their localities, especially when the government made the enlargement of the ting quota a ground for promotion. But popular resistance to such measures would have been strong, supported by custom and traditional ethics. The reported increment in the local ting population probably failed as a rule to reflect the extent of the actual growth of the local population. It was common practice for local authorities to report a small increment, ranging anywhere from three or four to one or two hundred, depending on the extent and size of the population of the locality.

A number of records show that during the late seventeenth and in the eighteenth century local officials were generally on the people's side. At the quinquennial ting assessment they did their utmost to prevent an enlargement of the existing quotas. In fact, to defend or to curtail the existing ting and land-tax quota was one way to win affection and fame in a locality. The biographies and epitaphs of those "model" early Ch'ing officials almost invariably contain brief mention of achievements of this kind. Lu Lung-ch'i (1630–1693), to name one of those kindly local officials, once petitioned the provincial authorities:

> At the time of ting assessment, no attempt should be made to enlarge the original quota . . . . In Chihli province the number of ting at the present time is already large enough. We should be content with keeping the quota as it is, but should never attempt to exceed it . . . . In brief, [if the superior authorities] could be somewhat more lenient toward the local magistrates, they would pass on the benefit to the poor people accordingly.[35]

This kind of political ethics, although it has become a headache to modern demographers, accounted in part for that rare period of contentment in the history of the Chinese people.

The movement to merge the ting tax into the land tax went on apace during the first eighty years of the Ch'ing period. The rationalization of the tax system, once started, could brook no limitation until the whole fiscal structure was thoroughly overhauled. Furthermore, if it was desirable to transfer the incidence of the labor services which had originally fallen on the household and the adult male so that they impinged on the latter only, it was even more reasonable and just to shift the incidence of the combined labor services to the land or the land tax. In the last analysis the male adult was assessed according to his property, and his property, in a predominantly agricultural society, was almost invariably land. The ultimate logic of this long series of multilateral fiscal reforms called the single-whip system, was nowhere better and more succinctly stated than by Li Fu, governor general of Chihli, who said in 1726: "The world has destitute ting but no destitute [cultivated] land."[36] Before the long K'ang-hsi reign ended in 1722, the merger of the ting payment into the land tax had been practically completed on a provincial scale in Kwangtung and Szechwan. In 47 of the 77 counties of Chekiang the ting payment had either been merged into the land tax or apportioned according to the acreage of the taxed land.[37] In 1696 the magistrate of Lo-t'ien county in Hupei, in a petition urging tax reform, stated that not only his neighboring counties but "half the empire" had virtually completed the merger of the ting and land taxes.[38] During the first half of the reign of the Yung-cheng Emperor (1723–1736) one province after another followed the example of Kwangtung and Szechwan and the merger of the ting and land taxes was completed in some seven-tenths of the counties in the whole country.[39] Thus ting in most parts of the country had become completely divorced from the adult-male population.

Ever since the sixteenth century, perhaps even earlier, the ting had begun to supersede the household and mouth (*hu-k'ou*) as the hard core of the registered figures. Although the national ting totals during the late Ming period cannot be exactly known because of the continuance of the time-honored household and mouth registration and the existence of very few late Ming *Fu-i ch'üan-*

shu, it is safe to say that the ting figures represented substantially less than one-half of the registered population. The Ch'ing government recognized the obsolescence of the household and mouth registration, for which the ting assessment was substituted. When the ting returns were collected by the Ch'ing government for the first time on a national scale in 1651, the registered ting amounted, in round numbers, to 10,600,000. As the government became more stable and peace was restored to most parts of the country, their number rose to 19,000,000 by 1660. Owing to the campaigns against the southern feudatories and against the Ming loyalists in Formosa, the ting totals sharply dropped in the 1670's. It was not until 1684, after the southern rebellion had been put down and Formosa conquered, that the total number exceeded 20,000,000. Thenceforth the number of ting slowly increased until it reached 26,400,000 in 1734.[40] The ting returns of this period, as has been explained, were neither population returns, nor returns of households, nor returns of tax-paying adult males. They were simply tax-paying units. For this reason, it is not to be wondered at that the ting return of a locality may read "15,232.68949664 ting."[41] It may also read "Increase of ting for the year: 0.7819 bushel."[42] An extreme case may give a decimal fraction of as many as fifteen digits.[43] The last quarter of the seventeenth and the first half of the eighteenth century was a unique period of peace, prosperity, and government retrenchment. When the government could manage comfortably on its revenues from the taxes, there was no need to increase the ting quota. Some modern writers have been greatly puzzled by the fact that thirteen of the so-called household or adult-male returns prior to 1712 in the official Ch'ing records are exact repetitions of those for the preceding year. They have even complained about the absurdity of the early Ch'ing population returns. The ting returns of the period 1651–1740 were never intended to indicate the population, and the Manchu government deserves kindlier treatment.

CHAPTER III

Population Data, 1741-1775

Some writers on Ch'ing population have thought that China's data since 1741 covered the entire population.[1] Rockhill in particular thinks highly of the registration for the year 1741, for the government was then "in strong and intelligent hands, its mandate executed with more faithfulness and precision than at any subsequent period, and the Empire enjoying perfect peace." He is, however, so puzzled by the sharp fluctuations in the population figures since that date that he accepts only the returns of 1741 and rejects those for all subsequent years.[2] His willingness to accept the 1741 figure and his readiness to reject all subsequent ones is in itself puzzling, and must be accounted for by his inability to understand the lack of uniformity in the methods of population registration during the period 1741–1775. Yet official documents, memorials, and particularly local histories, show the complexity of the system of population registration during this period.

The decree of 1711, which permanently froze the amount of the national ting payment, greatly reduced the importance of the quinquennial ting assessment. The obsolescence of the ting assessment became even more evident after the completion of the merger of the ting and land taxes, on almost a national scale, during the late 1720's and early 1730's. In view of the changes in the fiscal system, Li Fu, governor general of Chihli, suggested in a memorial in 1726 that the quinquennial ting assessment be abolished forever in Chihli, and that the male population above fifteen years of age be registered through the instrumentality of *pao-chia*, the traditional small self-governing units into which the people were organized for the purpose of local policing.[3] Although his suggestion was not immediately approved by the Yung-cheng Emperor because many other provinces were still in the process of completing the merger of the ting and land taxes, his memorial gained wide currency and his ideas were generally considered reasonable and feasible. After the tax merger was completed in most areas in the early 1730's, the time-honored ting assessment was temporarily suspended. Then, in the early winter of 1740, the young and energetic Ch'ien-lung Emperor (1736–1795) decided it was necessary to overhaul the system of population registration. His decree read:

In spite of the quinquennial [ting] assessment, provincial officials have hitherto been bound, in matters pertaining to the increase and decline [of the number of ting], by administrative formality. They have scarcely realized that this matter is indeed the foundation of good government. From now on the provincial officials should, in the eleventh month of each year, send in detailed reports as to the changes in the numbers of households and mouths and the amounts of grain stored in the government granaries within their respective jurisdiction. This will provide me with data for constant reference, in order to plan carefully in advance and to prepare for the eventuality of flood, drought, and famine. It is ordered that after the next quinquennial assessment is made the provinces should report the numbers of households and mouths. As to how detailed provisions will be made so as to be uniformly enforced in the provinces, the Board of Revenue should memorialize after due consideration.[4]

After brief consideration the Board of Revenue memorialized that "from now on the quinquennial [ting] assessment be carried out as of old and that the provincial officials be ordered, after the completion of the [ting] assessment in 1741, to enumerate from household to household all the ting of every prefecture and county, together with the adults and children of every household. There must not be any omissions and evasions."[5]

Had the recommendation of the Board of Revenue been adopted without qualification, China would have had a census system since 1741, however crude its methods. But the Board's recommendation, which constituted such a drastic departure from old practices, was subjected to further deliberate study by a group of ranking ministers who reached the following conclusion:

Minute of the Conference of Grand Secretaries and Nine Ministers

Censor Su Lin-po has memorialized that the recommendation of the Board of Revenue that the actual numbers of the population be annually enumerated is a suggestion which, though useful in ascertaining the actual increase of population, can hardly be feasible in real practice. This is because hitherto the principle of the quinquennial [ting] assessment has been based on the households and the distribution of public grain for famine relief has been based on on-the-spot surveys and not on the ting registers. Moreover, people are scattered all over the countryside. If they should be ordered directly to report to the local authorities, they could not comply. If the local officials should be ordered personally to enumerate, they could not bear [such an administrative burden]. Probably they would entrust such a task to their office underlings. Besides, some merchants are without fixed residence, as are the transients and those in services and on public errands. In the areas inhabited by the aborigines enumeration is hardly feasible. It is requested therefore that [the annual population enumeration] be stopped. But it is important that the amounts of grain stored in government granaries be strictly checked, in order to prepare for various eventualities. [In the light of Su's memorial], we believe that the population of the provinces is truly numerous and the annual enumeration would be troublesome. It is recommended therefore that after the completion of the [ting] assessment in 1741, when the numbers of households and mouths are ascertained, the provincial officials need only to report, annually in the eleventh month, the amounts of grain and the total numbers of households and mouths of their respective provinces, excluding the *liu-yü* (i.e., those who had taken up residence in localities which were not their officially registered ancestral counties) and the aborigines. There is no need to make a door-to-door enumeration. Approved [by the emperor].[6]

The importance of this recommendation of the ranking ministers, which has not been available to previous writers on Ch'ing population, is manifold. First, the ministers strongly rejected the system of direct annual official enumeration of the population proposed by the Board of Revenue. Secondly, their recommendation reveals the lack of even rudimentary demographic interest on the part of the high officials and the strength of the conception that "population" was of interest to the state primarily because of its fiscal significance. Thirdly, although in common usage the expression "households and mouths" meant the population in general, in the bureaucratic language of the middle eighteenth century it carried more fiscal than demographic connotation. The lack of clear definition of "households and mouths," coupled with the emphatic rejection of the idea of door-to-door enumeration, was bound to bring considerable confusion into the new system of population registration. In any case, the ranking ministers' lack of enthusiasm for the new system was known to, and shared by, some of the provincial officials.

The recommendation of the ranking ministers was embodied in the half-hearted directive of the Board of Revenue, which read:

> It would be too troublesome if the annual report on the numbers of the population were compiled according to the method of [ting] assessment. Since various counties of the provinces have already had their pao-chia door placards, the numbers of the natives and the liu-yü can readily be checked from such [pao-chia] registers. The actual number of the civil population can be obtained by deducting the liu-yü [from the pao-chia registers]. The provincial officials are ordered to report, annually in the eleventh month, the numbers of households and mouths and the amounts of grain stored. The regions inhabited by the vassal peoples and aborigines, where no ting assessment has ever been made, are to be excluded.[7]

Explicit and implied in this administrative order are two factors deserving special attention. First, the old method of ting assessment, whatever its drawbacks, had been based on the principle of direct assessment, carried out by the county magistrates and assisted by the li-chia personnel. Now, in 1741, the task of registering the entire population of a locality was transferred solely to the pao-chia machinery, because the li-chia system had been in decay since the late Ming period. Since the law did not require the magistrate personally to supervise the population registration, the new system became one of indirect registration, based entirely on the returns of the unpaid agents of the local government. Secondly, the pao-chia system had always been functionally concerned almost exclusively with local policing, only accidentally with population registration. Only at times of local disturbance had it ever been entrusted with the task of making an on-the-spot check of the truculent elements in the local population. Although there had been a long series of laws since the founding of the Ch'ing dynasty attempting to set up the pao-chia machinery in the whole country, it is extremely doubtful that it had ever been organized on a truly national scale as the li-chia system had been during the early Ming period.[8] In certain areas where the pao-chia system had not been functioning consistently, population registration was at once handicapped by a lack of experience and proper machinery. Statutes, memorials, and private testimonials indicate that the pao-chia system, whose primary function during the period 1741–1775 was still the maintenance of local peace, did not work uniformly in the provinces, least of all in performing its secondary task of population registration.

In 1736, for example, when the ting assessment was temporarily suspended and the pao-chia population enumeration was already being discussed, the government approved the request of the governor of Kiangsi that the pao-chia system should be set up on a household-to-household basis only in cities and populous towns, because the inhabitants of small scattered farmsteads and isolated villages could not easily be organized. Since this kind of difficulty was by no means confined to one province, the government took the occasion to confer optional power on provincial authorities in matters pertaining to the setting up of pao-chia apparatus in backward rural areas.[9] In 1742 the governor of Kwangsi memorialized that the pao-chia regulations were not strictly observed in many parts of the province. When the pao-chia registers were compiled for cities and suburbs, they lacked relevant breakdowns, with the result that they were almost useless.[10] In 1757 a censor of the Kiangsi circuit testified in a memorial that the pao-chia system had become a mere formality and its personnel was often made up of people without fixed occupation. The result was that although the pao-chia registers might be available, the population returns would not always be authentic.[11] In 1758 the acting governor of Kiangsi reported that the military rank and file had for years evaded the enumeration and assessment.[12] Wang Hui-tsu (1731–1807), the author of a famous guide for local officials, recalled in 1793 that during his long service as a staff member to various local officials of Kiangsu and Chekiang between 1752 and the early 1770's he had not been successful in influencing his patrons to enforce the pao-chia regulations energetically.[13]

The 1758 memorial of Ch'en Hung-mou, one of the famous provincial officials of the middle eighteenth century, throws interesting light on the working of the pao-chia population registration in Kiangsu. He suggested that as long as the pao-chia machinery helped to maintain local peace it should not be too concerned with an accurate enumeration of the entire local population. In his opinion the women and the children of both sexes, who played practically no part in fiscal administration and in the maintenance of local order, should be excluded from the pao-chia door placards.[14] It is unlikely, however, that the female and juvenile population of Kiangsu thenceforth ceased entirely to be registered. The complete lack of reference to Ch'en's 1758 memorial in the court annals and the collected statutes of the Board of Revenue indicates that his suggestion was not adopted by the central government, and that provincial authorities had to abide by the Board's 1740 directives.[15] Yet Ch'en's memorial reflects the emphasis that was placed on the pao-chia system and the possibility that portions of the female and juvenile population may have been overlooked in the pao-chia registration. Yüan Mei (1716–1798), the foremost poet of the time, in a long letter to Ch'en said that it was common knowledge that the pao-chia registration rarely covered the entire population and that in time of famine the number of people in need of public relief often exceeded the totals of the pao-chia registers. His advice was that the best government was the least government and that the people should not be relentlessly regimented into the pao-chia machinery.[16]

All this suggests that during the period 1741–1775 the provinces and counties were unable to harness the pao-chia machinery completely for the purpose of population registration. The system of pao-chia registration seems to have continued to work inefficiently. Although for the period 1741–1775 local histories generally contain only the data of the fiscal population because of the continuance of the ting assessment until 1772, the few that do give breakdowns for the entire local population are of considerable interest. The returns of Hsiao-shan county in Chekiang as shown in Table 10 may serve as an example. When the breakdowns are computed, the following results appear: (1) average size of the household: 4.5 persons; (2) sex ratio (number of males per 100 females): 132; (3) sex ratio among children under sixteen: 128; and (4) percentage of children in total populations: 44.5. On a prefectural scale we may cite the returns of Yung-p'ing in Chihli. In 1773 the seven counties of the prefecture altogether had a population of 1,375,645, not including salt boilers and their families who numbered 56,386. The following results are obtained: (1) average size of the house-

Table 10. Population of Hsiao-shan county.

	Fiscal Population in 1746 Urban	Rural
1711 quota	9,909.5	23,768.5
Increase since 1711	721	1,753
Total	10,630.5	25,521.5
	Total number of ting: 36,152	
	Actual Population in 1749 (46,461 households) Males	**Females**
Adults	66,807	49,302
Children	52,323	40,911
Total	119,130	90,213
	Entire populations: 209,343	

Source: *Hsiao-shan HC* (1751 ed.), 9.3b–46.

hold: 5.5 persons; (2) sex ratio (all ages): 121; (3) sex ratio (children): 115; and (4) percentage of children in total population: 34.7.[17] These ratios and percentages appear to be more reasonable than those of the late Ming period. A few other local histories which yield relevant information on the period 1741–1775 also indicate a degree of continuity in population registration.[18]

The problem is, however, that in a country as vast and varied as China exceptions were as important as the general rule, and local under-enumerations contributed to the errors in the annual national totals. The following are a few examples of local under-registration and other irregularities. Lin-shui county in Szechwan reported in 1750 a population of 7,688 households and 17,125 mouths. Two years later the registered population suddenly rose to 26,527 households and 58,910 mouths. The editor of the county history thoughtfully remarked that the returns of 1752 "differ drastically from those of 1750." [19] Modern readers may still doubt that the 1752 returns actually covered the entire population, as each household averaged only 2.2 persons. It seems likely that children were either entirely unreported or seriously under-reported.

Ch'i-shui county in Hupei represented a different kind of faulty registration, by duplicating earlier figures. This county reported in 1749 a population of 13,159 households and 118,771 mouths, an exact repetition of the returns for 1542. As the editor pointed out, the registered population of the county varied but very slightly during these 205 years. But in 1752 the new magistrate personally gave out door placards and made a thorough canvass, an example which was followed by subsequent magistrates. From 1752 on, the population returns show a reasonable continuity until they reached 51,856 households and 265,895 "male and female adults and children" in 1793.[20] Similarly, though without clear explanation, the population of P'ing-ting county in Shansi jumped from 2,014 households and 15,481 mouths in 1736 to 23,475 households and 102,749 mouths in 1751. It seems that only after 1751 did this county begin to have more reliable population returns.[21] Cases like these were therefore responsible for sharp increases in the annual national totals for certain years.

The population returns of Hsiang-yang prefecture in northern Hupei, which represented error of a more serious kind, are shown in Table 11. At first glance the returns appear to be remarkably

Table 11. Population Returns of Hsiang-yang prefecture in 1756.

	Ordinary households[a] (min-hu)	Military households[b] (chun-hu)
Ting who pay the ting tax	26,134	202
Ting, registered since 1712, not liable to the ting tax	6,371	185
Native male and female adults not classified as ting	264,670	9,961
Native male and female children	134,207	7,371
Total	431,382	17,719

Source: *Hsiang-yang FC* (1760 ed.), 12.1b–2a.
[a] Total number of ordinary households: 106,334.
[b] Total number of military households: 6,973.

complete and seem to indicate coverage of the entire population. A comparison of the 1756 returns with those of 1812, however, reveals that in less than two generations the prefectural population had jumped to 2,122,923, an improbable rate of growth.[22] This was because the law and directives of 1740 excluded the liu-yü from the pao-chia population registration, while after 1776 they were included. Northern and western Hupei, southern Shensi, and the whole province of Szechwan throughout the eighteenth century attracted the largest number of immigrants, mostly pioneer maize farmers of the mountainous areas. Prior to 1776 the number of people in the country in general and in the newly developed areas in particular who had been excluded from the pao-chia registration on this account must have been very considerable.

Errors of yet another kind occurred because some counties still compiled their population returns in accordance with older regulations that emphasized the fiscal population. Until its final abolition in 1772 the quinquennial ting assessment was carried out throughout the provinces; although the pao-chia population registration had been instituted, it was not yet uniformly enforced. An example, Lo-ch'uan county in north-central Shensi is shown by Table 12. The 1660 and 1711 returns are obviously those for

Table 12. Population returns of Lo-ch'uan county.

Year	Households	Male		Female		Total registered population
		Adults	Children	Adults	Children	
1660	788	....	....	....	....	1,644
1711	....	....	....	....	....	1,642
1735	....	11,414	6,243	9,178	4,257	31,092
1751	....	....	....	....	....	33,673
1756	....	....	....	....	....	33,876
1786	18,605	....	....	....	....	90,293
1802	19,233	32,820	19,750	26,290	15,154	94,014

Source: *Lo-ch'uan HC* (1944 ed.), 14, section on population.

the registered *ting*. Concerning the figures for 1735, 1751, and 1756 the editor commented that they were based on the Yellow Registers, as distinguished from the 1786 returns, which were based on the pao-chia registers and included, in addition to the natives, all the liu-yü, merchants from other places, officers and soldiers, Buddhist and Taoist monks and nuns, farmhands and employees, and men without fixed residence. We do not know exactly which portions of the population were covered by the returns between 1736 and 1756.[23] Although the liu-yü must have been excluded, this county was too poor and too far to the north to have been within the area of heavy immigration. What we do know is that since the figures were based on the local Yellow Registers, they covered the fiscal population in the broad sense, and the difference between the population returns before and after 1776 is at once qualitative and quantitative.

The worst cases of under-registration during the period 1741–1775 were found in certain counties of Hupei province. It is doubtful that they had the pao-chia registers at all; if so, they must have been compiled in ways similar to those for the ting registers. Since 1712 it had become customary for counties to report a very small annual increment in the number of ting. By 1775 it had come to the emperor's knowledge that one of the Hupei counties had reported an increase of eight persons annually for a number of consecutive years. Some others had annually reported an increase of from five to twenty persons.[24] It was not until systematic evasions were exposed that the emperor decided to reform thoroughly the system of pao-chia registration throughout the country.

These illustrations make it abundantly clear that although the population returns between 1741 and 1775 in theory represented the total population, in reality the nation's population was under-reported. The population of certain provinces, such as Szechwan, Hupei, Hunan, and Kwangtung, was very much under-reported. Owing to a more strict enforcement of the pao-chia population registration and because of the inclusion of the liu-yü in the report, the population of Hupei rose from 8,532,187 in 1771 to 14,815,128 in 1776. Likewise the population of Hunan and Szechwan respectively jumped from 9,082,732 and 3,068,199 in 1771 to 14,987,777 and 7,789,791 in 1776.[25] It was common knowledge that throughout the eighteenth century large numbers of people from Kiangsi and other parts of the southeast migrated first to Hupei and Hunan and then to Szechwan.[26] The hundred per cent increase in the registered population of Kwangtung in these few years is traceable to a similar reason and also to the administrative laxity for which the province was notorious. Small wonder, then, that the average rate of annual growth between 1771 and 1776 was an impossible 5 per cent. The basic reason for the confusion in the system of registration during the period 1741–1775 was the ambiguity of the law and directives of 1740 and the central government's lack of persistent interest in securing accurate population data. The latter may best be shown by the fact that, although the counties in general had their population registers, as distinguished from the ting registers, the local population registers were not required to be handed to the Board of Revenue until after 1776. Among the existing 15,000 volumes of the Yellow Registers originally kept in the Ch'ing cabinet storehouse none of the local population registers dates earlier than 1781.[27]

Historians have yet to discover documentary evidence definitely proving that provincial and local authorities purposely magnified the population figures after 1741, to satisfy the Ch'ien-lung Emperor.[28] This intelligent ruler had, in fact, been keenly concerned about the problem of a limited supply of land and rapidly increasing population ever since his accession in 1736. In view of the confusion and lack of uniformity in the system of pao-chia population registration, the data of the period 1741–1775 must be used with great caution and should not be taken as returns for the entire population. It has been a grave methodological mistake in previous studies of China's historical population to treat the data before and after 1775 as if they were comparable.

Defective as the population data of 1741–1775 are, they are not entirely useless to modern students in their attempt to reconstruct the population of eighteenth-century China, if they are properly understood. They provide a set of figures that are considerably below the truth. In the returns of 1775 and 1776, the difference is almost 20 per cent. Therefore, since the 1776 national total is likely to be still somewhat too low because of the inevitable confusion and minor omissions at the beginning of a new era of population registration,[29] we may assume that the actual population during the period 1741–1775 was at least 20 per cent more than the officially registered population. It must not be assumed, however, that the difference throughout these thirty-four years was a uniform one, nor can this approximate percentage difference serve as a solid foundation for reconstruction.

CHAPTER IV

Population Data, 1776-1850

I

The quinquennial ting assessment was finally abolished in 1772. This in itself reflected a change in the conception of population registration. The circumstances which led to the Ch'ien-lung Emperor's taking vigorous steps to ascertain the actual number of the entire population are interesting. In 1775 crop failure in eastern Hupei affected between 60 and 80 per cent of the people of nineteen counties. But those who were listed as being in need of relief outnumbered the "entire population" of those counties by over 100,000. After thorough investigation it was discovered that some of these counties had hitherto reported a small annual increment arbitrarily. The emperor was thus awakened to the fact that the system of pao-chia population registration had not been strictly enforced throughout the country. A series of decrees was therefore issued, the provincial authorities were sharply reprimanded, and the local pao-chia machinery fully mobilized to ascertain the true population of the empire. This time there was no ambiguity in the directives and the new regulations were to be uniformly and energetically enforced.[1] Realizing the complexity of the problem, the emperor instructed the provincial authorities to perform the task conscientiously and not to be too concerned with the usual deadline. Chihli province therefore did not send in its first detailed population returns until 1778.

In contrast to the halfhearted nature of the directive of 1740, the 1775 order of the Board of Revenue was clear and emphatic. It read:

> It is ordered that the provincial officials instruct their subordinate local officials to register the actual numbers of the population [within their respective jurisdiction]. The actual numbers of the population, together with the amounts of grain, should be compiled and sent in annually in the eleventh month. Should any falsification of the popula-

tion returns be discovered, the provincial officials shall be investigated and punished. When the county magistrates compile their population returns, their totals should be based on those of the pao-chia door placards and registers. They do not have to check the names of the inhabitants from household to household. . . . The quinquennial assessment and the compilation of the ting registers are abolished forever.[2]

The only technical drawback was the lack of a system of double-checking; the pao-chia machinery alone was responsible for the registration of the local population. Before analyzing the population data of the period 1776–1850 it is necessary to study the working of the pao-chia population registration system during these seventy years.

Despite the confusion in the pao-chia population registration during the period 1741–1775, the general political usefulness of the pao-chia system to the traditional state cannot be doubted. In an age of cheap government the pao-chia personnel served as the unpaid local police, helped to maintain peace and order and performed sundry duties that would otherwise have had to be carried out by regular local government personnel. The efficiency of the pao-chia system, however, varied with the current general political, economic, and social conditions and in proportion to the degree of pressure exerted upon it from above.[3] There was a twofold reason for the partial failure of the pao-chia population registration between 1741 and 1775. First, population registration was not made one of the primary functions of the system. Secondly, sustaining interest and pressure in the part of the central government, both indispensable to the success of the plan, were lacking.

In the winter of 1775, however, for the first time in the Ch'ing period, the registration of the local population was made one of the primary functions of the pao-chia system. The provincial authorities were repeatedly reminded of the consequences of falsifying population returns. It is difficult to imagine that, at a time when Ch'ien-lung's prestige and authority had reached their apex, he would fail to harness the provincial and local officials to work for accurate population returns which he now earnestly desired. During the second half of his long reign, despite its outward pomp and glory, the economic condition of the nation was beginning to deteriorate. It was precisely the deteriorating economy and signs of increasing unrest that made it necessary for his successor, the

Chia-ch'ing Emperor (1796–1820), to enforce the pao-chia regulations more strictly than ever before. Contemporaries as qualified as Chu Yün-chin (an experienced provincial official and author of the best concise geographic treatise of Honan province) and Wang Ch'ing-yün (vice-president of the Board of Revenue in 1851–1853 and author of the best financial history of the Ch'ing period) testified that the system of pao-chia population registration had been working satisfactorily by and large.[4] It is significant to note that Father M. Huc, who traveled extensively in China in the 1840's and was aware of the skepticism of some European authors about Chinese population returns, concluded that "the Chinese statistics are, nevertheless, kept with care." [5]

Moreover, the technique of pao-chia population registration was steadily improving. One notable feature was that the once effective method of compiling the *hsün-huan-ts'e*, or rotating registers, was reintroduced first regionally and then nationally. Yeh Pei-sun, financial commissioner of Hunan, revived it in 1781. The method, briefly stated, was first to make out duplicate pao-chia registers to be kept separately by the pao-chia personnel and by the local government. Once in every three months or in any reasonable length of time the pao-chia personnel, after making all revisions in accordance with actual changes in the population of their pao, exchanged their copy with the unrevised one of the local government. By the time the second register was brought back to the local government for exchange, county officials could check the results of the first period against those of the second. Further to check the accuracy of the pao-chia returns the local government might carry out verification of the books from time to time. The details of this method were submitted to the throne by Yeh's son in 1816, a year in which by imperial order pamphlets explaining this method were printed and circulated to county magistrates by various governors. The compilation of the *hsün-huan-ts'e* became a standard practice throughout the country.[6]

The forms of the pao-chia door placard still varied from place to place. Some emphasized only the number, age, and sex of the occupants of the household; some included brief information on occupation, property, and amount of tax payment. The general tendency was for the contents of the door placard to approach a bit closer to the questions asked in a census.[7] To harness the pao-chia headmen to work willingly on this thankless job, some thoughtful local officials would treat them with great courtesy by calling them elders and by feasting them on the occasion when the rotating registers were exchanged. To prevent omissions, special registers and door placards were prepared, in a number of counties at least, for the Buddhist and Taoist monks and nuns, inns and taverns, fishermen, beggars, and coolie carriers.[8] During the early years of the nineteenth century various methods and experiences in pao-chia organization were exchanged and propagated through repeated imperial exhortations and by officials who had made effective use of the pao-chia machinery. All these improved practices seemed to justify in part the high hopes entertained by the Emperor in 1775 that ascertaining the actual number of the population through the pao-chia machinery should not be too difficult.[9] S. Wells Williams, an American chargé d'affaires and a keen observer of pre-Taiping China, stated that the pao-chia system "was observed long before the Mongol conquest, and is followed at present; so that it is perhaps easier to take a census in China than in most European countries." [10] In any case, the system of pao-chia registration presumably worked much better during the period 1776–1850, when the central authority remained unimpaired, than it worked during the period of political decentralization following the Taiping Rebellion.

In spite of the principle that county magistrates did not have to supervise the population registration personally, some county officials were nevertheless instructed on special occasions to check from household to household the accuracy of the pao-chia returns. The famous etymologist and phonologist Tuan Yü-ts'ai, for instance, personally took part in the registration of the population while serving as acting magistrate of Fu-shun county in Szechwan in 1776.[11] Tsung Chi-ch'en (1788–1867), an able official, recalled how his father, while serving as magistrate in Hunan during the early nineteenth century, spent two years in making a door-to-door check of the pao-chia registers.[12]

Although the pao-chia population registration seems in the main to have been more faithfully executed during the period 1776–1850, regional under-enumeration in a country as vast and varied as China was still unavoidable. In the more backward parts of Kweichow, for example, Li Yen-chang (*chin-shih* of 1801, a conscientious official and expert on early-ripening rice) testified that during his personal supervision of population registration in En-ssu prefecture he discovered 1,421 villages hitherto unregistered. Consequently, the population of the prefecture, together with the subdistricts under the rule of tribal chieftains, was increased several times.[13] A selection from the 1812 edition of *Ta-Ch'ing i-t'ung chih*, the imperially compiled geographic treatise of the empire, reveals that in a number of prefectures of Yunnan, Kweichow, Szechwan, and Kwangsi, where the aborigines mingled with the Chinese, the local populations were either entirely omitted or seriously under-reported. While one of the most impressive local population returns is found in an 1845 edition of a Yunnan local history,[14] an early nineteenth-century governor general of Yunnan and Kweichow stated in a memorial that pao-chia registration in these provinces was as a whole rather unsatisfactory.[15]

In fact, there is good reason to believe that under-registration was most serious in the southwestern provinces, for the Board of Revenue directives of 1775 provided that the inhabitants of the national capital, all Manchus, the Chinese and Mongolian Bannermen and their families, diverse vassal peoples (such as Mongols in Inner and Outer Mongolia, Tibetans, and various tribesmen of Chinese Turkestan and Kokonor), and the aborigines of the southwest were to be excluded from the pao-chia registration. As the 1953 census-registration indicates, by far the most populous national minority groups are still in the southwest. The Chuang people alone, who concentrate in western Kwangsi, are now numbered at 6.6 millions. The various minority races of Yunnan are estimated by an atlas published in 1956 to be in the neighborhood of 7 millions.[16] The other large minority groups, such as the Miao, Yao, and Puyi, who inhabit the hilly districts of Kweichow, Kwangsi, Kwangtung, and western Hunan, also amount to several millions.[17]

It is not easy to estimate the extent of under-registration of the pre-1850 population due to the exclusion of national minorities. While the 1953 census-registration is suggestive, there have been significant changes in minority populations since the late eighteenth century. With the rapid growth of population Chinese migrant peasants systematically encroached upon the aboriginal districts in the southwest and the minority groups were pushed farther and farther into the hills and mountains. The Miao people of western Hunan, for example, were greatly reduced in number by the double onslaught of Chinese migrant peasants and repeated massacres by government forces, particularly during the late eighteenth and early nineteenth centuries. There can be little doubt that before 1800 the Miao population of western Hunan was much larger than the present estimated 320,000.[18] The Moslem population, reckoned as a minority in the 1953 census-registration, of many localities in Kansu and Shensi was nearly exterminated by Tso Tsung-t'ang's campaigns of the 1860's and 1870's.[19]

Outside the southwest, under-registration was fairly common in Kwangtung, where its distance from the national capital, the rugged individualism of its people, and a most lucrative maritime trade had contributed to corruption in the provincial and local administration. An official at the beginning of the nineteenth century observed that in Kwangtung the gentry often bribed local government underlings in order to evade the pao-chia registration.[20] More than local population data of other provinces of the period 1776–1850, population figures of some Kwangtung counties betray certain absurd features. The 1828 pao-chia return of Hsin-ning county in Kuang-chou prefecture, for example, yielded 128,863 males but only 68,109 females, with an improbable ratio of 189 males for every 100 females.[21] The 1786 and 1796 returns of Tung-kuan county consistently yielded a sex ratio of over 190 and the 1818 return omitted females entirely.[22] Other known cases of under-reporting were mostly connected with the rise of a restless or otherwise marginal and mobile section of the population, a sign that in itself may have indicated the growing pressure of population on the nation's land resources. As the plains and valleys of central and south China had been fully occupied in the course of the late seventeenth and early eighteenth centuries, there had been large-scale migration of people from the densely populated southeast to the inland Yangtze provinces, whence millions went further upstream along the Yangtze to Szechwan or trekked along the Yangtze tributaries to hills and mountains. Next to Szechwan the area that attracted the largest number of immigrants was the whole Han River drainage. Pioneering maize farmers lived in temporary shacks erected in the mountains and moved from one place to another every few years. Their proper registration presented considerable difficulty, which was sometimes aggravated by the absence of regular local government and pao-chia machinery.[23] Several new counties were created before and shortly after the turn of the eighteenth century and some unusually able officials were appointed. The same difficulty in registering the shack people also existed in a much smaller area along the Fukien-Kiangsi border.[24]

In some unusually depressed areas in the plains, like the parts of Anhwei and Kiangsi north of the Huai River where poor soil, natural calamities, and a growing population had created considerable vagabondage, even local officials as able and conscientious as the famous geographer Li Chao-lo (1769–1841) found it difficult to get all the local population properly registered.[25] Over the country as a whole, starting in 1796 revolts on an inter-provincial scale, like the uprisings of the White Lotus and other religious sects in the Hupei, Shensi, and Szechwan areas, disrupted pao-chia population registration in several provinces for a number of years.

Possibilities of exaggeration of population figures must also be briefly investigated. After the inclusion of liu-yü in the pao-chia population registration in 1776 there was a possibility of double counting merchants who carried on business in localities remote from their native counties. Double counting could be fairly extensive in areas like Hui-chou in southern Anhwei and in parts of Shansi, where long-distance merchants constituted a significant portion of the local population. Although there is no explanation of double counting in official documents and local histories prior to 1850, there is clear mention of it in some twentieth-century works. For example, Ningpo county, famous for its enterprising merchants and financiers in modern times, enumerated in 1932 a total of 700,481 "inhabitants," out of which 47,325 males and females had left the district and taken up residence elsewhere.[26] *The Economic Yearbook of Kwangtung Province* for 1941 likewise states that those who had migrated to Hong Kong, Macao, and inland southwestern provinces since the outbreak of the Sino-Japanese war in July 1937 were still retained in the provincial population registers.[27] For comparison some recent experiences are worthy of brief mention. After the completion of the 1953 census-registration the authorities are said to have carried out a sample check of the populations of 343 counties in 23 provinces, 5 cities, and one autonomous region covering 52,953,400 persons. The result they reported was said to show that duplications were at the very low rate of 1.39 per thousand and omissions 2.55 per thousand. The pre-1850 population counts were entrusted not to specially trained personnel but entirely to the unpaid pao-chia officers; omissions were probably much more serious than these claimed for the registration of 1953.

There was still another case of serious exaggeration. An examination of pre-1850 provincial population figures reveals that although the late eighteenth-century returns of Fukien appear reasonable, its figures for years after 1800 are highly suspect. In the first place, throughout the Tao-kuang period (1821–1850) Fukien failed to report its total population altogether nine times, a record unmatched by other provinces.[28] Secondly, excluding provinces of heavy immigration and of known serious under-registration, Fukien had a suspiciously high rate of increase. The earliest available complete unrevised provincial totals for the period 1776–1850 are those of the year 1787. In that year Fukien reported 12,020,000. But by 1850 it reported a population of 19,987,000, an increase of 65.8 per cent in sixty-two years. The rate is extremely high as compared with 40.0 per cent for Kiangsu, 38.2 per cent for Chekiang, 30.1 per cent for Anhwei, and 28.2 per cent for Kiangsi.[29] Kiangsu, Chekiang, Anhwei, and Kiangsi all suffered severe losses of population during the Taiping wars, and hence their combined population in 1953 was smaller than that of 1850.[30] Fukien, however, by and large escaped from the scourge of the Taiping and major twentieth-century wars, yet its population at the 1953 census was nearly seven millions less than its 1850 figure. In sharp contrast to the 1850 figure, which cannot by any means be justified, the 1787 figure and those of the late Ch'ien-lung period agree with certain historical facts. Moreover, the reason that population figures for Fukien were gradually compiled arbitrarily after the end of the Ch'ien-lung era was known to the central government. An imperial edict of 1829 read:

> According to various memorials . . . the pao-chia personnel in Fukien frequently evade their duties because of their fear of incurring people's wrath. From now on the arresting of criminals and the collecting of taxes should not be assigned as official duties of pao-chia personnel. Their sole duty should be population registration and detection of the lawless.[31]

When all the cases of under-registration and of exaggeration are weighed, it is difficult not to agree with the unusually well-informed Wang Ch'ing-yün that the population of China could only have been somewhat under-registered and not inflated.[32] It is interesting that Father M. Huc, while always aware of the suspicion of some Europeans that China's population figures might have been magnified, concluded after his extensive tours of China that the Chinese population returns "would seem rather under than over the truth." [33]

II

Population data were first collected on the local level and then reached the provincial and central governments. The quality and the authenticity of local data are therefore vital to any appraisal of the national figures.

Unfortunately there seems to have been little demographic interest among the pao-chia personnel and local officials during the late eighteenth and early nineteenth centuries. So far there has been no information on such critical aspects as birth and death rates, detailed age composition, or occupations. The summary accounts on population preserved in local histories frequently give only the number of households and of inhabitants. The better ones give the number of males and females and sometimes also of children under sixteen. It is from these fragments that we must attempt to evaluate the population data of the period 1776–1850.

Size of household. The most convenient source for figures on the size of the household is the 1842 edition of the gigantic imperial topographic treatise, *Ta-Ch'ing i-t'ung-chih.* This work, having originally been compiled shortly before 1812, contains population and household data for a majority of provinces as of 1820. It has by far the most complete returns on the numbers of households, which are available for fourteen of the eighteen provinces of China proper (the number of households for Kiangsu, Anhwei, Shensi, Kwantung, and peripheral areas such as Manchuria and Turkestan are not included). In addition to provincial totals of population and households it also gives breakdowns for prefectures and department counties. It thus enables us to check the totals with the details, as may be seen by a study of Table 13.

The national average of 5.33 persons per household in 1820 may be compared with the average of 5.68 persons per household in 1393. If the average household in 1820 had about five persons, this

Table 13. Households and population of fourteen provinces in 1820.

Province	Households	Mouths	Average number of mouths per household
Chihli	3,956,950	19,355,679	4.89
Shantung	4,982,191	29,178,919	5.86
Honan	4,732,097	23,598,089	4.99
Shansi	2,394,903	14,597,428	6.10
Kansu[a]	2,909,528	15,377,785	5.28
Chekiang	5,066,553	27,411,310	5.41
Kiangsi	4,378,354	23,652,029	5.40
Hupei	4,314,837	29,063,179	6.74
Hunan	3,234,517	18,523,735	5.73
Szechwan[b]	7,058,777	28,048,795	3.97
Fukien[c]	3,152,879	16,759,563	5.32
Kwangsi	1,279,020	7,429,120	5.81
Yunnan[d]	1,010,225	5,933,920	5.87
Kweichow	1,118,884	5,348,677	4.78
Total	49,589,715	264,278,228	5.33

Source: *Ta-Ch'ing i-t'ung chih.*

[a] Kansu included Ninghsia, part of Chinghai and Sinkiang; the number of *t'un-hu*, or households which held land under military tenure, which is missing in the provincial total, is added after checking with prefectural breakdowns.

[b] Mao-kung-t'ing, for which data are incomplete, is omitted.

[c] Formosa not included.

[d] The prefectures and counties that have only population figures but no household numbers and vice versa are excluded.

would be consistent with the results of field studies in modern China, both in the Nationalist and in the Communist periods. A pre-Communist study indicated an average of 4.84 persons per family.[34]

There were substantial differences in the size of households in various provinces in 1820. The difference between Hupei's 6.74

and Szechwan's 3.97 is 2.77 persons. The large household of Hupei could have been affected by the survival of an old custom of merging several families into one in order to lighten if not entirely to evade the burden of labor services and land tax. In Ming times this illegal practice was particularly rampant in Szechwan and Hupei.[35] Szechwan, however, underwent drastic changes during the second quarter of the seventeenth century, when its population in the rich Red Basin was nearly exterminated by the bandit leader Chang Hsien-chung and also by the brutal Manchu forces. Throughout the late seventeenth and eighteenth centuries Szechwan received the largest number of immigrants.[36] Its 1820 average of 3.97 persons to a household may have been affected by many of the immigrants being single adult males.

Sex ratios and age structures. We are less fortunate with regard to the data on numbers by sex and age groups. Official publications and provincial histories usually give only figures for the total population. Relatively few local histories yield specific information. The local histories of certain provinces, such as Kwangtung, Anhwei, Kiangsi, Chekiang, and Hupei, contain practically nothing concerning numbers by sex and age. As far as age structure is concerned, the best that could be done was to calculate in Table 14 the percentage of recorded children in the recorded total population, for in some local histories children under sixteen are grouped separately from the adults.

These data cannot, of course, be regarded as realistic records of the actual situation, but they go along with the common impression that in China the male population has outnumbered the female population. Generally speaking, the excess of males appeared as higher in the south than in the north. The ratios for Ch'eng-ch'eng and Ning-shan counties in Shensi indicated a surplus of more than 50 males for each 100 females. As to these particularly suspicious figures, the 1849 edition of the history of Shih-ch'üan county in southern Shensi, neighboring to Ning-shan, testified that the locality abounded in single male immigrants who were pioneering maize farmers of the virgin hills and mountains and that within the memory of the local people women had always been outnumbered by men two to three.[37]

As to children under sixteen in the total population of various sample areas, some Shansi counties and prefectures had the highest percentages. It seems not merely coincidental that Shansi, whose average household was given as the second highest in the country in 1812, should also record very high youth ratios.[38]

Evidence on infanticide. The existence of high male ratios in Chinese population can be better supported when the evidence on female infanticide is examined. The evidence is truly overwhelming in magnitude, but it is qualitative only. Local histories, particularly of southern provinces, have preserved various official exhortations prohibiting it and songs composed by local magistrates and scholars depicting its inhumanity. The 1864 edition of the history of P'u-ch'i county in Hupei gives a most vivid description of how female infanticide was carried out:

> The rustic people of Hupei and parts of Hunan customarily rear two sons and one daughter at the most. Any further birth is often disposed of. [The custom] is particularly against female infants. This is why in this area women are proportionately scarce and single unmarried men abound. When a baby girl is born, she is usually killed by drowning. Her parents, of course, cannot bear this, but none the less they close their eyes and turn their backs, while continuing to immerge her in the water tub until she ceases to utter her feeble cries and dies.[39]

Table 14. Sex ratios and percentages of children in Chinese populations, 1776–1850.

Year	Place[a]	Province	Population	Sex ratio (number of males per 100 females, all ages)	Children under 16 (percent of total population)
1778		Chihli	20,746,519	118.8	31.4
1773	Yung-p'ing P	Chihli	1,432,031	120.6	34.7
1777	Yung-ch'ing C	Chihli	196,576	108.3	38.3
1837	Chi-nan P	Shantung	4,086,511	111.5	36.1
1826	Chi-ning C	Shantung	400,237	115.7	39.0
1816		Honan	23,481,492	....	39.7
1776	Ta-t'ung P	Shansi	702,401	....	46.7
1748	Yü-tz'u C	Shansi	264,801	....	46.4
1784	Tai-chou C	Shansi	505,116	....	52.8
1843	Yang-ch'ü C	Shansi	354,340	....	40.6

Table 14 (*continued*).

Year	Place [a]	Province	Population	Sex ratio (number of males per 100 females, all ages)	Children under 16 (percent of total population)
1771	Fen-chou P	Shansi	1,277,834	....	43.9
1803	Lo-ch'uan C	Shensi	93,990	126.7	37.1
1784	Chou-chih C	Shensi	234,456	125.0	......
1783	Ch'eng-ch'eng C	Shensi	158,310	154.2	22.7
1829	Ning-shan-t'ing C	Shensi	115,392	156.4	34.2
1820	Su-chou P	Kiangsu	5,908,436	134.4	......
1793	Ch'ang-shu C	Kiangsu	525,617	135.1	......
1816	Sung-chiang P	Kiangsu	2,472,974	128.1	......
1816	Feng-hsien C	Kiangsu	261,898	131.1	38.2
1785	I-wu C	Chekiang	513,878	118.3	......
1826	She-hsien C	Anhwei	617,111	120.7	35.6
1829	Chang-chou P	Fukien	3,609,030	....	41.6
1829	Lien-chiang C	Fukien	174,406	....	39.5
1835	Lung-yen C	Fukien	356,121	....	40.0
1835	Hsün-chou P	Kwangsi	877,337	109.5	42.4
1845	Ta-yao C	Yunnan	95,451	110.9	......
1843	Ch'ung-ch'ing P	Szechwan	2,071,695	112.8	......
1814	San-t'ai C	Szechwan	184,679	107.2	......
1795	Ch'iung-chou C	Szechwan	135,788	101.1	......
1810	Pi-hsien C	Szechwan	134,488	111.5	......
1815	Ch'eng-tu C	Szechwan	386,397	125.7	......
1833	Shih-ch'üan C	Szechwan	113,963	116.5	......

Source: The 1778 returns of Chihli are printed in *Shih-liao hsün k'an*, no. 21, pp. 770a–770b; *Yung-p'ing FC* (1880 ed.), 45; *Yung-ch'ing HC* (1779 ed., printed in 1813), *Hu-shu* or "The Book on Fiscal Matters," 2.31a; *Chi-nan FC* (1839 ed.), 15; *Chi-ning CC* (1843 ed.), 3.5b; Chu Yün-chin, *Yü-ch'eng shih-hsiao lu*, 1.21a–21b; *Ta-t'ung FC* (1776 ed.), 13.3a; *Yü-tz'u HC* (1862 ed.) 6.4a; *Tai-chou chih* (1784 ed.), 2; *Yang-ch'ü HC* (1843 ed.), 7.20a–20b; *Fen-chou FC* (1771 ed.), 7.1a–1b; *Lo-ch'uan HC* (1806 ed.), 9; *Chou-chih HC* (1785 ed.), 4.1b; *Ch'eng-ch'eng HC* (1784 ed.), 5.4b; *Ning-shan-t'ing chih* (1829 ed.), 2.16a–16b; *Su-chou FC* (1877 ed.), 13; *Ch'ang-chao ho-chih* (1793 ed.), 3.3a–4a; *Sung-chiang FC* (1819 ed.), 28; *I-wu HC* (1802 ed., 1.34a–36b; *She-hsien chih* (1937 ed.), 3.3b–4a; *Chang-chou FC* (1877 ed.), 15; *Lien-chiang HC* (1927 ed.), 8.34a; *Lung-yen CC* (1835 ed.), 4; *Hsün-chou FC* (1826 ed.), 18.1a–3a; *Ta-yao HC* (1845 ed.), 3; *Ch'ung-ch'ing FC* (1843 ed.), 3; *San-t'ai HC* (1814 ed.), 1.47b–48a; *Ch'iung-chou chih* (1818 ed.), 17.5a; *Pi-hsien chih* (1813 ed.), 6; *Ch'eng-tu HC* (1815 ed.), 1.33a; *Shih-ch'üan HC* (1833 ed.), 3.

[a] C = county; P = prefecture.

That this is not an exaggerated account is proven by another nineteenth century Hupei local history which says: "The worst custom in Hupei in general is that people not only drown baby girls but sometimes also baby boys." [40] A Kiangsi local history testifies: "Although Ching-an is a secluded small county, it has twenty-eight pao and over 30,000 households. It should be reasonably expected that annual births of female babies would amount to several thousands. Yet those baby girls who are actually reared do not amount to ten or twenty per cent, while some 80 or 90 per cent are drowned." [41] Nor was infanticide confined to southern provinces. Li Tsung-hsi, *chin-shih* of 1847 and governor-general of Kiangsu and Anhwei in the mid-1860's, said of his native Shansi:

> I have learned of the prevalence of female infanticide in all parts of Shansi, but particularly in such southern counties as P'ing-ting, Yü-tz'u, etc. The first female birth may sometimes be salvaged with effort, but the subsequent births are usually drowned. There are even those who drown every female baby without keeping any . . . . This is because the poor worry about daily sustenance . . . and the rich are concerned over future dowries.[42]

A Hunan magistrate of the late seventeenth century, obviously a profound student of human nature who understood that honor exists only when bread is not a daily problem, appealed not to the poor but exclusively to the rich. He made it widely known that any well-to-do family that had reared two daughters was to be awarded a wooden tablet on which the virtue of the family would be extolled.[43] Father Huc during his tours in the 1840's saw a proclamation placarded before the office of the provincial judge in Canton which prohibited the practice of female infanticide.[44] It would appear that the practice must have been sufficiently extensive to call for a province-wide prohibition.

The repeated official exhortations could not halt this obnoxious practice. Even rich people in some areas like Amoy and certain northern Fukien counties continued to drown female infants, partly in fear of having to provide heavy dowries in the future and partly in dislike of rearing oversized families.[45] The poor regarded the practice as an almost legitimate means of maintaining their minimal standard of living and, in any case, as a dire economic necessity. Unless the custom of drowning female infants had been common throughout the country and usually tolerated by the public, Wang Shih-to (1802–1889), a gifted scholar doomed to poverty, would not have had the blunt courage to advocate female infanticide en masse as one means of checking the growth of population.[46]

The effectiveness of orphanages in lessening the incidence of female infanticide varied from place to place and from time to time. An orphanage adequately provided with funds in nineteenth-century Hsiao-kan county in Hupei was said to have saved the lives of some ten thousand female babies within the first three years of its founding.[47] The orphanage of Chiu-chiang (Kiukiang), the main port of Kiangsi, received successive donations from local and customs officials in the 1860's and early 1870's that well exceeded ten thousand taels. With ample funds at its disposal the orphanage was able to take care of an increasing number of baby girls who would otherwise have been abandoned by their poor parents.[48] An early nineteenth-century Hunan county schoolteacher bore testimony to the effectiveness of orphanages as institutions in saving the lives of baby girls of the poor, but he deplored that the benefit did not reach the rich who would continue to drown female infants unless the custom of providing substantial dowries was changed.[49]

Many local orphanages, after inauspicious beginnings, were eventually put on a sound financial footing. In T'ung-an county on the southern Fukien coast, for example, an orphanage was first established in 1737 on the initiative of the magistrate and did well for a number of years. It gradually became inoperative because of shortage of funds. By the 1850's the practice of drowning female babies so shocked the conscience of the local gentry that a permanent fund was collected for its support. It was said that by the early twentieth century the odious custom had almost entirely disappeared.[50]

The question as to whether the gradual decline of female infanticide in modern times may be somewhat connected with China's extensive contacts with the West cannot be clearly answered. But one century of Christian propaganda, charitable and educational work along the sea coast and inland must have had some effect on the educated Chinese. If the gradual disappearance of foot-binding and the abolition of the double sexual standard was due in part to the gradual permeation of Western ideas,[51] female infanticide could not have been an exception. Partly because of the inroads of Western ideas and largely owing to a series of major civil and international wars after 1851 which probably levied a much heavier toll of male than female life,[52] the sex ratio may have been brought down (it is given as 107.5 by the census of 1953). The traditional disproportion between male and female populations is borne out by some modern surveys[53] and particularly by comments in local histories.

Data of Chihli Province. The population data of a single compact area of considerable extent may also be appraised. Since among the provinces only Chihli yields more specific information, we may use Chihli as a sample. The population returns of Chihli in 1778 may be summarized as follows: households, 4,346,799; male adults, 7,650,231; female adults, 6,557,278; male children, 3,615,962; female children, 2,923,046.[54] These figures show the following facts: (1) size of the average household: 4.77 persons; (2) sex ratio (all ages): 118.8; (3) sex ratio (children under 16): 123.7; and (4) percentage of children in the total population: 31.5.

The size of the average household is slightly smaller than the provincial average in 1812. Children constituted only 31.5 per cent of the recorded total, which may be compared, on a purely suggestive basis, with a similar figure estimated in more recent times.[55] Among all the provinces Chihli had the lowest rate of increase between 1776 and 1850. Although the province's population record increased from 20,746,519 in 1778 to 22,819,000 in 1786, between 1786 and 1850, when the juvenile population reported in 1778 would have attained its majority and taken over the task of procreation, the net increase was a negligible 782,000 or a mere 3.4 per cent in 64 years. The general quality of the post-1776 Chihli population figures was reflected in the approbation that the Tao-kuang Emperor gave to the Chihli pao-chia machinery during his inspectional tour of 1829.[56]

Rates of growth. Having made the above checks on the population data, we are now in a position to look at the average annual rate of increase recorded for the entire population during the period 1776–1850. Writers on Ch'ing population have been prone to begin their study by computing the mean annual rate of increase of the whole population, a process that may be dangerous if they are not sufficiently equipped with knowledge of Chinese institutional history. Owing to China's vast extent and to regional and local uprisings which became increasingly frequent after 1795, the population returns sent in by the provinces were likely to contain omissions. It was not always possible for all the counties and provinces to report their populations punctually in the eleventh month every year. The omissions were not always clearly explained, in the national totals at least. So far the best sources for revision of the annual national totals are the archives of the Board of Revenue, which unfortunately are inaccessible to students outside China. Even the Board's archives do not seem always to contain clear mention of regional and local omissions and have serious gaps between 1792 and 1818, 1821 and 1829.[57] For these reasons Mr. Lo Erh-kang, formerly of the Institute of Social Sciences of the Academia Sinica, who has used the Board's archives to revise some of the annual national totals, has had only limited success. His revised data, together with my minor emendation based on *Ch'ing shih-lu*, are given in the Appendix.

In view of the factors that complicated the compiling of the annual national totals, it is not surprising that the population data of this period still show sharp fluctuations. The sharp fluctuations between 1774 and 1779, for instance, may have been connected with temporary disruption of pao-chia registration by Wang Lun's revolt in western Shantung in the latter half of 1774[58] and by the confusion that accompanied the effort at strict enforcement of a uniform system of population enumeration. The Ch'ien-lung Emperor, as we have seen, gave provincial authorities considerable time to improve the pao-chia registration and to compile their returns with care. It appears that the system of pao-chia registration did not become more orderly and uniform until 1779. It is advisable, therefore, to take the year 1779 instead of 1776 as the datum for any estimate of the average annual rate of growth.

The next sharp fluctuations occurred between 1795 and 1805. We have no reasons for the sudden drop in 1795, which marked the end of the Ch'ien-lung Emperor's long reign. The drop in 1796 was caused by the temporary omissions of Hupei, Hunan, and the Fu-chou prefecture of Fukien.[59] The ensuing drops, though not explained by official documents, were obviously due to more serious temporary omissions and perhaps also to some actual losses of population resulting from the uprisings of the White Lotus and other religious sects of the Szechwan, Hupei, and Shensi areas. The sharp increase in 1805 was presumably due to the inclusion of areas that had previously failed to report. We are not yet able to explain the fluctuations between 1813 and 1822, but they were undoubtedly caused by unexplained omissions rather than by population movements. From 1822 to 1850 the population data indicate a steady growth and Mr. Lo's revisions for this period seem more satisfactory.

For all these reasons 1779 is used as the datum and 1850 as the end. In our calculation we have to assume that the population of China was increasing at an even rate during these seventy-one years, since an attempt to compute the yearly differences is futile. By this method the population of the entire country increased from 275,000,000 in 1779 to 430,000,000 in 1850, an increase of 56.3 per cent, giving an average annual rate of growth of 0.63 of one per cent. If we compare the apparent rates of growth of 1779–1794 and 1822–1850, the difference is substantial. For the former the rate is 0.87 per cent and for the latter it is 0.51 per cent. Up to 1794 the economic and political conditions in general had been very favorable. After 1796 the rate of growth may have been slowed down by the operation of Malthus' positive checks. While these rates cannot be considered very accurate, because of the flaws in the annual population totals, their general order of magnitude nevertheless fits in with what we know of economic conditions of the time.

In conclusion it may be stipulated that one prerequisite to any correct understanding of the difficult question of the Ch'ing population is a correct periodization. The prerequisite to a correct periodization is a fairly intimate knowledge of Chinese institutional history. Any study that lumps the three subperiods together, or any two of them, is bound to lead to absurd results.

CHAPTER V

Population Data, 1851–1953

The early Republican and the Nationalist governments published a number of population figures which have been analyzed and revised by Chinese demographers. In the absence of true census returns these figures have been widely used by population experts with varying degrees of confidence. Few, if any, however, have taken pains to study the machineries with which such figures were collected. Thanks to more abundant government publications on the system of population registration during the first half of the twentieth century, it is possible to make a systematic study of the ways in which population data were collected and compiled. A knowledge of the administrative machineries for the collection of population data is a prerequisite to a critical appraisal of the official figures for the period 1851–1953.

Space does not allow a detailed discussion of why the highly centralized Manchu government was transformed by the events of the middle nineteenth century into an effete and decentralized government. Such a fundamental change in the power and efficiency of the central government can be indicated by the remarks of contemporaries. An official of the early nineteenth century described the pre-1850 government in the following hyperbole:

> For some 170 years our country has been blessed with such a prolonged peace and prosperity as to be free from even the slightest disturbance that would make cocks crow and dogs bark . . . . Central orders reach the officials through various stages and constitute a control as effective as that of the parent over his children. Unless sanctioned by the court, not even a minor matter can be decided upon by the officials. Political power is so centralized and government discipline so rigid that it is entirely without parallel in history.[1]

The basic factor that brought about political decentralization after 1851 was that the exigencies of the Taiping period forced the central government to delegate to provincial authorities the power to organize new armed units and to raise funds. The central government's loss of control over vital military and financial matters was testified to by the reformer K'ang Yu-wei (1858–1927): "A perusal of the memorials and correspondence of the famous post-1850 provincial officials shows that they all raised their own armies and funds and in the worst cases fought with one another over soldiers' pay." K'ang said of the equally vital changes in financial administration:

> The power of financial control is now vested not with the central government but with the provincial authorities. When the Board of Revenue needs money, it asks the provinces to apportion among themselves without bothering with the methods of fund-raising . . . . This is like the Son of Heaven's depending on tributes from feudal vassals. The Board of Revenue exerts no direct [financial] control over the nation. The provinces, therefore, have different methods of raising revenues which are decided upon and executed by governors general and governors. They need only to send [to the Board] a report which is a mere formality. The central government is forced to approve even such shameless improvizations as gambling tax and sale of offices. It is so obvious that provincial authorities exert financial control to the utmost and the Board of Revenue is utterly powerless.[2]

Institutionally the most significant change after 1850 was that the provincial financial commissioner and the provincial judge were reduced to mere subordinates of the governor general and the governor. Prior to the Taiping Rebellion the provincial financial commissioner, who was appointed by the Board of Civil Appointments and directly answerable to the Board of Revenue, and the provincial judge, who was directly responsible to the Board of Punishments, enjoyed an independent position vis-à-vis the governor general and governor. In fact, this kind of separation of powers within the provinces was one of the basic reasons for the central government's ability to exert effective control over the provinces. But the Taiping wars provided governors general and governors ample opportunities to curb the powers of the financial commissioner and the provincial judge until they were relegated to a hopelessly subordinate position.[3] Thus, by the end of the Taiping period the highest provincial officials emerged triumphant, with military command and financial and judicial powers in their hands. Effective central control was gone.

The events of the Taiping period made it necessary for the central government to delegate to provincial officials the power of organizing, financing, and training new army units. Since these new armies had a strong sense of solidarity and local patriotism and looked exclusively to their leaders for orders, they were virtually "personal" armies. It would not be an exaggeration to say that the origins of twentieth-century warlordism can be traced back to the Taiping period, for many early Republican war lords, including the first President of the Chinese Republic, Yüan Shih-k'ai, were members of the Anhwei clique.[4]

With the passing of the imperial system in 1912 and the death of a relatively strong political figure, Yüan Shih-k'ai, in 1916, China was ushered into an era of warlordism with which all students of modern China are familiar. Although after 1927 Chiang Kai-shek valiantly attempted to reestablish central control, it is common knowledge that his authority scarcely went beyond a number of central and lower Yangtze provinces. His task of political reunification was interrupted by his wars with the Communists and then by the Sino-Japanese war of 1937–45. Not until the establishment of the People's Republic of China in Peking in October 1949 did central authority reassert itself. It is important for our purpose to bear in mind that the central governments of the last century were in no such position as the Ch'ien-lung Emperor to harness the provincial and local governments. This changed character of the governmental system was fully reflected in the system of population registration.

I. POPULATION DATA, 1851–1902

The military situation during the Taiping period was so critical that an immediate shift of emphasis in the pao-chia function was clearly discernible. In the nation's capital in the 1850's, for example, the control and supervision of pao-chia was transferred from the Board of Revenue to the General Commandant of the Gendarmerie.[5] After a militia was organized in Hunan in 1852, the name pao-chia became almost identical with *t'uan-lien*, which literally means "grouping and drill."[6] It is true that in order to organize t'uan-lien there had to be some form of population registration, but the purpose of registration was mainly the detection of the lawless elements among the local populace and the enlistment of able-bodied adult males for militia service. This shift of emphasis in pao-chia function was nowhere better reflected than in a famous collection of political and economic essays and memorials of the late Ch'ing period in which pao-chia was classified under "military affairs." [7]

In the exigencies of the time, the idea of registering the entire local population by pao-chia personnel was inevitably pushed farther and farther into the background. In the Shanghai area, for example, the so-called population survey of the Taiping period excluded all "well-known gentry, the rich, prominent families and great clans" and covered only those segments of the local population who had little means of escaping from militia and sundry duties.[8] This being the case, the canvassing of all but the poor adult males could afford to be casual, so casual indeed that Shanghai county registered a perfectly stationary population between 1864 and 1881, despite an inflow of war refugees and giant strides in commercial development.[9]

As another example, Chü-yung county, near Nanking, proclaimed its pao-chia regulations in 1897:

> The names, ancestral counties, ages, occupations of adult males should all be filled out on the door placards and household certificates. Women and children under ten are to be exempted . . . . All households except those in government service should have certificates and door placards.[10]

This indicates the extent to which the system of population registration had been changed in most parts of China since the outbreak of the Taiping Rebellion in 1851.

Table 15. The 1872 population returns of two Chekiang counties.

County	Male		Female		Sex Ratio	
	Adults	Children	Adults	Children	Adults	Children
Wu-ch'eng	140,033	72,112	111,485	37,106	125.6	194.3
Ch'ang-hsing	44,425	18,776	22,817	4,347	194.7	431.9

Source: *Hu-chou FC* (1874 ed.), 39, *passim.*

Had these new regulations been uniformly and strictly enforced throughout the country, local population data collected after 1850 might still be of some use for analysis. But the confusion in population registration during this period was so great that sometimes even neighboring counties had different systems. This can be seen in Table 15, giving figures for two counties in northwestern Chekiang. Neither of these two local systems conformed to the regulations of Chü-yung and it is impossible to know why Wu-ch'eng covered proportionately more female population than Ch'ang-hsing.

Data on the provincial level for this period are even more confusing. The latest edition of the history of Hupei province comments:

> Although since the beginnings of the wars of the Hsien-feng period (1851–1861) population registration was carried out for the sake of organizing t'uan-lien, it was dictated only by the exigency of the time and by the local conditions and was never carried out in the whole province. Some local officials reported [local population figures], some did not. The documents in the financial commissioner's office were lost and destroyed [in war] so that it was impossible to check [such figures]. Population figures are consequently missing from local histories compiled during the T'ung-chih (1862–1874) and Kuang-hsü (1875–1908) periods.[11]

Hence Hupei failed to report population figures between 1852 and 1857. There can be little doubt that its annual provincial totals after 1858 were at best based on rough estimates or compiled entirely arbitrarily, because it was impossible to compute provincial totals without local returns.[12]

Other wartorn Yangtze provinces revealed similar or even stranger features in population registration. The 1885 edition of the history of Hunan province testified: "Our old provincial history was compiled in 1819, when all local governments kept their population registers and duly reported to the financial commissioner's office, despite the abolition of *ting* assessment. But lately [such local registers] have been entirely missing." [13] Yet Hunan provincial authorities were so concerned with an outward compliance with the law of 1775 that they went on reporting a very small and consistent annual gain until 1898. Kiangsi province missed only the year 1860, when the Taiping forces invaded the province on a large scale, and reported a perfectly stationary population for nearly four decades (24,516,000 in 1851 and 24,617,000 in 1890) although there is detailed evidence of severe population losses suffered by some prefectures during the Taiping period.[14]

Other provinces cared less about the law of 1775 because the war made it impossible for them to register the entire provincial population. Anhwei, a province that suffered very severely from the Taiping and Nien wars, never reported any population figures after 1852, when a peak population of 37,650,000 was registered. Kiangsu did not report between 1853 and 1873. Although the loss of population in southern Kiangsu during the Taiping period was very heavy, its 1874 reported population of a mere 19,823,000, as compared with 44,155,000 in 1850, can hardly be accepted as authentic. Chekiang, which was not seriously affected by the Taiping wars until 1860, showed a drastic drop from 30,399,000 in 1859 to 19,213,000 in 1860. In 1860 and 1861 the population of northern and parts of central Chekiang suffered exceptionally heavy losses. After a lapse of five years it began to resume its annual reports. Despite the scourge of the Taiping wars, it is extremely unlikely that its population could have been as small as 6,378,000 in 1866 and 11,900,000 in 1898. The confusion in pao-chia population registration in the last half of the nineteenth century was by no means confined to the lower and central Yangtze provinces. In Shensi "the local authorities regarded it [population registration] as a mere formality and it was extremely difficult to get at true population figures." [15] Figures released by the provincial government of Shansi did not agree with local figures.[16] In a province-wide exhortation for the setting up of t'uan-lien, Lu Ch'uan-lin, governor general of Szechwan, 1895–1897, openly deplored that the once effective pao-chia machinery had fallen so much into desuetude that nothing short of a door-to-door reenumeration could enable him to know the exact provincial population.[17]

It is interesting that Hsü Shih-ch'ang, a rapidly rising political figure who was made the first governor general of three Manchurian provinces in 1907 and later became president of the Republic, while taking considerable pride in the 1907 preliminary count of the Manchurian population carried out by his amply financed police department, expressed profound contempt for later pao-chia registration in general.[18] All this clearly indicates that the crux of the post-1850 registration problem was no longer how to harness the pao-chia machinery to work on population registration, as had been the case during the late Ch'ien-lung and early nineteenth century, but whether for most parts of the country such an apparatus still existed. The most eloquent testimonial to the obsolescence of pao-chia population registration is that the Veritable Records of the Ch'ing Dynasty, *Ch'ing shih-lu*, which had invariably given annual national population totals, abandoned the practice from 1874 onward, although provincial totals up to 1898 are available in the Board of Revenue archives.

From these facts and figures it becomes abundantly clear that, during the period when the empire was in the throes of a life and death struggle, the provincial and local authorities had little time or energy to work on population registration. In wartorn areas the combined and cumulative effects of wars, slaughter, property destruction, famines, and epidemics wiped out substantial portions of local populations and made a total collapse of the pao-chia machinery inevitable. In areas less affected by wars the energy of the officials was consumed in the more urgent matters of raising revenue and organizing militia for local defense. After the pacification of the Taiping Rebellion there were still the lingering Nien wars and campaigns against the rebellious Moslems of the northwest. The whole empire was so prostrated that neither the central nor the provincial governments had the will and resources to revive the once nation-wide pao-chia machinery. From time to time attempts were made by some provincial and local officials to strengthen pao-chia for local policing, but none actually restored the early pao-chia registration.[19] Not even the imperial exhortation of 1898 could salvage this ancient and once useful pao-chia machinery after a lapse of nearly half a century.[20]

The *coup de grâce* came in 1902, when Yüan Shih-k'ai, governor general of Chihli, memorialized:

> The evils of the pao-chia system are such that although it is incapable of preventing lawlessness it is more than sufficient to harass the people. The whole machinery has to be thoroughly overhauled and then replaced by a police system . . . . Since [the Boxer uprising of] 1900 Chihli has been depressed and unable to wipe out all lawless elements. Without an imperial decree sanctioning the prompt establishment of a police system, [the existing machinery] would be inadequate

to enforce the law and to know the actual conditions of the people. In the fourth month of this year I studied Western methods, drafted plans, and set up a police department with five sub-offices in the provincial capital, Pao-ting . . . . It is hoped that the police system will in course of time be extended to other prefectures and counties of the province.[21]

It did not take long for the imperial government to realize the futility of trying to salvage the pao-chia system. In the early fall of 1902 an imperial decree was issued:

Instruction to the Cabinet: Judging from the memorial and drafted plans submitted by Yüan Shih-k'ai, [it is clear that the police] will be a satisfactory instrument for maintaining local peace and order. It is thereby ordered that all governors general and governors should adopt the plans drafted by Chihli province and duly report on how they are to be undertaken. There must not be any delay or pretext for dodging the problem. Be this made known to all concerned.[22]

This was a belated recognition of a long accomplished fact and marked the formal abolition of the pao-chia system. In its stead emerged the modern police system, which was to be the sole agent for population registration until the partial revival of pao-chia after 1930. Misleading as the post-1850 national and provincial population totals are, in all fairness it must be said that there is one type of local data which are reasonably accurate and highly useful for our study of population losses, interregional migrations and resettlements. In localities where the population had been drastically reduced by wars, famines, and epidemics and had suffered extensive destruction of property, local officials were ordered to carry out a door-to-door population count in order to ascertain the true number of survivors upon whom the burden of taxation fell. In the greater parts of Anhwei, southern Kiangsu, and northern Chekiang, where war devastation was most extensive, local officials were sometimes assisted by surviving local scholars in counting the population and estimating the extent of originally cultivated land to be temporarily tax exempt. The number of immigrants who were attracted to settle in the wartorn places was also of great fiscal significance and hence carefully checked. The native survivors were as a rule entitled to some form of government aid which was generally distributed according to the size and needs of each family.

In those matters the mutual dependence of local officials and people and the sense of community facilitated the process of population registration.

II. POPULATION DATA, 1902–1927

The demand for the creation of a modern police system had become vocal after China's defeat by Japan in the war of 1894–95 and the failure of the "One Hundred Days of Reform" in 1898. In 1901, after the withdrawal of the allied expeditionary forces from Peking, the imperial government had already established a police officers' training school in the nation's capital, with Japan's help. As a result of Yüan Shih-k'ai's memorial of 1902 the provinces were ordered to set up a police department. By 1905 a new Police Ministry was created, whose name was changed to *Min-cheng pu*, or Ministry of Civil Affairs, in 1906. Whatever the later changes in the name of this Ministry, the police remained under its jurisdiction.[23]

Meanwhile, after 1905 the rapidly declining Manchu dynasty was trying to win the support of the nation by some halfhearted constitutional reforms. In anticipation of the impending congressional election, conscription, local self-government, and educational reform, the Directorate of Statistics was created in 1908, under the jurisdiction of the Ministry of Civil Affairs.[24] In that year the Directorate worked out a six-year plan for taking the first national census and sent standard census forms and detailed instructions to the provinces. Provincial authorities were required to enumerate people of both sexes and all ages and to specify the number of adult males and children of school age. Because of the immediate political exigency, this six-year plan was hurriedly "completed" in four. Theoretically, therefore, China had taken her first modern census by 1911.

The so-called census of 1908–1911 is significant not only for its genuine demographic emphasis but also for its practical consequences. To some writers on China's population the 1911 figure of 341,913,497 was almost a godsend with which they could substantiate their view that China's population could not have been as large as official Ch'ing figures showed. Walter Wilcox consistently used the 1911 figure as one of the major justifications for his estimate of 342,000,000 as the peak Chinese population for the nineteenth century. This view has many admirers in China who believe that China's population since the middle of the nineteenth century has been stationary.

The practical consequences of the 1911 census were such that the census machinery and its efficiency merit study. Among Chinese writers on the subject Wang Shih-ta has made the most laborious study of national totals. Although he had no access to local and provincial returns for 1911, he has exhaustively studied provincial reports submitted to the central government. He concludes that the census of 1911, whatever its defects, was a true census which "had penetrated deep into the populace" and which cannot have been a fabrication.[25] Yet it was precisely the lack of information on the local level that led him to such a positive conclusion. While the wording of the provincial reports could easily have been manipulated by experienced officials and clerks, the reliability of the census figures depended on the efficiency, honesty, an resourcefulness of the local census machineries.

The police would have been the sole census-takers between 1908 and 1911 but for the fact that a number of provinces, due to lack of funds, had not yet created a police department. The statutes of this period provided that the provincial police department, where such a department already existed, should serve as the director of provincial census. In any province that still lacked a police department, the financial commissioner was to direct the census. Even on the provincial level, therefore, there was a lack of uniformity. On the local level the statutes provide:

> Census duties are vested in the elected officers of lower self-governing bodies or in the heads of *hsiang* [a sub-division of a county] . . . . In localities where such local self-governing bodies have not yet been created, the [county] director of census should supervise the police and select some local gentry of proven integrity to undertake [the census].[26]

It is obvious from a critical perusal of the law that the whole census machinery from the provincial down to the local level was patchwork apparatus at best.

Although detailed and candid discussions on late Ch'ing police are scarce, the draft history of Anhwei province compiled during the 1930's gives data on the system which cannot be found in most official publications.[27] Anhwei, to comply with the imperial decree of 1902, transformed the headquarters of the old listless provincial pao-chia into a police department in 1903. A police officers' training school was also established in the provincial capital, but it was seriously handicapped from the beginning by a shortage of funds; consequently, even by the 1920's, it had but few graduates. The inadequacy of the newly created police system was further aggravated by the fact that constables were generally recruited from the "mean and low classes who did not even know what police were about."

The capacity of the newly created police of Anhwei is indicated by the total number of county police offices and suboffices. Out of 68 counties in the province only 44 had police offices. The distribution of county police offices and suboffices is shown in Table 16.

Table 16. Police offices and suboffices of forty-four Anhwei counties up to 1911.

Number of offices and suboffices	Number of counties
1	16
2	6
3	4
4	5
5	9
6	2
11	1
13	1
Total 141	44

Source: *Anhwei t'ung-chih-kao* (latest uncompleted ed.), "Book on Civil Affairs," chs. on "Police," *passim*.

Since the average county comprised a considerable area, the number of police offices and stations was adequate neither to maintain peace and order nor to carry out the more sophisticated business of census-taking. As to the number of police constables, Fu-yang, one of the largest counties in Anhwei with a reported population of 1,379,692 in 1930,[28] had but 150 constables, which far surpassed the number in any other county in the province. Smaller counties like I-hsien and Ch'i-men had but 8 and 10 respectively. In a province whose area exceeded 150,000 square kilometers there were only 1,939 constables who, notwithstanding their lack of proper training and pay, were the main census-takers during the years 1908–1911.

While other provinces were not so candid and voluble about the sad state of police administration, there is reason to believe that the police situation in Anhwei was probably not too much worse than in other provinces. In a period when the provinces vied with one another in claiming progress in modern reforms, only Fengtien, the southern Manchurian province, consistently prided itself on its police and census administration, owing to the able administration of Hsü Shih-ch'ang, an unusually ample allotment of funds for the expansion of the police force, and the docility and relative material prosperity of its people.[29]

Mr. Wang's own evidence indicates that the execution of the census of 1908–1911 undoubtedly bristled with difficulties. Kiangsi province, for example, reported in 1910:

> When the preliminary census was carried out, the foolish rustics were incited by rumors that taxes would be increased. Consequently they sometimes offered resistance. The census procedures had to be simplified. After considerable explanation and persuasion by the gentry the people gradually understood [the aim of census]. Not until then was it possible to get the number of adult males and school children which was originally unobtainable. These figures were amended afterward and are hereby submitted.

The 1910 report of Kwangtung province is even more revealing:

> From the beginning of the census-taking disturbances have repeatedly broken out in Kwangtung. In localities like Hsin-an, Ta-p'u, Lien-shan, etc. the agitation developed into open riots which were put down only by the provincial army . . . . [Serious disturbances] were due partly to the ignorance and misunderstanding of the rustics who, upon seeing the mounting of door-placards and the new population registration, believed that adult males would be drafted and taxes increased. They therefore became tumultuously agitated . . . . Local Officials would have liked to request that [the census] be put off but for their fear of government discipline and punishment. In trying to carry it out they were afraid of more serious riots . . . . As a result, in most cases the census registers so compiled contain omissions and are un-

authentic. Local officials for fear of punishment had to manipulate skillfully in order to comply [with regulations] and higher authorities had to forbear in order to save face. There cannot have been more self-deception and deception of others.

In view of the various difficulties, the governor general of Kwangtung and Kwangsi Yüan Shu-ku memorialized that "the census be postponed until the police system has been extended to *hsiang* and *chen* (all sub-divisions of a county)."

The case of Kwangtung is most instructive. In the first place, the unusually candid statements of the governor general soon cost him his official post. Other provincial officials were keenly aware of the consequences of a frank self-exposé and had to resort to a manipulation of words. Understanding this, it is possible to read between the lines of those provincial reports and conclude that the census of 1908–1911 was hurriedly and confusedly carried out by most provinces. Secondly, this ill-starred official hit upon the crux of the problem by pointing out that there was no census apparatus that could actually reach the villagers. Even down to the heyday of the Nationalist regime, the farthest that the police could reach was the heart of a county centering around the walled city and some large and populous towns or boroughs. After the collapse of the pao-chia system there was no administrative organ, let alone a specially trained census organ, that could bridge the local government and the broad mass of the people.

Therefore, the so-called census of 1908–1911 and those of the early Republican period were in most localities carried out — or rather manipulated — by the county government and local gentry. The unusually informative and unfinished history of Ningpo testified:

The drafting of the constitution and the preparation for national election [during the last years of the Ch'ing period] provided the impetus for census-taking. The people, so accustomed to traditional government exactions, often thought mistakenly that the purpose of the census was to increase taxation. They therefore purposely underreported. During the early Republican period there were some people who, vying with one another for more electoral seats, inflated population figures. There were also those who, remembering the adult male's fiscal burden, evaded the census. The local officials in charge, disgusted with all this, compiled population figures arbitrarily. The figures were all unauthentic because of both exaggerations and omissions. Up to the present [the late 1930's] there have been few reliable figures.[30]

The 1934 edition of the history of Hua-yang county in the heart of Szechwan is equally revealing. Its 1908–1911 returns show that the county had 119,201 households and a total population of 481,192. Among 270,350 males only 49,584, or 18.3 per cent of the male and 10.3 per cent of the total population, were adult males between 20 and 45. This unreasonably low percentage for adult males indicates serious evasions by the segment of the population most liable to taxation and sundry services. What is more, from the detailed returns of various t'uan, or militia units, it is discovered that four consecutive t'uan reported households in exactly the same round number of 500. Small wonder, then, that the editor commented:

During the last years of the Ch'ing dynasty and the early years of the Republic the emphasis was placed on election and population census was carried out for that purpose. At the beginning people were deeply suspicious of the census and refused to report true numbers. Then, owing to the scramble for electoral seats, the figures were sometimes exaggerated. All those figures were unauthentic.[31]

Recalling the Szechwan population figures of the early Republican period, a modern geographer has said that they were compiled and copied entirely arbitrarily, although local officials seldom failed to appear to comply with orders from higher authorities.[32] The same thing was also recorded by a modern Fukien local history.[33]

The danger in accepting the results of the 1908–1911 census and the subsequent figures of the early Republican period is fully shown by the fact that, although many counties of Fukien had neither witnessed a census nor gathered any population figures between 1908 and 1927, Fukien managed to submit its provincial "census" figures to the Ch'ing and early Republican governments without delay.[34] So hurriedly submitted and so incomplete were the 1911 returns for Kiangsu that this most densely populated province reported only 4,995,495 households and a population of 9,356,755. Shanghai, however, reported a population of 5,550,100 as compared with the probably more reliable figure of 3,417,497 for 1932–33.[35] The early Republican government, which knew the inner workings of the 1908–1911 census system, concluded that "it failed to secure good results and its figures were falsified and trimmed as before." [36]

In retrospect, what the early Republican government said of the 1908–1911 census may be even more legitimately applied to the censuses under its own sponsorship, for the machinery, or rather the lack of machinery, with which to collect population figures remained practically the same and the country was approaching greater political confusion. The early Republican statutes provided that the police force, wherever one existed, was to be the sole census-taker. In localities that had no police but had *pao-wei-t'uan* or local defense corps, census taking was to be their duty. In places with neither police nor *pao-wei-t'uan* the county magistrate was to use his discretion in enlisting the assistance of "local gentry of proven integrity and administrative experience." [37] The results shown by some local histories were even more random compilations. The so-called census of 1912, which was belatedly released in 1916–17, gave a total of 419,640,279.[38] Apart from the emendation of the Kiangsu figure, no reason is given for the sudden increase of 78,000,000 over the 1911 figure.

With the passing of the emperor, for two thousand years the symbol of common allegiance, and after the death of Yüan Shih-k'ai in 1916, China reached the zenith of political chaos. A number of regional warlord governments were either beyond the control of, or openly defied, the nominal national Peking government. No census was again attempted. Altogether, it would not be too unfair to say that all the official population figures between 1902 and 1927 were the result of governmental self-deception.

III. POPULATION DATA, 1928–1949

Although the establishment of the Nationalist government in Nanking in 1927 signified a new era, it is well known that its authority was accepted reluctantly and gradually by the provinces. At first the area under its effective control comprised only the lower and central Yangtze provinces. Even in Kiangsi, Anhwei, Hupei, and Honan there were Communist strongholds up to 1934. Whatever its political significance, the Nationalist government was confronted with a gigantic task in seeking to unify the country and renovate the governmental system. Inevitably there was considerable discrepancy between its goals and achievements. This was particularly true in the system of population registration and the compilation of population figures.

Before the end of 1927 the *Nei-cheng pu*, or Ministry of the Interior, which superseded the Ministry of Home Affairs of 1906–1927, had already notified the civil affairs departments of various provinces that the population of the entire country was to be counted. New census forms were issued in 1928 and the provinces were twice instructed to carry out this task through the county governments. Only sixteen provinces reported before the end of 1928, and up to the summer of 1931 twelve provinces had still failed to report. The Ministry's first national total, 474,787,386, which was based substantially on estimates, was subsequently revised downward by the Directorate of Statistics and by population experts outside the government.[39]

The Nationalist population figures have been widely used and widely questioned at home and abroad. So far as is known, no study has ever been made of the Nationalist machinery for population registration. The machinery was far more complex than the 1927–28 directives of the Ministry of the Interior stated. To begin with, the county government was ordered to get the local population figures with the help of the county police bureau, if there was one. The police bureau was therefore one of the census agents. Secondly, the laws of 1928 provided that each county should create a number of subdivisions, such as *hsiang*, *chen*, *lü*, and *lin*, in order to prepare for local election and local self-government, which was one of the late Dr. Sun Yat-sen's ideals. The officers of these county subdivisions, when elected, should carry out a population count.[40] Thirdly, in July 1929 the Ministry of the Interior issued further orders for the creation or expansion in each county of *pao-an-tui*, or peace-preservation corps, which, though not directly entrusted with census duties, were to be especially concerned with the registration of adult males between 20 and 45.[41] Last but not least, the pao-chia system, which had fallen into oblivion since 1902, again attracted the attention of Chiang Kai-shek, who was determined to wipe out the Communists. It was first revived in some particularly troublous areas before it was legally revived on a national scale in 1934. From 1934 the pao-chia machinery once again became the sole agent for population registration.

There had been some improvement in the police system since its humble and ludicrous beginning in 1902. But the improvement was slight because of fund shortages and particularly because of the ever shifting policy of the Nanking government. While the total amount allocated for police by the provinces of China Proper reached $26,000,000 (Chinese) in 1932, it was drastically curtailed to $16,000,000 in 1934, when the pao-chia system was statutorily revived. By 1934 only four provinces in China Proper had a police budget of over $2,000,000 and ten provinces had less than $500,000. So large and densely populated a province as Szechwan allocated a mere $31,604 for its police administration.[42] Up to 1935 in the whole country only 2,751 graduated from the police officers' training schools and 29,295 from the constables' training classes.[43] The quality of the police personnel may be partly inferred from their salaries. In large cities like Shanghai and Tientsin a police constable might receive $20 a month; in some rural areas a police officer got only $8 a month and a constable one or two.[44] A fairly large number of rural districts had no police at all. These facts and figures testify eloquently that the county government itself could never have been an effective census agent in the late 1920's and early 1930's.

The records of local self-governing bodies which served as census agents were hardly more encouraging. There were two main reasons for the government exhortations of 1927–28 to create local self-governing bodies. Sun Yat-sen, who was so disheartened by the sham local self-government of the late Ch'ing and early Republican periods, which was virtually a monopoly of "local magnates and evil gentry," had insisted on a genuine and direct participation in local administration by the people. Secondly, the Nationalist leaders were aware that by far the greatest problem in local administration and population registration was the lack of local subadministrative units that could link the county government and the people. Owing to this lack, "the government and the people have been separated by an unbridgeable gulf." [45]

By August 1932, when the anti-Communist campaigns were in full swing, Chiang Kai-shek, as commander-in-chief of the Nationalist forces, had reviewed the problem of local militia and its prerequisite population registration in detail. His own words provide the aptest explanation for the failure of population registration between 1927 and 1934:

> It is generally known that population registration should be faithfully conducted by the county director of the bureau of militia (*ch'ing-hsiang-chü*) with the assistance of the elected officers of such sub-units under the county as *ch'ü*, *hsiang*, *chen*, *lü*, and *lin*. Since such officers cannot have been elected [within a short period], there has been a lack of a responsible organ to carry out the task of population registration . . . . The underlying difficulty is that for a long time there have been no chains of sub-divisions under the county that can reach the bottom [i.e., the people]. At a time of population survey the county government either dispatches a few of its personnel [to get the numbers] or entrusts the gentry with forms and registers to be filled out. The result is that *population figures are absolutely inaccurate* . . . . The fundamental cause [for this failure] is that while current laws and provisions recognize the importance of population registration, they fail to understand the necessity of establishing the pao-chia system. They are concerned with the election of officers for various local self-governing bodies but not with the appointment of pao-chia headmen for local defense. They fail to realize that the establishing of a pao-chia system and the task of population registration are so intimately related that they are one and inseparable. For, from historical times the basic unit in the pao-chia system has always been the household; as every household has its head, so does every chia, which consists of ten households, and every pao, which consists of ten chia. The fixing of a head for each household and the appointment of a headman for each chia and pao is already the beginning of population registration. The actual numbers and changes of the local population depend on pao-chia headmen's periodic counts and reports. Functionally speaking, the pao-chia is virtually a rural police force. It is clear therefore that in order to fulfill the important task of population registration we will have to put off for the time being the creation of such local self-governing units as *hsiang*, *chen*, *lü*, and *lin* and find an entirely different path. It is especially important that [population registration] should be entrusted to pao-chia and carried out simultaneously with the establishment of the pao-chia system.[46]

This is undoubtedly the most authoritative appraisal of the efficacy of local self-governing bodies as agents for population registration. Indeed, it clearly testifies that before local self-governing bodies could be elected they were indefinitely postponed and that between 1927 and 1934 there was no proper machinery to register the population. It further strengthens the belief that in studying official Chinese population data the machinery for the collection of the data is quite as important as an analysis of the data themselves.

The regulations on the formation of local peace-preservation corps were first proclaimed in July 1929 and subsequently amplified. Although at first the group liable to this local militia duty was adult males between 20 and 40, there were many exceptions. Kiangsu fixed the age range between 20 and 45, and this soon seemed to become the majority practice. Kiangsi's lower age limit was 18 because its need for militia was greater during the famous campaigns against the Communists in the early 1930's.[47] Within this age range male adults in the following categories were exempted from militia duty: university and high school graduates, people who served in the regular armed forces while temporary residents in their native counties, the sole male adults of the households, civil servants, shopkeepers and entrepreneurs who could find no substitutes to manage their enterprises, the sick and maimed, and convicts.[48] It would appear that those who could not escape the service were the ordinary hapless peasants.

The inner working of this peace-preservation system was aptly reviewed by Yang Yung-t'ai, one of Chiang's ablest and most trusted associates. As the secretary general of the Generalissimo's Nan-ch'ang headquarters, he candidly concluded, at the second interprovincial conference on peace preservation, held in 1935 at Nan-ch'ang, that the peace-preservation corps still remained so fragmented that they were in the hands of "local magnates and evil gentry." Such corps had proven useful only in Hunan and Kiangsi, where they had been brought under the control of the provincial governments.[49] No special review was made of the population registration by the peace-preservation authorities, but there is ample evidence that the militia duty of adult males almost invariably accounted for their being seriously under-registered.

This under-registration of adult males was reflected in the pao-chia population registers. For reasons explained by the Generalissimo himself, the pao-chia system was first revived in the provinces where Communists had established their bases — Kiangsi, Hupei,

Anhwei, and Honan. By November 1934 it had been extended to the whole country, and became the sole agent in ascertaining the population between 1934 and 1949. A brief analysis of the available pao-chia figures will show their quality. Kiangsi reported an adult male population of 3,477,039 in 1928, before the pao-chia revival. Its 1935 figure (under the pao-chia system) was only 2,665,065, a decline of 801,974 or 23 per cent in six years.[50] It is true that five years of intensive civil war between the Nationalists and the Communists effected some real decline in the adult male population, but it is hardly believable that the adult male population constituted only about 15 per cent of the total provincial population. Honan, as another example, reported in 1935 a total of 5,749,768 households as against only 4,836,021 adult males.[51] It is highly improbable that nearly 16 per cent of the households had no adult males between 20 and 45. Similarly we find that the adult males constituted only 16 per cent of the total reported population of Kweichow and 16.8 per cent of the total reported in Hupei.[52]

Table 17. The 1931 and 1933 Pao-chia returns of ten representative Anhwei counties.

County	1931			1933		
	Males	Females	Total	Males	Females	Total
Huai-ning	355,033	319,116	674,149	348,067	305,214	653,281
Wu wei	416,914	332,056	748,970	400,122	325,302	725,424
Ho-fei	751,759	553,887	1,305,646	736,150	538,202	1,274,352
Shou-hsien	426,665	255,501	682,166	423,737	275,623	699,360
Ho-hsien	159,017	119,948	278,965	160,896	135,484	296,380
Ssu-hsien	328,957	272,223	601,180	291,534	256,134	547,668
Fu-yang	753,250	626,442	1,379,692	970,882	833,758	1,804,640
Kuei-ch'ih	158,527	129,177	287,704	151,706	127,999	279,705
Hsüan-ch'eng	321,558	224,371	545,929	281,184	204,968	486,152
She-hsien	181,729	162,331	344,060	153,625	134,899	288,524
Total	3,853,409	2,995,052	6,848,461	3,917,903	3,137,583	7,055,486

Source: *Nei-cheng nien-chien* (1935), Vol. II, Chs. 4 and 5. These ten counties, excluding the capital city Huai-ning, were the most populous among the ten administrative areas (*ch'ü*) into which the province was divided.

Sometimes even within the same province the results of pao-chia registration showed conflicting tendencies. In Table 17 Anhwei, which gave more details than other provinces, serves as an example.

Among these ten counties only Fu-yang reported a substantial gain in male and female populations between 1931 and 1933. With the exception of Ho-hsien, which reported a very small increase in male population in 1933, the other eight counties all reported a decline of male population in two years. It may be said that while the general tendency for most counties was to report fewer adult males, occasionally a more energetic enforcement of pao-chia regulations could substantially increase the registered population of both sexes, as was the case in Fu-yang and all other counties in the same administrative area. These conflicting tendencies make an overall appraisal of the provincial population figures impossible.

There is no evidence that the pao-chia system was ever established on a truly provincial scale outside a number of lower and central Yangtze provinces, which were under Nanking's effective control.[53] It took the highly centralized Ch'ing government decades to revamp the traditional machinery and to establish a nation-wide pao-chia network, and this was made possible only by the prolonged domestic peace and prosperity of the eighteenth century. With our knowledge of the political conditions of the Nationalist period it seems inevitable that there should have been a wide discrepancy between regulations and practice. Most provinces had scarcely had time to establish an extensive pao-chia network when war with Japan broke out in 1937. After 1945 civil war again broke out and proved to be the undoing of the Nationalist government. For most provinces, therefore, the population figures of the Nationalist period were almost pure guesswork. Even in certain Yangtze provinces the sharp fluctuations in the number of pao and chia indicated that the pao-chia system was far from stable and orderly as Table 18 on Hunan illustrates.

By far the worst feature in the Nationalist population administra-

Table 18. Number of Pao and Chia in Hunan, 1937–1941.

Year	*Pao*	*Chia*
1937	41,652	457,711
1938	39,138	1,046,452
1939	22,159	270,912
1940	19,783	251,322
1941	20,087	286,597

Source: *Hu-nan min-cheng t'ung-chi* (Department of Civil Affairs, Hunan Provincial Government, 1941).

tion was not the lassitude of the provinces, which could not be expected to establish an efficient province-wide pao-chia network overnight, but the manipulation of national and provincial population totals by the central government. The Ch'ing Board of Revenue was at least honest in keeping the provincial returns as they were originally submitted, but the organs of the Nationalist government constantly revised and trimmed the provincial figures without explanation. For instance, the results of the 1928 count supervised by the Ministry of the Interior and computed in 1931 gave a total population of 474,787,386. This figure, admittedly a rough estimate, was not only rejected by the Directorate of Statistics; it alarmed the Ministry of the Interior itself. After considerable trimming the following figures shown in Table 19 seem to have been accepted by both these organs.

Table 19. Official population figures for all China.

Year	Households	Mouths	Source
1928	83,865,901	441,849,148	Ministry of the Interior
1933	83,960,443	444,486,537	Directorate of Statistics
1936	85,827,345	479,084,651	Ministry of the Interior

Source: *Chung-kuo jen-k'o wen-t'i chih t'ung-chi fen-hsi* (Directorate of Statistics, 1946), p. 11.

There was still another series of population figures, given annually in the *China Yearbook*, which differed from those of the two government organs. Its editorial board, nominally independent of the government but actually consisting of members of the government and those closely affiliated with it, relied on neither of these government organs but on experts from the Institute of Social Sciences of the Academia Sinica. This was because its figures were for foreign consumption and had to be further polished. It would be tedious and unnecessary to compare various sets of figures compiled by government organs and nongovernment experts of the Nationalist period. It is enough to point out that the national population totals during the two decades of the Nationalist rule ranged between 430,000,000 and 480,000,000 and all of them were guesswork, since no figure could be authentic without a nation-wide machinery to collect the data.

IV. THE CENSUS OF 1953

Since we lack details of the 1953 census returns which are likely to remain classified material for years to come, it is not possible to make a systematic appraisal. In discussing the organization, procedures, regulations, and methods of the census, it is important to bear in mind the character of the Chinese Communist state, which differs fundamentally from that of any previous regime and which must have had an important bearing on the collection of population data.

For all its oriental despotism, the traditional Chinese state at its most efficient could seldom reach the broad masses of the people directly. The individual, after paying taxes and performing certain services required by law, had little direct contact with the state. He was cushioned in Ming and Ch'ing times by such subcounty organizations as the li-chia and pao-chia, whose personnel were mostly selected and appointed by the county government, and by village and clan authorities chosen by the people themselves. While the government remained authoritarian in form as well as in practice, the traditional saying contained a good deal of truth: "The sky is high up and the emperor is far away." The Nationalist regime at its height, for all its pretensions, had no administrative machinery with which to bridge the gulf between the county government and the masses.

The Communist regime, on the other hand, had perfected its technique of mass control and mobilization in areas under its control even before its rise to power in 1949. After 1949 the Red Army, the various types of police force, the party cadres and nonparty activists, who were all inspired by a militant ideology and high ideals and who went out in millions to the peasants and workers, the numerous mass organizations which the people were encouraged and almost forced to join, the residents' teams and residents' committees which kept vigilance over possible counterrevolutionaries and the activities of the inhabitants, have all made the existence of individuals as individuals extremely difficult. The agrarian revolution and nation-wide redistribution of land, the institution of food requisitions based on individual family needs and production, the formation of rural mutual-aid teams which have been rapidly superseded by agricultural producers' cooperatives have further enabled the state to establish direct contact with the individuals. In the last analysis it is the genius of the Communist regime for controlling the masses at the grass-roots that has made it unique in Chinese history. A Soviet census adviser said [54] that under the new regime the people no longer had any incentive to evade census authorities. But it is also an irrefutable fact that in 1953 the people had no place to hide from the omnipotent and omnipresent state.

The population figures released officially and unofficially between 1950 and July 1953 have given rise to criticism and skepticism of the 1953 census itself. Not all the officially released and privately estimated national population totals between 1950 and 1953 are available. The first official figure given by the Ministry of the Interior in May 1950 was 483,869,678, including Formosa. Two months later the headquarters of the People's Liberation Army gave the figure as 492,530,000.[55] These figures are obviously adapted from the commonly accepted notion that the population during the late Nationalist period was around 475,000,000. The two available figures for 1951 — 483,000,000 given by the Economic and Financial Commission, and 486,600,000 by the newspaper *Ta-kung-pao* of Shanghai — differ very little from earlier ones.[56] The 1953 census figure of 582,600,000, excluding Formosa, exceeded all these figures by about 100,000,000 and this has been one of the main reasons why some question its authenticity.

That most, if not all, of the mainland-released figures up to the end of 1952 should have been a continuation and adaptation of previous Nationalist figures was indeed inevitable. During the early stage of Communist power consolidation a total rejection of the late Nationalist figures would have meant a complete statistical vacuum. Even the present Communist land statistics are essentially an adaptation of better estimates made during the Nationalist period. In the light of the circumstances and of the dubious quality of the Nationalist population figures the contradiction between the Communist figures of 1950–1952 and the census figure of 1953 constitutes no valid reason for the rejection of the latter.

There were two important figures computed by the State Statistical Administration for 1951 and 1952, which were based on local and regional administrative reports and some sample government surveys and which were not made known until after the completion of the 1953 census.[57] The figure for the end of 1951 was 563,194,227 and that for the end of 1952 was 575,286,076, both presumably including estimates for Formosa. Probably based on some of the same administrative reports released in newspapers, one of the two atlases compiled in 1953 gave a total of 545,156,042 and another was content to say that the total national population, including Formosa, was over 540,000,000.[58] From these facts and figures it is clear that after the release of some initial necessarily stopgap figures in 1950–51 the Communist government organs became fairly cautious and reluctant to release further figures, presumably because they were amazed by the substantial differences shown by more responsibly collected information in local and regional administrative reports. It is important to bear in mind that the 1953 atlas figures were patchwork private estimates assembled from fragmentary newspaper reports and that all the post-1949 and pre-census figures, except the two unreleased figures arrived at by the State Statistical Administration, were guesswork.

Although detailed information is lacking, the machinery for ascertaining the size of the population seems to have undergone significant changes between late 1949 and the time of the census. Immediately after the establishment of the People's Republic in October 1949 the Nationalist system of reporting local populations by the pao-chia machinery was continued out of necessity. The pao-chia reports lacked accurate information on sex, age composition, and nationality, which impeded the new regime in carrying out economic planning and educational and health reforms. Through checks carried out in a number of rural and urban areas the new regime discovered the untrustworthiness of pao-chia figures in general and for urban areas in particular.[59] Before the end of 1950 the pao-chia system was abolished and interim population reports were made by local administrative organs. It seems more than coincidental that this change in the machinery for population registration should have resulted in the vastly different totals computed by the State Statistical Administration for 1951 and 1952.

By 1952 there was a greater need for authentic population figures, due to the impending general election of deputies to the National People's Congress and preparations for the launching of the first Five-Year Plan. Using the 1939 Soviet census as a model, the State Statistical Administration began to develop a total organizational plan for taking the first census—by designing forms and working out census procedures, training census personnel, and dispatching directives to administrative organs on various levels. The census procedures and the decision to take the census in mid-1953 were widely publicized in newspapers and radio networks.

Originally the census authorities planned to count both the permanently resident and the mobile populations. Since the counting of people in transit would have to be carried out within a very short period of time in trains, railway stations, hotels, and many other places, and since such a plan would require an enormous number of census-takers, it was later decided that the census should undertake only an enumeration of the resident population. A special census sheet was designed to avoid the dangers of double-counting. Special pains were taken not to offend the family, which traditionally regarded certain members as temporarily absent from home even though they had lived in another locality permanently or over a long period. The general family registration schedule (Form A) [60] was therefore divided into two parts. All members who at the time of the census lived in the family were registered on the right-hand side, including those who were by census definition "temporarily absent" (for not more than six nmoths). Those who had been living away from the family permanently or for over six months were registered on the left-hand side.

It was originally planned that the census should register sex, age, nationality, literacy, occupation and place of work, and social class. Partly because of technical difficulties and partly because of the extra work involved, the census authorities decided in the late stage of planning to retain the first three items only. The census was therefore concerned merely with sex, age, nationality, and the permanent resident population.

Although Form A had two parts, in actual practice only the part that dealt with members living permanently at home was used; the other part was designed merely to avoid offending members of the family and to prevent the double-counting. Persons who were by census definition permanently absent from home were mostly students living in dormitories, workers living in factories or doing construction work away from home, certain categories of government employees, members of armed forces, long-term hospital inmates, etc. They were all registered at their institutions or places of work. Seasonal workers, students away for short-term classes, temporary patients in hospitals and sanatoriums, and persons in hotels were registered on the right-hand side of Form A and regarded as permanent residents of their home localities. Special efforts were also made to count the boatmen of the southeastern coast and those in transit on and after the national census day.

FORM A

Location of Town	Location of Rural District
Province	Province
Town	Administrative Area
County	County
Street	Village *
Lane	Hamlet *
House Number	

Absent population (family members living permanently in another place)				Resident population (persons living in the place in question)									Category of population
Total number of persons living in another place ——, Men ——, Women ——				Total resident population ——, Men ——, Women ——								Head of family	Relation to head of family
													Surname and given name
													Sex
													Age
													Nationality
													Additional remark

Person completing census sheet: —— Date of filling form: 1953, —Month —Day.

* In the absence of original Chinese terms it is difficult to know what the lowest rural administrative units are. They are probably *hsiang*, which means a group of villages, and *ts'un*, which means a village.

Form B was designed especially for single persons without families. It covered mainly students, members of military units, and others who lived a communal life. Since the registrant was a single person without family, Form B contained no left part and omitted the item of relationship to the head of the family. Additional space was provided for the name of the organization or institution to which the registrant belonged. Otherwise the two forms were alike.

A great deal of thought had been given to proper methods of listing family and given names, classifying sex, and determining age. In a country like China even these simple aspects of a census could present serious problems. As it turned out, many women were given personal names for the first time. In view of the general skepticism about the high sex ratio for males, the census authorities insisted that the Chinese character *nü* (female) must be used on all occasions concerning female persons. No abbreviations or ditto marks were allowed in the case of several female entries in one family. The determination of age was by far the most troublesome. For illiterate persons or for those who could remember only the year in pre-Republican imperial chronology and the date in the lunar calendar, census-takers were equipped with conversion tables. The common Chinese custom of counting *sui* was in every case to be converted to actual age.

Careful as the census plan was, it was not without flaws. The first drawback was that the term nationality was not strictly referred to ethnic affiliation. According to the census regulations, the census-taker was only to ask the registrant of what national group he considered himself to be a member; there was no objective criterion for determining nationality. The cases of *Hui* (Chinese Moslems) and Manchus were both misleading. The Chinese Moslems, though classified by census definition as *Hui*, were in reality mostly Chinese in race. It is common knowledge that since the Manchu abdication in 1912 most of the Manchus have adopted Chinese surnames and become thoroughly Chinese.

Secondly, not all of China's population was directly counted during the census. Geographic inaccessibility, linguistic and other difficulties made it impossible to carry the census to the whole of Tibet, and to sizable parts of Sinkiang, Chinghai, and Sikang, where the population consisted exclusively or predominantly of national minorities. For these areas census authorities had to rely on reports sent in by tribal chieftains or local governments. Altogether a total of 8,397,477 persons were indirectly registered.

A more serious flaw was that the census failed to observe the critical-moment concept so vital to any census. The critical moment was fixed at the twenty-four-hour period between midnight of June 30 and midnight of July 1, 1953, a far from happy choice because most of the peasants were out working on the farms. The size of the population and of the country made it impossible to fulfill the original goal, which was to have provincial population figures available by November 1953. By March 1954 the census was only 30 per cent completed.[61] For some remote patrs of Kwangtung it was not completed until April 1954. Only by simplifying the original procedures were the final tabulations completed and results announced in June 1954. As it turned out, the critical moment was applied only to births and deaths and to the determination of age. By regulation babies born after the critical moment were not to be counted, but persons who died between the critical moment and the time of actual enumeration were to be included. The age of each registrant was determined as of the critical moment irrespective of when the actual count took place.

There was yet another possible flaw of a grave nature. Since the census personnel, before the census began, were given certain reference material, compiled from local and regional administrative reports and showing the general character of the local population to be covered, the possible relation between the pre-census estimates and the census results is worth speculating on. In 1956 two writers expressed the opinion:

> It would seem probable that local census enumerations or registrations which agreed with the figures given at the outset to the census personnel would be assumed to be correct; that, contrariwise, divergent results would be assumed to be in error and either verified or adjusted. The checking of census figures against the estimated figures may well have assumed massive proportions in the last few months when the census was being completed swiftly along with the registration of voters.[62]

All in all, therefore, the nationwide enumeration of 1953 was not a census in the technical definition of the term. On the other hand, never in the history of China had a similar project been undertaken on a comparable scale; two and a half million workers, students, school teachers, civil servants, and activists were trained and mobilized as census-takers between late 1952 and early 1954. Even if the two pre-census estimates made by the State Statistical Administration exerted undue influence on the final tabulations, the census results seem likely to be closer to the truth than any previous Chinese population figures.

The preliminary tabulations, completed in June 1954, yielded a total population of 582,584,859. As a check, a sample survey was carried out in 343 counties in 23 provinces, five cities, and one autonomous region. This check was said to have covered 52,953,400 persons or 9 per cent of the total population. It was reported that duplications were found to have been 1.39 per thousand and omissions 2.55 per thousand. A revision was therefore made of the figures for 98.5 per cent of the total population reported as directly counted, and this yielded a total of 582,603,417. The 1953 population figures for various provinces are given in Table 20, together with the 1947 Nationalist provincial totals.

Table 20. Official population figures by provinces,[a] 1947 and 1953.

Province	A Population (1947)[b]	B Population (1953)	B as a Percentage of A
Kiangsu	37,089,667	41,252,192	111.2
Chekiang	19,942,112	22,865,747	114.7
Anhwei	21,705,256	30,343,637	139.8
Kiangsi	12,725,187	16,772,865	131.8
Hupei	21,784,415	27,789,693	127.6
Hunan	26,171,117	33,226,954	127.0
Szechwan	48,107,821	62,303,999	129.5
Sikang	1,651,132	3,381,064	204.8
Hopei	28,529,089	35,984,644	126.1
Shantung	39,425,799	48,876,548	124.0
Honan	28,473,025	44,214,594	155.3
Shansi	15,025,259	14,314,485	95.3
Shensi	10,015,672	15,881,281	158.6
Kansu	7,671,106[c]	12,928,102	168.5
Chinghai	1,346,320	1,676,534	124.5
Sinkiang	4,012,330	4,873,608	121.5
Fukien	11,100,680	13,142,721	118.4

Table 20 (*continued*).

Province	A Population (1947) [b]	B Population (1953)	B as a Percentage of A
Kwangtung	29,101,941	34,770,059	119.5
Kwangsi	14,603,247	19,560,822	133.9
Yunnan	9,171,449	17,472,737	190.5
Kweichow	10,518,765	15,037,310	143.0
Liaoning	18,673,664	18,545,147	99.3
Kirin	14,212,369	11,290,073	79.4
Heilungkiang	5,298,235	11,897,309	224.6
Jehol	6,109,866	5,160,822	84.5
Tibet	1,000,000	1,273,969	127.4
Shanghai	3,853,511	6,204,417	161.0
Peiping	1,602,234	2,768,149	172.8
Tientsin	1,679,210	2,693,831	160.4
Chahar	2,114,288 [d]	....	....
Suiyuan	2,166,513 [d]	....	....
Inner Mongolia		6,100,104 [d]	
Total	454,880,279	582,603,417	128.1

Source: The 1947 figures are from *China Yearbook* (1948).
[a] Taiwan figures for 1947 and 1953 are omitted.
[b] The populations of 1947 special municipalities which were abolished by the People's Republic of China before 1953 are added to the 1947 totals of the provinces to which they now belong.
[c] Includes the 1947 population of Ninghsia, which has been merged with Kansu.
[d] Not comparable.

The unavailability of full census returns enables us to discuss only a few demographic aspects of China's 1953 population. Shortly after the completion of the census the State Statistical Administration announced that the urban population as of June 1953 was 77,257,282 or 13.26 per cent of the total population. The so-called urban areas included *chen* (townships), which are under the jurisdiction of the county, the industrial and mining *ch'ü* (administrative districts), provincial cities and cities under the jurisdiction of the central government.[63] This division of rural and urban areas is arbitrary and misleading. The average chen is a comparatively populous shopping center surrounded by farms rather than factory sites. It is essentially a rural township. The ordinary provincial cities which account for the bulk of the "urban population" are in reality the areas within the city walls, centers of rural counties. Since the definition of urban areas is based not on the size of the local population and only partly on the type of the local economy, the figure on urban population is at best of partial reference value.

Based on a 1954 survey of "29 large and medium cities, the whole province of Ninghsia, 10 counties of other provinces, as well as one representative ch'ü (administrative district), 2 representative chen (rural township), 58 representative *hsiang* (group of villages), and 7 representative *ts'un* (village) of 35 counties, covering a population of 30,180,000, the present birth rate of China averages 37 per thousand; and the annual rate of population growth 2 per cent." [64] Lacking detailed information as to the localities surveyed and registration procedures used, it is difficult to say how representative this survey may be. There is reason to believe that both the birth and the death rates are too low, because not until the adoption of new regulations on population registration in January 1958 did the government formally admit to a hitherto comparative neglect of, and lack of facilities for, the registration of births and deaths, especially in the rural districts.[65] The two-per-cent annual increase must await future verification, although it is likely that even rudimentary sanitary reforms and health campaigns must have contributed substantially to lowering the traditionally high death rates.

The census abstract shows a total of 297,553,518 males as against 276,652,422 females, excluding areas under indirect registration. The males constitute 51.82 per cent and the females 48.18 per cent of the total population, giving a sex ratio of 107.5. When this ratio is interpreted against the traditional cultural, economic, and sociological background and in the light of major modern civil and international wars which all brought about more male than female deaths,[66] it should confirm the common notion that China has had a substantial excess of males over females. There is some evidence that female infanticide has been in process of decline in modern times. The laws of the Communist government prohibit it, and amid the legal and social campaigns to exalt the status of women it is likely that the sex ratio for children, if available, would conform more to the natural pattern of the modern West.

More is known about the age composition, although the information is far from complete. Out of a total population of 574,206,440 directly enumerated, infants through four years old constitute 15.6 per cent, children of five to nine years, 11 per cent, and children of ten to seventeen, 14.5 per cent.[67] The total juvenile population up to age seventeen is therefore 41.1 per cent of the total population. It should be pointed out, however, that the age distribution of the youth may not be regarded as typical of the past. The effect of war and peace on the age distribution of the youth population is clearly discernible. Children under four, constituting the highest percentage group in the total youth population, were born after the end of the Nationalist-Communist war in 1949–50. Children between five and nine, who account for the lowest percentage group, were born in the war years of 1944 and 1948. Children aged ten to seventeen, who account for 1.1 per cent less than the infant group, were born between 1936 and 1943. But for the prewar carry-over and the comparatively stable economic conditions during the early phase of the Sino-Japanese war, the percentage of this age group might have been somewhat lower.

To sum up the official Chinese population data of the past five centuries, it may be said that the more useful ones are those of the Ming T'ai-tsu period (1368–1398), of the period 1776–1850, and of the census year 1953. The data of the period 1741–1775, while highly defective, may be of some help. The century between 1851 and 1949, despite the availability of various figures, is practically a demographer's vacuum. To appraise the probable pattern of population movements during the past half-millennium, it is necessary to know the historical circumstances peculiar to each period and to investigate the various economic and institutional factors that were related to population changes.

PART TWO

FACTORS AFFECTING POPULATION

CHAPTER VI

The Population-Land Relation: Ming, Ch'ing, and Modern Land Data

The following chapters concern matters which have a bearing, directly or indirectly, upon China's population and its history. Here again, the figures recorded in China's voluminous records are very numerous, but the verifiable facts are few. Yet the written record deserves careful study, for it remains our chief avenue of approach to these problems.

Although modern official Chinese land figures have been widely used and often criticized by economists, geographers, and agricultural experts, the nature of traditional Chinese land returns and the influence of traditional principles of land assessment on modern land figures have not been systematically studied. Failure to understand these two points has not only impeded research on certain major problems in the economic history of China but has given rise to misleading interpretations of China's historical population-land relationship. The influence of traditional land data on modern figures can be shown by Table 21. With the exception of the 1602 return, there was no consistent change in the reported area during the 469 years from 1398 to 1867. Yet official population figures would indicate that China's population increased from approximately 65,000,000 in 1398 to 430,000,000 in 1850.

It is clear that this highly improbable proportion between population and official land figures cannot be a reflection of the historical relation between population and land, although some scholars have worked out population-land ratios for the Ch'ing period based on official figures. The traditional land returns were, in the last analysis, numbers of units in landtax payment. They were returns of fiscal rather than actual acreage under cultivation; consequently, they were invariably too low. It is the aim of this chapter to explain the nature of traditional land figures as a basis for criticism of all official land figures of Ming, Ch'ing, and modern times.

I

The traditional Chinese principles of land assessment were complex but based in general on considerations of economic justice. Since land varied in fertility, equal amounts of land in different locations could not be assessed at the same rate. With the disappearance of public ownership of land and the complete triumph of private ownership after the middle of the eighth century, land had come to be classified for taxation purposes into various grades. There were three main grades — upper, medium, and lower — each further classified into three subgrades. This was known as *san-teng chiu-tse* or "three main grades and nine sub-grades." Upper-grade land

Table 21. Official Chinese land data, 1398–1932.

Year	Reported area (in *mou*)	Reported area (in acres)
1398	813,187,917	133,900,000
1502	422,805,881	69,600,000
1578	701,397,628	115,600,000
1602	1,161,894,881	176,000,000
1645	405,690,504	66,800,000
1661	549,357,640	90,500,000
1685	607,843,001	100,000,000
1724	683,791,427	112,600,000
1753	708,114,288	116,600,000
1766	741,449,550	122,000,000
1812	791,525,100	130,000,000
1833	737,512,900	121,000,000
1867	815,361,714	134,000,000
1887	936,090,273	153,700,000
1932	1,248,781,000	205,700,000

Source: All figures except those for 1602 are conveniently assembled in Wan Kuo-ting, "Chung-kuo t'ien-fu niao-k'an chi ch'i kai-ke ch'ien-t'u" (A Bird's-eye View of China's Land Taxation and the Prospect for its Reform), *Ti-cheng yüeh-k'an* (Journal of Land Economics), vol. IV, nos. 2–3. The 1602 figures are from *Ming-shih-lu*, reign of Shen-tsung (Kiangsu Sinological Library Photostat ed.), 379.14b–15a.

bore a higher tax rate than lower-grade land so that land was taxed roughly according to its productivity.

Because of the technical difficulties and the time and labor involved in such a complex assessment system, a simpler principle gradually emerged without sacrificing the basic conception of economic justice. While historical accounts on the evolution of the principles of land assessment are extremely scarce for the early periods, there is definite evidence that by Sung times (960–1279) medium- and lower-grade land in some areas was converted to upper-grade land at substantial discounts. For example, in 1113 a Shansi official stated in a memorial that the land in Shansi was commonly classified into ten grades, with ten mou of the tenth-grade land equated to one mou of the first-grade land. Not satisfied with this ratio, he requested more lenient rates of conversion for marginal and submarginal land.[1] This practice was by no means confined to North China, for in Ch'ang-shu county in the southern Kiangsu delta the medium and lower grades of land had been converted to upper grade at the rates of 1.5 to 1 and 2 to 1 ever since the twelfth century.[2]

By the Ming period, this practice of mou conversion was common. T'ang Shun-chih, a famous scholar of the early sixteenth century, said:

> The difficulty of land survey lies not in measuring but rather in properly grading the land. This is because land varies in fertility and cannot be subjected to a uniform rate of taxation. [An equitable principle is that] the land of a county should first be graded into three general categories according to productivity. After this the land should be measured and converted to actual [i.e., fiscal] mou. This is like the basic principle of land allotment stated in the *Chou-li* . . . . I have studied the method used at the beginning of the reigning dynasty whereby the number of mou was first converted before the tax rates were fixed. In a fertile county the size of the mou was invariably small and in a poor county the size of the mou was always large. This practice is quite in keeping with the ancient [Chou-li] concept.[3]

While there can be little doubt that this was correct as a general principle, the Ming system of land assessment was very complex. Local histories show that during the Ming and Ch'ing a considerable number of communities still employed the earlier system of assessing land into "three main grades and nine sub-grades." Barring possible large-scale evasions of land taxation and registration, the land returns of such localities should in theory be closer to true acreage returns than those of the areas where mou conversion was extensively practiced. In view of the fact that mou conversion has contributed much to under-registration of utilized land in Ming, Ch'ing, and modern times, it is necessary to know the areas in which cultivated land was extensively converted to fiscal acreage.

II

The conversion of actual mou into fiscal mou was practiced particularly in the low plain areas of north China during Ming times. It may be said that one of the basic reasons for the existence of various sizes of mou, which are deviations from the standard mou of 240 *pu* (5 square feet), was the various rates of mou conversion. Ming works clearly explain that natives of the low plain areas of north China had from Mongol times occupied large-mou land. In other words, the land had already been converted to fiscal mou at substantial discounts. After the population decline caused by wars, rebellions, and calamities toward the end of the Yüan period, people from more peaceful loess high plain areas were ordered by the early Ming government to migrate to the low plain areas, where they were settled on small-mou land, which bore heavier rates of taxation. This economic injustice gradually gave rise to widespread complaints and made reassessment inevitable. It is interesting that, although the new assessment discovered large areas of land hitherto unregistered, the guiding principle was not to increase the original local quotas of taxed land but to allow the small-mou land to be converted at reasonable rates to the large-mou land. It is not surprising, therefore, that in many parts of Hopei, Shantung, and Honan 1.8 or 2 actual mou of upper-grade land and sometimes 7 or 8 mou of poor-grade land were customarily equated with a single fiscal mou.[4]

Ming and Ch'ing local histories of Hopei, Shantung, and Honan that yield information on mou conversion are too numerous to be listed in full, but two cases deserve our special attention. The first is Yü-chou in southwestern Honan, whose 1,532 assessment received wide publicity and often served as a model for other localities. The second is Li-ch'eng county, or modern Tsinan, the capital city of Shantung, which may be regarded as having reached an advanced stage of rationalization and simplification in land assessment. In the 1532 assessment of Yü-chou, land was classified into five grades. The uppermost grade was given no discount, the other four grades were equated at 2, 3, 5, and 10 mou to a single tax-paying mou.[5]

In Li-ch'eng various grades of land were gradually converted to a standard fiscal mou called the *i-li ta-liang-ti*, or the "uniform large tax-paying land," which is defined and described as follows:

> Among privately owned land, the uppermost grade is called golden land, 240 pu of which constitute a [large tax-paying] mou. The second grade is called silver land, 280 pu of which constitute a mou. The third grade is called copper land, 360 pu of which constitute a mou. The fourth grade is called tin land, 600 pu of which constitute a mou. The fifth grade is called iron land, 720 pu of which constitute a mou. As to land under military tenure, the uppermost grade corresponds to privately owned golden land, 240 pu per mou. The medium grade corresponds to privately owned copper land, 360 pu per mou. The poor grade corresponds to privately owned iron land, 720 pu per mou.[6]

The customary conversion rates for five grades of privately owned land are therefore 1, 1.166, 1.5, 2.5, and 3 respectively, and those for the three grades of land under military tenure are 1, 1.5, and 3. Li-ch'eng is a level and fertile district and its conversion rates were less generous than those of poorer districts. In Yung-ch'eng county on the western tip of the Shantung promontory, for example, 1,200 pu or 5 mou were equated with one fiscal mou during the reign of Ming T'ai-tsu, although later on the conversion rates were slightly reduced.[7] To give another example, when military tenure was established in the marshy flats south of Tientsin at the turn of the sixteenth century, usually 5 or 10 mou of newly cultivated land were reckoned as one tax-paying mou in order to attract settlers.[8]

These cases illustrate the trend in land assessment in the low plain provinces of north China, but inevitably there were deviations from the general practice. In localities where the original conversion rates were unusually generous or at times when the need for local government revenue was great, upper and medium grades of land were converted to lower grade, thus increasing the local quota of taxed land.[9] Such cases were comparatively few, however, and were almost entirely confined to the unique tax-raising period of Wan-li (1573–1619).

Although the general tendency was for medium and poor grades of land to be converted to a uniform upper grade, which was equated with the fiscal mou, there were cases where even the uppermost grade was still converted to the fiscal mou with a certain discount. In Chi-yang county in Shantung the uppermost grade was equated to the fiscal mou at the rate of 1 to 0.5917.[10] In still other cases mou conversion and the earlier principle of assessing land into "three main grades and nine subgrades" were merged. Despite the conversion of land of all grades to the fiscal mou, tax rates were far from uniform. Table 22, giving figures for Wen-hsien county in

Table 22. Conversion of land in Wen-hsien county.

Original grade of land	Rate of conversion to fiscal mou	Tax rate per fiscal mou (in shih of grain)
A	1	0.105
B	2.1	0.05
C	3	0.035
D	3.5	0.0087
E	11.06	0.0087

Source: *Wen-hsien chih* (1746 ed.), 10.2b–3a.

Honan serves as an example. There were also cases like Sung-hsien in Honan where about one-half of the total utilized land was converted to the lowest grade and the other half was assessed according to "three main grades and nine subgrades."[11]

The local histories of the loess high plain provinces do not usually yield similar information on mou conversion. But this does not mean that the land returns of Shansi and Shensi were more accurate. The 1609 edition of the history of Fen-chou prefecture in southern Shansi gives more specific information than other Shansi local histories. In Ming times the land of the prefecture, much as in Li-ch'eng in Shantung, was classified into five grades—golden, silver, copper, tin, and iron. While there is no mention of the conversion rates for the upper three grades, four mou of the lowest two grades were equated to one fiscal mou.[12] In fact, mou conversion must have been fairly common in Shansi, because the practice in certain Shansi counties dates back to the early twelfth century.

The system of mou conversion must also have been sufficiently ancient in Shensi as to have almost binding force. In north central Shensi attempts were made by some officials in the 1650's to register the acreage of newly cultivated land without traditional discount. So strong were local customs and the people's resistance that from 1660 the time-honored generous conversion rates were not only firmly upheld for old developed land but were also applied to newly tilled land. For the entire north central Shensi area the conversion rates ranged from 3 or 4 to 8 or 9 mou for one fiscal mou.[13] In 1735 Shih I-chih, a native of Shensi, concurrently governor of Shensi and president of the Board of Revenue, testified in a memorial that similar conversion rates prevailed in other parts of Shensi and suggested that one way to encourage large-scale agricultural development was to allow medium-grade land to be converted at the more generous rates hitherto given exclusively to the poorest grade of land.[14] In Kansu, the most northwestern province of China Proper, the "large-mou land" remained common up to the twentieth century.[15]

In the densely populated lower Yangtze area land was converted to fiscal acreage less extensively than in north China, partly because of more careful assessments in the early Ming period and partly owing to the prevalence of highly productive irrigated paddies. In contrast to the piecemeal land assessments carried out by local authorities in other areas during the Ming T'ai-tsu period, the land of southern Kiangsu and Chekiang was ordered to be carefully measured before tax rates were fixed. Northern Chekiang was surveyed and assessed twice during the reign of the Ming founder, once in 1368 and once shortly before 1398. Although the *Yü-lin t'u-ts'e* were compiled by most provinces in the early Ming, none was compiled with greater care than those of Chekiang and southern Kiangsu, the richest parts of China. The true nature and scope of the famous early Ming land surveys, which have often been misunderstood to be on a national scale, are best described by the early sixteenth-century scholar Huang Tso:

On March 2, 1398 the *Yü-lin t'u-ts'e* were completed. Formerly, the emperor had ordered the Board of Revenue to assess the land of the country. The rich people of Su-chou and Ch'ang-chou prefectures of Kiangsu had evaded labor services by cunningly arranging to have the title deeds of their properties registered under names of their kinsmen, neighbors, tenants, or servants. This had become a vogue and given rise to various illegal practices, with the result that the rich had grown richer and the poor poorer. Upon hearing this, the emperor dispatched the student of the Imperial Academy, Wu Ch'un, and a number of his colleagues to go to those areas, which were to be subdivided for purpose of land-tax payment. For each such land-tax subdivision four headmen were appointed who were to assemble the li-chia personnel and elders to carry out a land survey. The areas of holdings were computed and given registered numbers with the owners' names attached. All these were written [for convenience] at the four corners of the registers which were officially coded. They were called *Yü-lin t'u-ts'e* because the maps look like fish scales. By now [March 2, 1398] Chekiang province and Su-chou and some other prefectures of Kiangsu had submitted their registers. The emperor was pleased and rewarded Wu Ch'un and his colleagues with silver and paper money.[16]

This account, based on archives of the Imperial Academy in Nanking whose members played such an important role in land survey in the late fourteenth century, is much more accurate and specific than the brief account given in the official history of the Ming dynasty, which contains errors.[17] The important thing we learn from Huang's account is that the so-called land survey of the Ming T'ai-tsu period was confined to Chekiang and southern Kiangsu; the rest of the country had never been surveyed with equal care. It is interesting to observe that Chekiang, which was twice assessed and surveyed in the early Ming, returned a total of 51,705,151 mou, a figure that exceeds any subsequent official return and J. L. Buck's estimate for the 1930's. From the figures and from Ming accounts there is some reason to believe that among traditional Chinese land statistics the 1393 return of Chekiang was perhaps the closest to the truth.

Yet, despite the relatively thorough and careful surveys, it cannot be said that this was a true acreage return. Although Chekiang local histories are usually reticent about mou conversion, occasionally there are references to it. For example, in Chia-shan county, on the south of the T'ai-hu Lake, one of the best irrigated districts in China, the 1583 assessment allowed certain categories of medium- and poor-grade land to be converted at the rates of 1.5 or 2 actual mou to one fiscal mou, "according to well-established precedents." [18] In defending such a practice the local scholars cited cases of neighboring Chekiang counties and those of southern Kiangsu. Similarly the 1582 assessment of Chiang-shan county in the hilly southwestern corner of Chekiang fixed the following equations shown in Table 23. These 1582 conversion rates may be different

Table 23. Land assessment of Chiang-shan county.

Category of land	Amount of fiscal mou per actual mou
Irrigated land	0.9257
Nonirrigated land	0.793
Hilly land	0.9984
Ponds (fish)	1.211

Source: *Chiang-shan HC* (1623 ed.), 3.2b–3a.

from those of the Ming T'ai-tsu period, but from our knowledge of the strength of local customs it is highly probable that the early Ming assessment was based on a similar principle.

The system of mou conversion is more clearly recorded in southern Kiangsu local histories. Ever since the twelfth century, land of medium and poor grades had been equated to fiscal mou at the rates of 1.5 and 2 to 1 in Ch'ang-shu county. This is a strong indication that, careful as the late-fourteenth-century surveys were, they could not but abide by old local customs. The 1819 edition of the history of Sung-chiang prefecture, of which Shanghai was a leading county, has preserved detailed customarily fixed conversion rates. Table 24 gives the equations of Hua-t'ing county, a part of modern Shanghai. The first four categories were all irrigated land and the last two were marshy land where peasants gathered fuel.

I-hsing county of Ch'ang-chou prefecture, using a reverse process of conversion, had the conversion rates shown in Table 25. It is particularly important to bear in mind that even in this richest agricultural area mou conversion was common, although there are cases where the local customary mou is smaller than the standard mou of 240 pu.[19]

In contrast to the small allowance given to mou conversion in rich southern Kiangsu, northern Kiangsu has been one of the most

Table 24. Land assessment in Sung-chiang prefecture.

Category of land	Amount of actual mou per fiscal mou
A	1
B	1.5
C	2
D	2
E	3
F	3
G	6
H	6

Source: *Sung-chiang FC* (1819 ed.), 22, *passim.*

notoriously under-registered areas in China. The famous scholar Yen Jo-chü (1636–1704) commented on land assessment in Yang-chou prefecture:

Chiang-tu county, with some 1,700,000 mou of farm land, bears a total land tax of over 50,000 taels. Kao-yu county, with some 2,500,000 mou, bears a total land tax of 40,000 taels. T'ai-chou county, with some 900,000 mou, bears a total land tax of about 44,000 taels. It is not that T'ai-chou, with approximately one-third as much land as Kao-yu, bears three times as much tax [per mou] as Kao-yu. It is rather that the mou of T'ai-chou is large and the mou of Kao-yu is small. For another example, Hsing-hua county, with some 2,400,000 mou, bears a total land tax of about 20,000 taels. Pao-ying county, with only about one-tenth

Table 25. Land assessment in I-hsing county.

Category of land	Amount of fiscal mou per actual mou
Best level irrigated	1
Higher level irrigated	0.8595
Low level irrigated	0.7778
Hilly bamboo	0.4395
Highest level	0.4395
Tea growing	0.3334
Sandbars, ponds, and marshes	0.2528

Source: *I-hsing HC* (1799 ed.), 3.15a–16a.

as much land as Hsing-hua, bears ten times as much tax [per mou] as Hsing-hua. This is because the mou of Pao-ying is large and the mou of Hsing-hua is small. For the small mou, which is usually equated with the fiscal mou, the rate of taxation has to be light; conversely, for the large mou, which is the result of conversion of several actual mou into one, the rate of taxation has to be heavy. This is not clearly stated in the *Fu-i ch'üan-shu*.[20]

Further north in the lower Huai River valley of Kiangsi, as in Huai-an prefecture, a late seventeenth-century official testified that there had been two systems, of large and small mou, ever since the late fourteenth century, if not earlier. In general, 4.2 small or actual mou were equated with one large or fiscal mou. All landtax computations were based on this conversion rate.[21] In Hai-chou, in the extreme northeast of Kiangsu along the coast, the original land quota of the sixteenth century was 1,146,000 mou. It was raised at the beginning of the Ch'ing period to 2,050,000, on a uniform principle that 100 actual mou be equated with 31 fiscal mou. Since then the population of the area had been increasing and much land reclaimed, but the reclaimed land was converted at the rate of 8 to 1.[22] As a result of extensive use of mou conversion in northern Kiangsu, the historical and modern Kiangsu land returns are as a rule considerably below the truth.

Perhaps the most systematic and revealing information on the disparity between traditional land statistics and the actual extent of cultivated land is in the various editions of the history of Anhwei province. It may be said that the system of mou conversion was extensively practiced in the whole province throughout early modern and modern times. Even the local histories of some exceptional districts which yield no specific conversion rates give clear indication that the total acreage of the locality is the converted fiscal acreage. Down to the twentieth century there were usually two kinds of land registers, one containing the returns for actual mou and one the returns for fiscal mou. Only the fiscal acreage is entered into the official land registers.

The samples given in Table 26 are probably fairly representative of the whole province. The smallest conversion allowances are found in the trading and densely settled Hui-chou area, of which She-hsien, the capital city of the prefecture, is typical, as may be seen in Table 27. In view of the truly province-wide mou conver-

Table 26. A sampling of land conversion rates in Anhwei province.

Locality	Category of land	Amount of actual mou per fiscal mou
Hsüan-ch'eng	Nonirrigated	3
	Hilly	12
	Sandbars	24
Shou-chou	Medium grade	1.5
	Lower grade	2
T'ai-ho	Medium grade	2
	Lower grade	5
Ho-chou Pref.	Medium grade	1.306
	Lower grade	1.935
	Grass	3.618
	Newly reclaimed hilly	4.680
	Publicly owned ponds	2.247

Source: *Anhwei TC* (1829 ed.), 51–53, *passim*; 1877 edition, 69, *passim*; and the uncompleted draft history of the province compiled in the 1930's, *Anhwei t'ung-chih-kao*, Book on "Finance."

sion, there is strong reason to believe that the land returns of Anhwei, one of the important food-surplus areas in China, have been unreasonably low.[23]

Local histories of other southern provinces only occasionally yield information on mou conversion. In Nan-ch'eng county in Kiangsi the best land was equated to fiscal mou without discount, but medium and lower grades were customarily converted at the rates of 1.5 and 2 to 1 fiscal mou.[24] Scholar-gentry of P'ing-hsiang, the important modern coal-mining district near the Kiangsi-Hunan border, petitioned for the maintenance of the traditional landtax

Table 27. Conversion allowances in Hui-chou area.

Category of land	Amount of fiscal mou per actual mou
Nonirrigated	0.561
Hilly	0.434
Ponds	1.191

Source: *Anhwei TC* (1829 ed.), 51–53, *passim*; 1877 edition, 69, *passim*; *Anhwei t'ung-chih-kao*, "Finance."

quota at the beginning of the Ch'ing period by citing certain precedents whereby twenty mou were equated to one fiscal mou.[25] It is known that up to the end of the Ch'ing period in some Hupei counties one large fiscal mou consisted of 7 or 8 to 11 actual mou.[26] In Chia-hsiang county in Hupei, for example, all grades of land had been consolidated into one uniform rate of taxation. Even without knowing the specific equations it is obvious that the land of the district had been converted to fiscal mou.[27]

Even in Fukien, where the mountainous typography and limited cultivated land forced up land values, the examples of mou conversion shown in Table 28 have been found.[28] Although the exact con-

Table 28. Land conversion rates in Fukien.

Locality	Category of land	Amount of actual mou per fiscal mou
Ning-hua	Upper grade irrigated	1
	Medium grade irrigated	1.4
	Lower grade irrigated	2.5
	Upper grade nonirrigated	2.1
	Medium grade nonirrigated	6
	Lower grade nonirrigated	8
	Upper grade ponds	2.5
	Medium grade ponds	3.4
	Lower grade ponds	6
P'u-t'ien	Upper grade irrigated	1
	Lower grade irrigated	1.03
	Upper grade nonirrigated	1
	Lower grade nonirrigated	1.028494
Hsien-yu	Upper grade irrigated	1
	Lower grade irrigated	1.52
	Upper grade hilly	1
	Lower grade hilly	1.82

Source: Liang, "Single-Whip System"; *P'u-t'ien HC* (1758 ed., 1879 reprint), 5.9b–10b; *Hsien-yu HC* (1770 ed., 1873 reprint), 4.4a–4b.

version rates are not given, some Ming editions of Fukien local histories clearly indicate that all grades of land were converted to fiscal mou.[29]

The accounts of fiscal matters in Kwangtung local histories are usually of poor quality. It is known, however, that in some counties in Hui chou prefecture the 1582 assessment allowed irrigated land to be converted at the rate of 300 pu or 1.25 mou to one fiscal mou, and nonirrigated land at the rate of 600 pu or 2.5 mou.[30] The extent to which land was converted to fiscal acreage in Kwangtung may best be indicated by the fact that the official Kwangtung land returns throughout the Ch'ing period, instead of being simply mou figures, were usually prefaced by the key word *shui* (tax).[31] In other words, the mou figures were clearly explained not as true acreage returns but as another way of expressing the amount of land tax. The figures on newly reclaimed land were prefaced with the same word.[32] The governor of Kwangsi testified in a memorial in 1737 that in Kwangsi it was fairly common to convert cultivated land into fiscal mou at rates of 2 or 3 to 1.[33]

These examples from various provinces indicate that mou conversion was extensively practiced in most parts of China, particularly in the regions north of the Huai River as well as in Anhwei. The extreme brevity of fiscal accounts in many of the local histories and the purposeful reticence about mou conversion on the part of gentry-scholar compilers of local histories in order to safeguard their favorable landtax quotas, make it very difficult to estimate the extent to which mou conversion has contributed to national underregistration. Modern students must accept this quantitative disappointment in their attempt to reconstruct historical land statistics. A description is given by Liu Kuo-fu, *chin-shih* of 1682, in his 1687 memorial:

> Having had the opportunity of working in the Board of Revenue, I have noticed that the regulations on land assessment differ from province to province. The large and small mou systems have existed along with the system of classifying land into upper, medium and lower grades. The number of sub-grades ranges from 1 or 2 to 3 or 4, or even to several tens. Not only does one province differ from another but within each province one prefecture or county may differ from another, because of the varied topographical conditions and different fertility of the land. Ancient people fixed the rates of taxation according to the quality of the soil because it was impossible to make all rates uniform. While among the *Fu-i ch'üan-shu* [of various provinces] many give the specific rates of conversion, those that omit such information are by no means less numerous.

After describing the complex systems of land assessment in his native Yang-chou, he goes on to say:

> The rate of taxation has to be light when one actual mou is equated with one fiscal mou and conversely it has to be heavy when several mou are converted to one fiscal mou. [The main difficulty is that] all this is often not clearly explained in the *Fu-i ch'üan-shu*. Having had the opportunity to go through the *Fu-i ch'üan-shu* of all provinces, I have come to realize that the failure to state specifically the systems of large and small mou is not confined to my ancestral area. Personally I think that although tax rates and principles of land assessment should be made as simple as possible, the systems of large and small mou particularly should be made clear. It is requested that at the time of the next compilation of the *Fu-i ch'üan-shu* edicts be issued to all authorities concerned [to the effect] that the systems of large and small mou and of land grading should be specifically explained.[34]

It is obvious from this memorial that the lack of specific reference to mou conversion in *Fu-i ch'üan-shu* or local histories is not proof of its absence in those localities. In fact, the local gentry's fear that land taxation would be increased if the extent of mou conversion were clearly stated was a common and potent reason for editors of local histories to make fiscal accounts little more than summaries of the locality's obligation to the state. For example, while the histories of Su-chou prefecture and its counties, with the single exception of the 1783–1793 edition of the history of Ch'ang-shu, are entirely reticent about mou conversion, the fact that it was fairly common in Su-chou was cited by some northern Chekiang gentry in defense of their local landtax quotas.[35]

Probably the main reason for the entire lack of reference to mou conversion in such provinces as Hunan, Szechwan, Yunnan, and Kweichow is that they were relatively lately developed and their land assessment was based not on mou but on rough estimates of seeding area or yield. In Li-ling county in Hunan, land has never actually been measured in mou but only by the amount of seed sown. Consequently, the land has always been measured and assessed by the *shih*, or Chinese bushel, of seed sown. To allow further discount, the area sown to one shih of seeds was entered into official land registers as only four-fifths of a shih of land area.[36] In I-yang county in the same province a shih of land was customarily reckoned as equivalent to 6.25 mou.[37] In many parts of Szechwan and Kwangsi similar customs have survived to the present and the size of the shih as the basis for land taxation varies roughly according to the fertility of the land.[38] Yunnan and Kweichow, the two provinces inhabited substantially by aborigines in Ming and Ch'ing times, were yet to develop customary mou systems. Because of their ludicrously small landtax quotas, much of their cultivated land was unregistered. The percentage of unregistered land in these two provinces was definitely higher than that in the areas where mou conversion was extensively practiced.

III

Other major factors having a deflating effect on official land returns have been the large amount of unregistered land and the traditional principle that national tax quotas, once fixed, should not be substantially enlarged. This principle provides perhaps the best explanation to the puzzling fact that, with the sole exception of 1602, there was no increase in officially registered land between 1398 and 1867. The prevalence of many and varied local mou systems, though a further complicating factor, may be regarded as a corollary to and result of mou conversion and hence will not be greatly elaborated. Some factors, on the other hand, have had an inflating effect on land returns. The first of these has been the existence in certain localities of customary mou systems based on a mou smaller than the standard of 240 pu. The second has been found in a number of rice-rich lower Yangtze localities where a portion of the fiscal burden, originally borne by the ting, has been shifted to the land, thus accounting for an exaggeration of local land returns. However, when all conflicting and compensatory factors are weighed, it will become evident that inflating factors are far from offsetting the deflating factors.

There have been two main categories of unregistered land. The first consisted of land that was legally permanently exempt from tax. At the beginning of the Ming period large areas in the low plains of north China which had been devastated by wars attending the downfall of the Yüan dynasty were made permanently tax-exempt, in order to attract agricultural settlers.[39] Although this category was subjected to taxation from the second quarter of the fifteenth century on, Ming local histories provide ample evidence that the aim in assessing was not to increase local quotas of taxed land but rather to relieve the burden of the hitherto tax-bearing land. The bulk, if not all, of this category therefore did not contribute to the increase of land returns for the low plains of north China. In the second quarter of the eighteenth century, when the Ch'ing empire was reaching the apex of peace, prosperity and governmental benevolence, a long series of imperial decrees permanently exempted certain categories of newly reclaimed land from taxation. Depending on the location, the tax-exempt ceiling for small plots of newly reclaimed irrigated land was one or two mou and that for parcels of nonirrigated land was as high as ten mou.[40] Since the growing population of the eighteenth century forced peasants to the increasingly minute division of holdings which has since characterized Chinese farms, the aggregate acreage of such tax-free small plots in various provinces must have reached a significant size. A substantial portion of such land has probably remained tax-exempt and unregistered in modern times.

The second main category of unregistered land has been the property of powerful local gentry and their clans. Historical and modern accounts of the gentry's evasion of land taxation are voluminous. Most local histories contain some reference to it and deplore the unfair distribution of the local fiscal burden. It is impossible to know even the approximate extent to which gentry properties have successfully evaded land registration, but there is ample evidence that particularly serious land under-registration was due to this factor. Alarmed by the fact that the total registered land of the country had dwindled by almost one-half between 1398 and 1502, the famous official Ho T'ao (1480–1540) testified in his 1529 memorial that evasions of land registration had been particularly rampant in Hu-kuang (Hupei and Hunan), Honan, and Kwangtung. In Hu-kuang and Honan the titled imperial clansmen and some powerful gentry had managed to remove their land from local land registers. In Kwangtung, where there were no fiefs of imperial princes, much of the land belonging to the gentry had effectively evaded the assessment.[41] Indeed, the loss of government revenue due to these serious regional evasions of land tax provided the most important incentive for the nationwide land reassessment that began with the second quarter of the sixteenth century.

From historical accounts and from modern figures it is reasonable to think that the amount of unregistered land in Kwangtung has been large. The governor of Kwangtung memorialized in 1732 that the province as a whole had had no *Yü-lin t'u-ts'e* and its land registration had been scandal-ridden.[42] An imperial edict of 1853 stated that millions of mou along the Kwangtung coast which had been turned into highly productive paddies had never been assessed and registered.[43] A sampling of the Yung-cheng Emperor's rescripts reveals that newly cultivated land in various provinces was not properly registered and sometimes even large portions of land within the traditional local quota were no longer recorded in official land registers, through the combined efforts of gentry and local government clerks. Although the backward districts of Szechwan, Yunnan, and Kweichow had been steadily developed, new assessments in these partly aboriginal districts were little more than a formality and seldom based on measurement. The discrepancy between the actual cultivated and the reported areas in these southwestern provinces was great. When the famous scholar Hung Liang-chi, "the Chinese Malthus," said in a poem written while serving as educational commissioner of Kweichow in 1792–95, "on the green furrows people till tax-free land,"[44] he merely pointed out a fact well known to Ch'ing officials, but seldem mentioned by them.

The fact that the land properties of twentieth-century warlords, generals, and officials, especially those of Szechwan, were mostly unregistered is so notorious as to need no elaboration. Professor J. L. Buck, in his monumental study of land utilization in China in the 1930's, carried out a field survey of unregistered land in all northern and southern localities. His findings show the unregistered land in north China to have been 39.2 per cent of the total cultivated area and that of the southern provinces 20.9 per cent. In four Kwangsi counties surveyed the percentages of unregistered land in the total cultivated area were as high as 37.3, 58.3, 64.3, and 79.4.[45] If his data are fairly representative of the whole country, unregistered land must have made a major contribution to the deflating of traditional and modern land figures.

A factor that was related to the persistence of land underregistration was the traditional fiscal principle of *liang-ju i-wei-ch'u* (measuring expenditures against revenues).[46] Usually after the total government expenditures had been roughly estimated at the beginning of a dynasty, the national tax quotas were fixed so as to give the government a small surplus. Theoretically, any good government lived within the bounds of its revenues, although necessities not infrequently forced it later to enlarge the tax quotas. The strength of the fiscal ethic that tax quotas should be maintained as far as possible was most evident in a number of Hopei localities, of which Kuang-p'ing prefecture is typical:

> There are large and small mou systems. At the beginning of the Ming period there was land that was permanently exempted from taxation in order to attract settlers. There was also poor and alkaline land which was not taxed. Later all this land was surveyed, with the result that the newly discovered acreage far exceeded the original quota. Local officials, for fear that this large increment might astound the central government and bring misfortune to the people, reported the total land area in large mou instead of in small actually surveyed mou so as to meet the original quota. In all subsequent returns the numbers were all those of large mou, in order to agree with [the quota] in the Yellow Registers. At a time of special levy the whole amount was apportioned on the basis of small [actual] mou so that fiscal equity could be maintained. One large mou varied from 1, 2, or 3 to 7 or 8 small mou. For medium-grade land the conversion rates were relatively small and for poor-grade land the rates were large, but all grades of land were converted to large *mou*. All this was designed to meet the original registered amount of land in the county and to distribute fairly the original local landtax quota.[47]

While all land returns between 1398 and 1867 show no increase in taxed land, the 1602 return gives a total of 1,161,894,881 mou or 176,000,000 acres, which amounts to 86 per cent of the lowest, or 75.8 per cent of the highest, estimates of the total cultivated area in China Proper by Professor Buck in the 1930's. It is generally agreed that Ming Shen-tsung, commonly known by his reign title Wan-li, was one of the greediest emperors in Chinese history. His need for revenue was a result of increasing extravagance, Korean expeditions, campaigns against southwestern aborigines and Japanese pirates, and his court's increasing awareness that silver was the most concrete expression of wealth.[48] Unable to make ends meet, he ordered fresh land assessments in the country throughout his long reign, with the result that by 1602 the total amount of taxed land registered an increase of 14.2 per cent over the traditional ceiling of 813,187,917 mou fixed in 1398.

It is true that in certain localities the landtax quotas during the Wan-li period were so enlarged that they exceed even Buck's estimates for the 1930's. But such localities, as far as can be attested from local histories, may have been a very small minority. Wen-shang county in Shantung is an illustration. Its 1581 assessment converted all its land into small fiscal mou and yielded a total of 1,745,867 mou, as compared with the original Ming quota of 450,163 large mou. In 1591 another assessment boosted the total to 1,945,342 mou, which exceeds Buck's estimate for the 1930's by about 300,000. But it is most interesting to observe that the 1591 quota was later greatly reduced and by 1764, after a long period of peace, population growth, and agricultural development, the assessed area was only 1,491,900 mou.[49]

The majority of late Ming and early Ch'ing local histories, however, suggest that even during the Wan-li period the combined efforts of local people and compassionate officials succeeded in defending original local landtax quotas. Some local officials, while reporting a net gain in taxed acreage over that of 1398, purposely kept the increment low; it was considered sound ethic that the amount of land newly discovered in a fresh survey should go mainly toward the relief of such land as had hitherto exclusively borne the incidence of taxation, instead of being fully reported to higher authorities. After the Wan-li period the national landtax quota reverted to a level below 800,000,000 mou, even though the mounting state expenditure was met by a number of surcharges.

The traditional fiscal principle was well observed by the early Manchu government, which fixed the national landtax and ting quotas on the basis of the 1570's, when "extortionate" assessments had not yet been made. Consequent upon large-scale devastation and slaughter in Shensi, Shansi, Szechwan, parts of Hupei, Anhwei, and other northern provinces during the second quarter of the seventeenth century, regional and local landtax quotas were greatly reduced by the new dynasty. The sharp drops in the national land returns between 1645 and 1685 may have partially, though of course not accurately, reflected the extent to which cultivated land and population had declined since the peasant rebellions of the 1630's and 1640's and the Manchu wars of conquest. But the gradual and moderate recovery in national land returns during the late seventeenth and early eighteenth centuries failed to indicate the rate of population growth and the expansion of cultivated area, for the K'ang-hsi, Yung-cheng, and early Ch'ien-lung eras were a period of benevolent despotism par excellence. The Manchu government managed to maintain the traditional Ming landtax quota till late in the nineteenth century, regardless of mounting government expenditures.

Some minor inflating factors must also be examined. In some cases in southern Kiangsu where the land is very fruitful the customary mou was only seven-tenths or eight-tenths of a standard mou.[50] In Wu-hsi county a modern agricultural expert has found 173 sizes of mou in use, ranging from 2.683 to 8.957 *are*, as compared with the standard mou of 6 2/3 *are*.[51] In 1951, when the holdings of a Sung-chiang peasant who broke the record for per-acre rice yield were measured, it was found that what had been customarily regarded as 17.5 mou was in fact only 16.1 mou.[52] In hilly and congested southwestern Chekiang, sometimes six-tenths or seven-tenths of an actual mou of the best irrigated land was equated to one fiscal mou.[53] In She-hsien in southern Anhwei one mou of fish pond has customarily been converted to 1.191 fiscal mou. There are certain *kung* or *pu* in north China which, instead of being 5 square feet, are only 4.5 or 3.2 square feet. Unlike the mou of special localities in the southeast, however, those of such northern districts are not necessarily smaller than the standard mou because they consist of from 260 to 720 such small kung or pu.[54] All in all, these mou which are smaller than the standard are very special cases. Speaking of the country as a whole, the amount of mou deflated by conversion seems likely to have been far in excess of occasional local inflated mou figures caused by the existence of small mou systems.

Another inflating factor has been the inclusion in the registers, as taxed land, of certain fiscal elements that were originally not borne by the land. In Ch'ih-chou prefecture in southern Anhwei the 1582 assessment consolidated the taxes borne by the ting and all grades of land into a single payment, whereby the original ting quota of 30,120 was converted outright into taxed land at the rate of 5 mou for each ting.[55] It is extremely unlikely, however, that this artificially increased fiscal acreage exceeded the actual cultivated area because, as we have seen, in the whole province of Anhwei agricultural land has been traditionally and extensively converted into fiscal acreage at a substantial discount. Many southeastern localities, like Ningpo and Shao-hsing prefectures in Chekiang,[56] simply merged the ting payment with the land tax without affecting the quotas of taxed land. This simple method thus increased only the tax of each fiscal mou and did not result in any increase in the total fiscal acreage.

There were still other localities in the southeast, such as Wu-chin county in southern Kiangsu, where the amalgamation of ting and land tax even resulted in a small net reduction of the total fiscal acreage because, in addition to the conversion of ting into land, the amounts of certain categories of land were reduced by new conversions. The complex process used in the 1538 assessment in Wu-chin is shown in Table 29. By adding 124,398 ting, which were

Table 29. 1538 land assessment in Wu-chin.

Category of land	Original quota (in mou)	Number of mou of originally taxed land per new fiscal mou	New quota (in mou)
Public	139,662	5	27,934
Private	1,295,881	1	1,295,881
Hills and marshes	75,478	10	7,548
Total	1,511,021		1,331,363

Source: *Wu-chin HC* (1605 ed.), 3.63a–63b.

converted outright into 124,398 mou of new tax-paying land, to the new land quota of 1,331,353 mou, we get a total new fiscal area of 1,455,761 mou, which is 55,260 less than the original total land quota.

These different local samples make it appear obvious that, although the inclusion of ting in the taxed land was theoretically a factor that could inflate the local land return, in practice it probably had comparatively little effect on the fiscal acreage, which seems to have remained below the actually cultivated area. It must also be mentioned that hills, mountains, marshes, bamboo, and woodlands — though providing secondary income for peasants are not cultivated land in the strict sense of the term — were nevertheless converted to fiscal mou. The conversion rates were very generous, however. In Shao-hsing, for example, originally 100 mou, and since the late sixteenth century 50 mou, of nonagricultural hilly land were taxed at the rate of a single ting.[57] Even up to the present the unit in measuring nonagricultural hilly and mountainous land in Kiangsi has been the human voice. That is, the area that can be reached by the voice of a shouting human is said to be reckoned as a single fiscal mou.[58]

It seems evident that the deflating factors far outweigh the inflating factors, for the customary mou smaller than the standard mou have existed only as exceptions. Even in southern Kiangsu, where the small mou are found in some villages, the bulk of the agricultural land has been converted to fiscal acreage at significant discounts. Even the inclusion of ting in some southeastern localities into fiscal acreage could hardly offset the combined effects of mou conversion, evasion of land registration, and the usually strong resistance to a substantial increase of the quota for taxed land. The concept that the state should live within the bounds of its revenues was so deeply rooted that local histories are full of praise for those "model officials" who successfully defended the original quota of taxed land, despite the actual expansion of the cultivated area. When we take into account the prevalence of large mou systems in north China, northern Kiangsu, and the whole province of Anhwei and the great quantity of unregistered land almost everywhere, and particularly in the southwest, it seems likely that the traditional, and in some degree also the Nationalist land statistics must have been considerably below the truth.

IV

The influence of traditional land returns on the Nationalist land statistics can be seen in Table 30.

It should be pointed out that in theory and by official definition the Nationalist government's land statistics differ fundamentally from Ming and Ch'ing land returns. For one thing, the Ming and Ch'ing returns were figures on fiscal acreage, while the Nanking returns by definition were figures on cultivated land. Secondly, the Nanking government established a Directorate of Statistics for the collection of land statistics; previously local governments had been responsible for the returns on taxed land. Thirdly, the unit of land measurement was changed to the new standard shih-mou, which is equal to 1.085 old mou. Last but not least, it was said that the Directorate had local agricultural correspondents who were to report annually on local cultivated areas and crops.

However, there was considerable discrepancy between the new government regulations and actual practice in the realm of land statistics. A comparison between the land returns of various counties given in the Directorate's *Statistical Monthly* (the combined issues of January and February 1932) and the traditional returns on taxed land in some local histories reveals a strange continuity in a fairly

Table 30. Official land returns by provinces, 1393–1932.

Province	1393		1502	
	Reported area as a percentage of total area of the province	Area in '000 mou	Reported area as a percentage of total area of the province	Area in '000 mou
Kiangsu	35.20	60,515	32.59	56,026
Chekiang	31.43	51,705	28.72	47,234
Anhwei	13.12	28,838	10.68	24,992
Kiangsi	15.75	43,119	14.69	40,235
Hupei, Hunan	34.03	200,218	34.56	233,613
Szechwan	1.17	11,203	1.64	10,787
Fukien	7.42	14,626	6.86	13,517
Kwangtung	6.51	23,734	1.99	7,232
Kwangsi	2.86	10,140	3.01	10,785
Yunnan	......	....	0.06	363
Kweichow	......	....	......	....
Hopei	25.47	58,250	11.79	26,971
Shantung	28.94	72,404	21.70	54,293
Honan	52.08	144,945	14.95	41,610
Shansi	15.89	41,864	14.84	39,081
Shensi, Kansu	3.03	31,525	2.51	26,066

Table 30 (*continued*).

1578		1887		1932	
Reported area as a % of total area of the province	Area in '000 mou	Reported area as a % of total area of the province	Area in '000 mou	Reported area as a % of total area of the province	Area in '000 mou
29.95	51,499	64.46	100,825	53.33	91,669
28.39	46,697	28.43	46,771	25.05	41,209
11.07	25,895	17.56	41,113	22.29	53,511
14.65	40,115	17.29	47,342	15.20	41,630
34.25	221,620	39.58	117,323	20.58	61,010
		9.94	34,874	13.01	45,612
2.52	13,483	7.07	46,416	14.65	96,272
6.81	13,423	6.80	13,400	11.82	23,290
7.05	25,687	9.53	34,731	11.65	42,452
2.63	9,402	2.50	8,964	6.09	29,840
2.77	1,799	1.44	9,320	4.18	27,125
0.18	517	0.96	2,765	8.01	23,000
21.54	49,257	30.30	69,305	45.22	103,432
24.68	61,750	50.34	125,931	44.23	110,662
25.28	74,185	25.75	71,675	40.59	112,981
13.97	36,804	21.44	56,477	22.99	60,560
		9.63	30,591	10.55	33,496
2.81	29,292				
		2.71	16,775	3.79	23,510

Source: Wan Kuo-ting, "Chung-kuo t'ien-fu niao-k'an chi ch'i kai-ke ch'ien-t'u." It should be noted that Shensi and Kansu in Ming times were one province. The percentages of their reported areas for 1393, 1502, and 1578 in the total and area are herein revised.

large number of cases. In a number of cases the figures for the so-called cultivated areas of 1932 are even lower than the traditional ones on taxed land. On the other hand, the Directorate did make a large number of substantial revisions which seem to have been based on its local agricultural correspondents' reports. In view of the limited number of provinces under the Nanking government's effective control at the time and of the Directorate's limited funds, it is doubtful whether the Directorate had special correspondents in all of the country's nearly two thousand counties. The official 1932 returns were therefore a mixture of new estimates of local cultivated areas and traditional local fiscal acreages. This is shown by the fact that the 1932 returns of Chekiang, Kiangsi, and Honan are lower than the corresponding figures for 1393, and the 1932 returns of Kiangsu, Hupei, and Shantung are lower than the corresponding figures for 1887. In fact, the Directorate did not seem to have confidence in its own data, for after the publication of Professor Buck's monumental *Land Utilization in China* in 1937 the Directorate warmly endorsed Buck's findings and practically discarded its own.

In general there can be little doubt that the findings of the University of Nanking's land utilization survey, under Buck's directorship, which were based on a field study of some 55,000 farms in 22 provinces during the period 1929–1933, are of far better quality than the Directorate's figures. It should be mentioned, however, that although in localities directly investigated by the survey Buck's revisions of the Directorate's corresponding figures are quite substantial, his data for the majority of localities were still based somewhat on the Directorate's figures. This was indeed inevitable because no privately initiated survey could possibly cover all the counties in so vast and varied a country as China. But the fact remains that Buck's study is still the best of its kind ever undertaken in China and that any critique of modern land statistics should be based on his findings.

Fortunately, although the Directorate failed to compile authentic statistics on the cultivated area, the Nationalist administration has left two types of land data which are useful for an appraisal of Buck's estimates. The first consists of data collected by a number of counties which experimented with "land self-reporting" in the early 1930's. Self-reporting meant that the owners were asked by the local government to report the size of their holdings. The method of land self-reporting started with the merger by the local government of a number of *pao* (usually 100 households) into a *lien-pao* (merged pao). Since a lien-pao consisted of at least several hundred households, the aggregate area of the holdings belonging to it was considerable. Trained personnel began by surveying the aggregate cultivated areas of all the lien-pao of the county, without mesauring individual holdings. This enabled the local government to know approximately the total cultivated area of the county before landowners were asked to report on their own holdings. After the approximate total local cultivated area was known, the county government enlisted all pao headmen to make a house-to-house call inquiring about the amount of land owned by each household. To avoid evasion, government personnel further traced the ownership of every piece of land on the field. All owners thus located were required to report the size of their holdings. The process therefore provided a double check. The sum total of acreages reported by owners was finally checked against the total cultivated area of the entire county obtained from the initial general surveys. Moreover, the land self-reporting movement was reasonably carefully prepared.[59] Months before it was launched, special personnel were trained and its purpose widely publicized. Special pains were taken to explain to the people that the purpose of land self-reporting was not to increase the land tax but to bring about greater fiscal equity. The method proved fairly successful in one Anhwei and seven Kiangsu counties.[60]

The other useful land data left by the National Administration result from land surveys by means of air photography, a technique that received great attention from the Ministry of the Interior and the Departments of Civil Affairs of various provinces in the 1930's. Air survey was carried out in a number of Kiangsi and Hupei counties. It was completed in twelve Kiangsi counties before the outbreak of war with Japan in July 1937. But the air survey of several Hupei counties was suspended by the war. Air-survey figures for Kiangsi counties are by far the most accurate extant and even the preliminary findings of air surveys in Hupei are revealing.

A comparison of the results of "self-reported" areas and Buck's estimates is made in Table 31. It should be emphasized that land self-reporting was by no means free from errors. As the reports of Hsiao-hsien, Chiang-tu, and Tang-t'u clearly explained, the main aim of land self-reporting was to bring about a landtax reform. This being the case, the provincial governments of Kiangsu and Anhwei were content with the reported figures, which were invariably larger than the original quotas of taxed land. The governments did not bother with discrepancies between the reported areas and the total local cultivated areas obtained by preliminary county government surveys. For this reason the self-reported areas,

Table 31. A comparison between Buck's estimates and "self-reported" areas.

Province	County	A Buck's estimate (in mou)	B Self-reported area (in mou)	B as percentage of A
Kiangsu	Shu-yang	3,375,000	3,145,561	93.2
	Hsiao-hsien	2,258,000	2,453,568	108.7
	Chiang-tu	2,139,000	2,326,889	108.8
	Li-yang	1,392,000	1,426,177	102.5
	I-hsing	1,160,000	1,296,533	111.8
	Chen-chiang	754,000	1,102,087	146.2
	Chiang-yin	1,243,000	1,242,141	99.9
	Total	12,321,000	12,992,956	105.5
Chekiang	(entire province)	44,213,000	56,500,000	127.8
Anhwei	Tang-t'u	845,000	1,197,804	141.8
Shensi	Hsien-yang	704,000	637,000	90.5
	Nan-cheng	256,000	580,000	226.6
	Pao-ch'eng	354,000	635,000	179.4
	An-k'ang	227,000	339,000	149.3
	Total	1,541,000	2,191,000	142.2
Yunnan	K'un-ming	600,000	431,877	72.0
	Chin-ning	50,000	123,415	246.8
	Ch'eng-kung	59,000	253,261	429.3
	K'un-yang	71,000	64,191	90.4
	I-liang	120,000	274,544	228.8
	Ch'eng-chiang	61,000	133,273	218.5
	Fu-min	56,000	66,807	119.3
	An-ning	140,000	234,544	167.5
	Sung-ming	346,000	363,242	105.0
	Yü-hsi	95,000	202,571	213.2
	Total	1,598,000	2,147,725	134.4
Grand Total		60,518,000	75,029,485	124.0

Source: *An-hui-sheng Tang-t'u-hsien t'u-ti ch'en-pao kai-lüeh; Chiang-su-sheng Chiang-tu-hsien t'u-ti ch'en-pao kai-lüeh*; *Chiang-su-sheng Hsiao-hsien t'u-ti ch'en-pao kai-lüeh* (all published as reports of the Local Taxation Reform Committee of the Ministry of Finance, 1935–36); also *Chung-kuo t'u-ti wen-ti chih t'ung-chi fen-hsi* (Directorate of Statistics, 1941); *Yün-nan-sheng nung-ts'un tiao-ch'a* (Executive Yüan, 1935); and Fang Hsien-t'ing, ed., *Nan-k'ai ching-chi yen-chiu* (CP, 1936), pp. 976–977.

though probably much closer to the truth than either the *Statistical Monthly* figures or Buck's estimates, were still somewhat less than the true cultivated areas. The difference between Buck's estimates and the actual cultivated areas of seven Kiangsu counties should therefore be somewhat more than 5.5 per cent. The self-reported area of Tang-t'u county in Anhwei exceeds Buck's estimate by 41.8 per cent. Although this single case cannot be regarded as typical of the rest of Anhwei, it seems more than coincidental that such a large difference should be found in a province where mou conversion was most extensively practiced.

While Buck's estimates for seven Kiangsu counties were remarkably close, his total figure for Chekiang province was undoubtedly much too low. As early as 1393 the province had reported an area of 51,705,151 traditional mou, or 47,650,000 shih-mou, which exceeds Buck's figure by 3,400,000 shih-mou. Since Ming times the population of Chekiang is recorded as having more than doubled itself. As compared with the 1393 figure, the total 1930 reported area of 56,500,000 shih-mou appears to be reasonable. There should be no danger of exaggeration in the 1930 figure because no sane person would willingly pay more tax by purposely magnifying the size of his holding. Experience suggests that self-reported areas could only be below the truth. For example, in Fukien and Szechwan, which in the early 1930's were still beyond the effective control of the Nanking government, cultivated areas were so underreported that they are not worth analysis.[61] Although the self-reported areas of Shensi and Yunnan counties exceed Buck's estimates by 42 and 34 per cent, the reported areas of Hsien-yang (a county neighboring the provincial capital of Sian) and K'un-ming (the capital city of Yunnan), places where officials and influential people held large amounts of landed property, are considerably less than Buck's estimates.

In any case it seems probable that the margin of error in Buck's estimates for the northwestern and southwestern provinces is fairly great. Kweichow, for example, under the able administration of Wu Ting-ch'ang, the financier who became an able governor, carried out land self-reporting between 1937 and 1940. The self-reported area for the taxed land of the province was 18,216,313 shih-mou, as compared with Buck's estimated cultivated land of 11,325,000 shih-mou. But as Governor Wu pointed out, the area actually cultivated by 1940 should be in the neighborhood of 31,500,000 shih-mou, or 12 per cent of the total provincial area.[62] Wu's estimate appears to have been an intelligent one because Kweichow traditionally has been one of the most under-reported provinces. Its population, which was formerly believed to be slightly over 10,000,000, was 15,000,000 according to the 1953 census. Since the number of farm households was one of the basic factors in Buck's estimation of cultivated areas, his figures on cultivated areas for northwestern provinces may have been much too low. The populations of Shensi and Kansu, which were generally believed to be around 9,500,000 and 6,900,000, respectively, were 15,880,000 and 12,930,000 according to the 1953 census. One gets the impression that, owing to the lack of reliable population and farm household numbers, Buck's estimates in general inevitably had a downward bias, but under-estimation was particularly serious in the peripheral areas where he had to rely heavily on the *Statistical Monthly* figures, for these figures for northwestern counties were heavily influenced by traditional figures on taxed land. When the extensive mou conversion in Shensi and the unusually large amount of unregistered land in Kansu[63] are borne in mind, the reasons for a serious under-estimation of the cultivated areas in northwestern provinces can be easily understood.

Since air-survey records for twelve Kiangsi counties are available, they should be compared with Buck's estimates. This is by far the most exact comparison that can be made between Buck's estimates and relatively accurate survey figures. In an over-all comparison (Table 32), Buck's figures are exactly 30 per cent lower.

In the mid-1930's the Department of Civil Affairs of Hupei province carried out an air survey of the cultivated areas of six counties. By the summer of 1937 the bulk of the land in three counties had been surveyed and photographed and that of the others partially measured. On the basis of incomplete air surveys the Department released the anticipated total cultivated areas for six counties. Though not as accurate as Kiangsi's figures based on completed air surveys, these should at least be accepted as by far the most responsible estimates. In Table 33 they are tabulated in comparison with Buck's figures. Buck's figures for T'ien-men and Chiang-ling, which are taken from the *Statistical Monthly* without revision and without field study, are even lower than the amounts of taxed land. When it is remembered that about three-fifths of a total land area of 303,400,000 shih-mou consists of level plains and valleys with abundant water supply and that the province, in addition to feeding its population of 36,000,000 (1953 census figure), still has a significant food surplus, Buck's estimate of 70,000,000 shih-mou, or 23 per cent of the total land area, as the cultivated area of Hupei is obviously wide of the mark.

Table 32. Cultivated areas of twelve Kiangsi counties in the 1930's (in shih-mou).

County	Total land area	Cultivated area A (Buck)	Cultivated area B (Air survey)	B as percentage of A
Nan-ch'ang	2,474,754	1,211,000	1,464,060	120.9
Hsin-chien	5,229,668	1,013,000	1,465,722	144.6
An-i	935,930	205,000	330,505	161.2
Chin-hsien	2,761,316[a]	1,076,000[a]	1,110,103	102.5
Tung-hsiang	1,642,500[a]	493,000[a]	641,435	130.1
Ch'ing-chiang	1,878,837	578,000	948,387	164.2
Hsin-kan	1,839,266	388,000	581,961	150.0
Feng-ch'eng	4,191,406	1,349,000	1,961,431	145.3
Chin-hsi	2,095,830[a]	461,076[a]	633,030	137.0
Chia-chiang	1,913,709	321,000	436,282	135.9
Chi-shui	4,124,556	650,000	893,472	137.4
Lin-ch'uan	3,428,258	1,272,000	1,376,226	108.2
Total	32,516,030	9,019,000	11,842,614	130.0

Source: *Chiang-hsi-sheng ti-cheng kai-k'uang* (Department of Economic Reconstruction, Kiangsi Provincial Government, 1941).

[a] Buck's estimates of total land areas are substantially different from the total land areas computed from air surveys. His cultivated areas are revised accordingly, but the revisions are still based on his percentages of cultivated areas in the total land areas.

We must point out, however, that Buck made altogether four estimates of the total cultivated area of China Proper during the 1930's and local figures add up to his lowest total of 205,000,000 acres. The difference between his lowest and his highest estimate (232,000,000 acres), which he believed to be the most accurate of the four, is 13 per cent. In the light of other available evidence Buck's estimates seem remarkably close for Kiangsu only and his figures for other provinces seem considerably below the truth. Since the difference between his estimates for other provinces and self-reported areas and air-surveyed areas is, percentagewise, at least twice the difference between his highest and lowest estimates, there is reason to believe that even his highest estimate is likely to have been somewhat too low.

Table 33. Cultivated areas of six Hupei counties in the 1930's.

County	Amount of taxed land (shih-mou)	Cultivated area A (Buck) (shih-mou)	Cultivated area B (estimate by Hupei gov't.)	B as percentage of A
Wu-ch'ang	1,057,696	1,781,404	2,360,000	132.5
Han-yang	675,000	1,121,692	1,791,000	159.7
Han-ch'uan	739,065	1,562,648	1,852,700	118.5
Sui-hsien	1,118,162	3,479,036	5,687,000	163.4
T'ien-men	1,776,172	532,000	3,106,670	583.8
Chiang-ling	2,873,461	2,240,000	5,018,870	224.0
Total	8,239,556	10,716,780	19,816,240	184.9

Source: *Ti-cheng yüeh-k'an*, vol. IV, nos. 4, 5.

V

A knowledge of the direction in which Buck's estimates err and the probable range of his errors is vital to a proper understanding and evaluation of the land statistics of the People's Republic of China. The official figure for the cultivated area in all China for 1952, the year that marked the completion of postwar rehabilitation and serious preparation for the first Five-Year Plan, is 1,618,-770,000 mou or 266,672,000 acres. It must be emphasized that this total is coincidentally almost exactly the sum of Buck's highest estimate of 232,000,000 acres for China Proper and the official 1942 "Manchukuo" return of 34,446,000 acres.[64] There seem to be only two possible explanations for this extraordinary coincidence.

First, if the Communist 1952 figure is accurate and independently arrived at, it would appear that Buck's highest estimate of the cultivated area of China Proper in the 1930's was remarkably accurate. Despite the lapse of one and a half decades, it is against common sense to suppose that within three years of Communist victory China's total cultivated area could have surpassed the pre-1937 peak. According to a scholarly and conservative estimate, up to 1943 a total of 71,000,000 acres, or more than a quarter of China's total cultivated area, had been seriously affected by the Sino-Japanese war. Out of these 71,000,000 acres some 50,000,000 had become battlefields.[65] Although repatriation and rehabilitation began shortly after the conclusion of war with Japan, they were delayed by full-scale civil war between the Nationalists and the Communists. But for the marked improvement in the administrative efficiency of the new regime, repatriation and rehabilitation would have been much slower. This being the case, the 1952 land return should roughly correspond to the pre-1937 peak area and Buck's highest estimate should be regarded as unusually accurate.

This explanation, though plausible, is difficult to accept. Our knowledge of the nature of traditional Chinese land statistics and their lingering influence on modern official and even unofficial figures, together with our criticism of Buck's estimates, has suggested that Buck's highest estimate was probably considerably under the truth. Although the lack of detailed official explanation as to how the 1952 land statistics were collected and compiled makes it impossible to prove or disprove them, one cannot but suspect that the 1952 official figure is a readaptation of Buck's data. At the very least we know that the Communist land figures are not necessarily bona fide land statistics. Any doubt about this statement should be dispelled by the following eyewitness account of the land reform in a northern Hunan village:

> Owing to the difference in soil and location, two pieces of land of approximately the same size could vary a great deal in value. Moreover, the rice fields of Hunan were mostly of irregular shapes and didn't lend themselves to exact measurement. Since measuring the land itself involved so many complications, the only possible standard was the average productivity. Thus each peasant was asked to "appraise" his own land. The correct procedure was for each peasant to write down the productivity of each plot of his land on a flag. Land that he owned himself had a red flag while a white flag signified land rented from landlord or rich peasant. Distribution officers would come around and make the assessment.[66]

This eyewitness went on to explain that although theoretically a peasant received a certain number of mou, actually the mou number was guesswork based not on measurement but on average yield.

The fact that the nationwide and historic land reform was not accompanied by a nationwide land survey is undeniable. It can be further said that never in Chinese history has the cultivated land of the entire country been subjected to a scientific survey. The so-called land surveys of the Ming T'ai-tsu period were confined to Chekiang and some prefectures of southern Kiangsu and they were more in the nature of land assessment than land survey. This was indeed inevitable because the technical difficulties involved in a scientific land survey are greater than those involved in census-taking or crop-reporting. Despite the remarkable improvement in administrative efficiency, mainland China is not yet able to undertake a nationwide land survey. While for political reasons and for purposes of economic planning there was urgent need to ascertain the size of the population and the amount of agricultural product, a survey of the cultivated area of the entire country, which would require hundreds of thousands of trained personnel and take years if not decades to accomplish, had to be left for the future. The 1952 land return, therefore, remains an intelligent guess at best.

As to the accuracy of the 1952 land figure, two factors must be considered. First, any series must have a datum. In the lack of better data for the pre-Communist period, the Peking government is compelled to use previous statistical series until direct and more reliable data can be collected. Land figures are not the only Communist statistical series which show continuity with those of the Nationalist period. The pre-1953 Communist population figures, which bear such strange resemblance to figures of the previous era and which have given rise to so much skepticism of the 1953 census total itself, are one of the most eloquent proofs. Despite the low quality of official and unofficial statistics compiled during the Nationalist period, they are selected and embodied in a recent reference work edited by a group of social scientists in mainland China.[67] A complete rejection of the figures of former years would mean a statistical vacuum.

Secondly, many statisticians of the new Peking government have served the previous regime and most of them are Western trained.

Hence they are less acquainted with the subtleties of Chinese institutional history than with Buck's estimates, which have long been acclaimed the best available. It is true that all this does not definitely prove that the 1952 land figure was not independently arrived at, but in substantial parts of China, as in the northern Hunan village mentioned earlier, local cultivated areas have always been estimated on the basis of average yield rather than on customary mou systems which, though confusing, are theoretically possible of being converted to the standard twentieth-century shih-mou. Considering the great complexity of the mou systems, it is doubtful whether such a conversion on so vast a scale had indeed been made before the release of the 1952 land figure. Therefore, it seems a major probability that the official 1952 figure is a readaptation of Buck's highest estimate for the 1930's.

If my reasoning is sound, the 1952 official land figure should be treated on an equal basis with the pre-1953 Communist population figures, which are a continuation and minor revision of figures of the Nationalist period. It should not be accepted as an authentic land return for the simple reason that there was no land survey on a national scale. The much publicized expansion of cultivated area in recent years may thus be more apparent than real. In the first place, rural rehabilitation after prolonged war and devastation was necessarily a slow process. In the second place, even up to the present the Peking government's agricultural policy has been concerned more with various immediate means of increasing food production than with the opening of virgin soils.[68] In the light of this historical study of official land statistics and our analysis of Buck's estimates, there is reason to believe that when a true nationwide land survey comes to be undertaken, the total cultivated area of China will be somewhat larger than is usually believed by Western experts and Peking statisticians.

CHAPTER VII

The Population-Land Relation: Interregional Migrations

An examination of official land statistics makes it clear that the changes in China's agricultural area throughout the Ming, Ch'ing, and modern periods must be studied from sources other than official figures. A great deal of useful if nonquantifiable information is available on the dissemination of various new food crops and on migrations from congested to sparsely settled areas. While it is impossible in most cases to approach quantitatively the problem of the expansion of the agricultural frontier, a fairly detailed marshaling of local history records on interregional migrations will at least offer an explanation in broad terms as to how the growing population was accommodated.

Because of the comparative scarcity of relevant information in pre-Ch'ing local histories, major interregional migrations in Ming times can be traced only through government-supervised undertakings. The prolonged turmoil attending the downfall of the Mongol Yüan dynasty brought about a drastic reduction in the registered population of the low plain region of north China and the Huai River area. Under the energetic administration of Ming T'ai-tsu the government made a systematic effort to move people from congested to war-devastated areas. Before the end of the fourteenth century nearly 150,000 households of landless tenants from southern Kiangsu and northern Chekiang were sent as settlers to the Feng-yang area along the Huai River in northern Anhwei, the birthplace of the Ming founder. An unspecified number of peasants from Tse-chou and Lu-an prefectures in southern Shansi, which had escaped the war scourge of the late Yüan period, were settled on the low plains of Hopei, Shantung, and Honan. To solidify the northern defenses nearly 70,000 Chinese and Mongol households were reshuffled in the area north of Peiping. After Ming Ch'eng-tsu moved the capital from Nanking to Peking in 1421, further efforts were made to transfer peasants from Shansi to the new metropolitan area. To curb the influence of the powerful landed gentry, Ming T'ai-tsu forcibly moved 45,000 gentry and other wealthy households [1] from southern Kiangsu and northern Chekiang to Nanking, where they were under closer government surveillance. All but the last category of government-enforced migrations had some direct bearing on agricultural development.[2]

In the interest of national defense Ming T'ai-tsu organized military colonization in various strategic areas, from Yunnan in the extreme southwest to the northern region of the Great Wall and beyond. In the late fourteenth and early fifteenth centuries military colonies were established even in parts of southern Manchuria and the vast northern vacuum in the region of the lower Amur River. It was said that in Yunnan alone Prince Mu Ying, the famous general and adopted son of Ming T'ai-tsu, was instrumental in developing more than one million mou of land held by his garrison forces under military tenure. Garrison soldiers along the Great Wall, as well as in inland posts, were all ordered to cultivate land and to become self-sufficient in time of peace. In time many of the northern frontier posts became prosperous towns and cities.[3] In addition to military colonization, civilians were encouraged to establish agricultural colonies in remote parts of the empire to which convicts were also sometimes sent as tenant farmers. After Ming Ch'eng-tsu's death in 1425 the government's interest in supervised migrations slackened, but military colonization remained an important government project till the very end of the Ming period.[4]

There can be little doubt that many if not all of these government-sponsored migration and colonization schemes left some impact on the economy of the underdeveloped areas. The scale and impact of the government-enforced migration from Nanking to Yunnan must have been very considerable; Hsieh Chao-che, a famous early seventeenth-century scholar and governor of Kwangsi, testified that in the Kunming area the economy was very prosperous and the culture, dress, dialect, and social customs in general had become almost identical with those of the Nanking area.[5]

The summary of interregional migrations during early Ming times given in the official history of the Ming dynasty is necessarily very sketchy. Among other things, it fails to mention migrations voluntarily undertaken by the people. The privately initiated migrations, mostly piecemeal and unrecorded, probably contributed even more to the expansion of the agricultural frontier than government-sponsored migrations. The fragmentary nature of the sources makes it difficult to reconstruct the history of voluntary private migrations during Ming times. The case of Hunan province, however, illustrates the importance of such Ming population movements. The exceptionally thoughtful chapters on the origins of prominent local clans in a few late Ch'ing Hunan local histories have led a modern historian to discover that by far the greater part of modern Hunan's population came originally from other provinces, mostly from Kiangsi during the Ming period.[6] This conclusion would seem to be substantiated by the fact that most of the available biographical material on prominent late Ch'ing officials from Hunan reveals that they were of Kiangsi stock. The local histories of some of the Szechwan counties that escaped the scourge of the mid-seventeenth century bandit leader Chang Hsien-chung testify that practically all their "old native" clans seldom date back beyond the early Ming period and most of them came to Szechwan from Hupei during the Ming period.[7] These facts, added to our knowledge that one of the major achievements of the Ming period was the extension of Chinese civil administration in the hitherto predominantly aboriginal southwestern provinces, give some reason to believe that Chinese migrations to, and colonization of, the southwest played an important role in the development of that region. That government-sponsored migrations were inadequate to develop the low plain area of north China is attested to by the early Ming laws which aimed at attracting new settlers by offering them permanent land tax exemption. Voluntary migrations from the loess provinces to the low plain provinces of north China must later have reached a significant scale; by the middle of the fifteenth century so much tax-exempt land had been settled that the government was forced to revoke the exemption laws in order to appease the bitter resentment of the old natives over the fiscal inequity.[8]

Although voluntary migrations during the Ming period seem to have been sizable, they are overshadowed by the dimensions of interregional migrations in the Ch'ing and modern periods. Voluntary migrations were motivated mostly by economic factors. We may assume that in Ming times the total national population was relatively small and arable land plentiful, and the incentive for migration less than during the subsequent period of rapid population growth. It was well known in Ming times that the vast Huai River region and the Hupei lowlands were comparatively sparsely populated, with large areas of arable land unused.[9] Hsieh Chao-che's descriptions of the geography of Ming China in his famous cyclopedia *Wu-tsa-tsu*, gave the impression that the whole of the mountainous areas of the southwest, much of Kwangtung, Hunan, and Hupei, the Huai River region, and parts of north China had arable land which had yet to be systematically developed. The only congested areas were the lower Yangtze, northern Chekiang, the Hui-chou area in southern Anhwei, and the province of Fukien.[10] The abundance of relevant information in Ch'ing local histories makes it possible to discuss migrations regionally.

I. MIGRATIONS TO SZECHWAN

The large-scale peasant rebellion led by Chang Hsien-chung during the second quarter of the seventeenth century, which is unparalleled for its bloodthirstiness, created a certain vacuum in the Red Basin of Szechwan. Although official Ch'ing accounts exaggerated Chang's cruelties, which were probably no greater than those of the conquering Manchu forces, the reduction of Szechwan's population seems to be a fact.[11] The depopulation in Szechwan made it necessary for the young Manchu government to proclaim in 1653 that the province should be thrown open to garrison soldiers and civilians who wanted to settle as farmers and that the provincial government should provide them with draft animals and seeds.[12] In 1671 the governor general of Szechwan, Hupei, and Hunan stated in a memorial, "there is an abundance of cultivable land in Szechwan but there are not enough people to cultivate it." Consequently, an imperial edict ordered that those who were willing to settle in Szechwan were to be tax exempt for a period of five years and that any local official who could attract three hundred immigrants would be promoted immediately. Further to stimulate agricultural settlement the edict provided that any expectant official or degree-holder would be immediately appointed an acting county magistrate upon bringing in a certain number of immigrants.[13] The incomparable economic opportunities thus offered by this fertile province must have brought immediate immigration from such neighboring provinces as Hupei and Shensi, for two edicts, in 1712 and 1713, found it necessary to provide regulations concerning property disputes between natives and immigrants.[14]

Ch'ing and Republican editions of Szechwan local histories which contain accounts of immigrants are too numerous to be listed in full. It is necessary to mention only a few which yield exceptionally revealing material on the geographic composition of the modern Szechwan population. Hsin-fan in the heart of the Red Basin specifically states that its population is composed of people from Hupei, Kiangsu, Fukien, Kwangtung, and Shensi.[15] Chien-wei county, along the Min River on the southwestern fringe of the Red Basin, takes special pains to analyze the components of its population and lists all the guildhalls established by immigrants according to geographic division. The Hupei guildhall has the most members; it is followed by those of Pao-ch'ing and Ch'ang-sha prefectures of Hunan, Kiangsi, Kwangtung, Fukien, Kweichow, and Kwangsi. The members so outnumbered the natives that during the Taiping period each geographic group elected its own executives, who were responsible for the collection of special taxes or monetary contributions for the government. From the beginning of the Ch'ing period to the early nineteenth century Chinese Moslems in smaller numbers also came from the northwestern provinces of Shensi and Kansu.[16]

That immigration took place continuously for more than two hundred years after the founding of the Ch'ing dynasty is borne out by various modern local histories, particularly by the 1932 edition of the history of An-hsien, in north-central Szechwan:

> Szechwan suffered severely from Chang Hsien-chung during the late Ming period. Between 70 and 80 per cent of An-hsien's cultivated land was laid waste, with the result that large numbers of immigrants were attracted to settle here. Hupei, neighboring to Szechwan, sent out the most. It was followed, in order, by Kwangtung, Shensi, Fukien, and Kiangsi . . . . Even in late Ch'ing times the immigrants still preserved their native dialects. There were Kwangtung dialects, Shensi dialects, Hupei dialects, and the dialects of Pao-ch'ing and Yung-chou prefectures of Hunan, which were all impure. It is only during the last several decades that their pronunciations have gradually become pure, owing largely to improvement in communications.[17]

There are, of course, a number of counties where the descendants of pre-Ch'ing natives still constitute a significant portion of the local population. Tzu-chung county, midway between Ch'eng-tu and Chungking testifies:

> Tsu-chung does not have any native clans that can be traced back more than six hundred years. [Among the old clans] six- or seven-tenths came from Hupei during the early Ming period. Those originating in Fukien, Kiangsi, and Kwangtung all came during the Ch'ing period. Those of the local population who are descended from Ming immigrants are now called natives of Szechwan. The rest are still called by their places of origin. The multifarious dialects have not yet entirely changed under new environments. The most commonly spoken is the [Szechwan] mandarin.[18]

A modern geographer summarizes the geographic composition of the modern Szechwan population by saying that, while descendants of pre-Ch'ing natives are found in a number of localities, they are comparatively few and are concentrated mostly in the extreme southeastern part and a few places in the western part of the province. People of Hupei and Hunan origin predominate in eastern, western, and southern Szechwan; people of Honan, Anhwei, and Kiangsi origin are found mostly in southern counties; a considerable number of people of Shensi and Kansu origin are in northern and some western counties; and people of Kwangtung, Fukien, Kiangsu, and Chekiang ancestry live mainly in large cities like Ch'eng-tu and Chungking. Chinese Moslems are scattered in northern and northwestern cities and descendants of Manchu garrisons are found chiefly in the Ch'eng-tu area.[19] This summary and much of the detailed local history evidence, all of which cannot be presented, indicate a wide range of immigration into Szechwan during Ch'ing times.

Although modern Szechwan local histories testify to the waves of immigration that started in the middle of the seventeenth century, few have mentioned the approximate date when such extensive immigration came to an end. There is reason to believe that for fully two hundred years between 1650 and 1850 Szechwan was the largest recipient of immigrants and that after the Taiping Rebellion it relinquished its leading position to other regions. In the first place, a selection from modern local histories of Hupei, which probably sent more people to Szechwan during the Ch'ing than any other province, suggests that formal accounts, poems, and folk songs on emigration to Szechwan without exception refer to the pre-1850 period.[20] In the second place, during the second half of the nineteenth century some Szechwan counties, particularly those of the northern mountainous fringe, became areas of emigration, sending out poor peasants to Shensi. Thirdly, the major factor that may have brought about a basic change in migration trends was the depopulation of the lower Yangtze area caused by fourteen years of Taiping wars. From 1864 the rich irrigated lowlands of southern Kiangsu and northern Chekiang could offer to land-hungry peasants elsewhere better economic opportunities than Szechwan.

The redevelopment of Szechwan in Ch'ing times was not reflected in official land statistics, however. In 1659 the province returned only 1,888,350 mou of taxpaying land, which was barely comparable to that of an average large county of lower Yangtze.[21] Despite the continual influx of immigrants, which must have reached large proportions after 1700, the province still returned a ludicrously low figure of 21,445,616 mou of taxpaying land in 1734.[22] The so-called land surveys carried out in Szechwan from 1730 on were but minor tax readjustments, and right down to the end of the Ch'ing the discrepancy between the reported taxed area and the actually cultivated area in Szechwan was one of the widest in the country.[23] The only figures with which to gauge Szechwan's function as a leading absorbent of the rapidly growing populations of central Yangtze and other southern provinces are its population returns between 1776 and 1850. The registered population of Szechwan increased from 8,429,000 in 1786 to 44,164,000 by 1850.[24] These figures suggest that, despite two centuries' continual influx of settlers, Szechwan acquired the bulk of its population during the end of the eighteenth and the first half of the nineteenth centuries.

Space does not allow a detailed discussion of agricultural development in Szechwan during the first two hundred years of the Ch'ing. The 1814 edition of the history of San-t'ai county offers a general illustration:

> Our soil is not poor and our people are not lazy. The innumerable immigrants have brought with them every conceivable food plant or

product, all of which have been extensively propagated here. Many things that were unknown in the past are now our staple products. Our locality is so full of life and vigor that prosperity surpasses that of any previous period.[25]

From detailed local evidence it may be said that rice was extensively grown on the lowlands and basins and also on terraced hillside fields where rainfall was sufficiently abundant. All northern dryland crops were found in the province and its fertile red earth was particularly suited to maize and sweet potatoes, of which Szechwan was, and still is, the leading producer in the country. Peanuts, although not famous for their quality, were extensively grown on the innumerable sandbars along rivers and streams. Even the Irish potato was fairly commonly grown on the lofty peripheral mountains before 1850. By 1850 land utilization in Szechwan seems to have assumed its modern pattern and the province ranked foremost in all-over agricultural production. Even in the 1870's Baron von Richthofen, the famous geologist and traveler, observed: "under normal circumstances, a degree of ease and well-being as regards the sustenance of life, not common in other provinces of China, appears to prevail in Sz'-chwan [Szechwan]." [26]

In brief, in terms of numbers and length of time, Szechwan may have served an even more important function in accommodating the growing population of China than did twentieth-century Manchuria.

II. THE DEVELOPMENT OF THE YANGTZE HIGHLANDS

The well-known nineteenth-century historian and geographer, Wei Yüan, summarized the main migration trends in Ch'ing China by saying that people of Kiangsi poured into Hupei and Hunan, and the people of Hupei and Hunan filled up Szechwan.[27] He is echoed by T'an Ch'i-hsiang, a modern historian of migrations and expert in historical geography, who concludes in his interesting case study of Hunan that this province changed from a recipient of immigrants to an emigrating province in Ch'ing times.[28] These generalizations need qualification. Since Mr. T'an's study of Hunan is by far the most detailed and valuable of its kind, a brief criticism of its sources and methods will be helpful to a general understanding of the interprovincial migrations in the vast inland Yangtze region during the greater part of the Ch'ing.

The sources on which Mr. T'an's study is based are the chapters on prominent clans given in a few nineteenth-century Hunan local histories, which, being based on genealogies, provide precise information as to the periods in which such clans immigrated and their places of origin. Precise as clan data are, modern researchers are apt to be deluded by them for two reasons. First, it usually took a considerable time for an immigrant family to become well established in a new locality. Most of the immigrant families that settled in Hunan in the Ch'ing, particularly during the late eighteenth and early nineteenth centuries, are unlikely to have become prominent soon enough to be listed in Ch'ing local histories. Secondly, a sampling of such chapters suggests that about one-half of these prominent clans were descended from officials, civil and military, and degree-holders. The percentage of poor peasant immigrant families that could become prominent in a new locality within the span of three or four generations was certainly very small, yet it was just such peasants that formed the bulk of immigrants. These Hunan clan data, therefore, tend to distort proportions through omission.

Given the historical circumstances and the biases of local history editors, which were probably against relatively new immigrants, the very fact that a significant number of Ch'ing immigrant families were listed as prominent clans in some of those local histories seems to suggest that Hunan in Ch'ing times on balance was far from a net emigrating province.[29] This is borne out by other Hunan local histories which do not have special chapters on prominent clans but contain clear references to immigration. For example, Ling-hsien, in southeastern Hunan, had in 1871 a total of 91,160 immigrants as compared with only 27,221 natives.[30] Wu-ling hsien, capital city of Ch'ang-teh prefecture west of Tung-t'ing Lake, received more than 10,000 immigrant households during the Taiping period and the geographic composition of its population may be partially reflected by its public cemeteries for people from various geographic areas, namely, Fukien, Shansi, Shensi, Kiangsi, Su-chou of Kiangsu, and Ching-hsien of Anhwei.[31] Yu-hsien, near the Kiangsi border, testified that during the third quarter of the nineteenth century immigrants from Fukien and Kwangtung were still pouring into the county and the neighboring area.[32] Ju-ch'eng county, in the extreme southeast bordering Kwangtung and Kiangsi, took every effort to attract immigrants during the post-Taiping period because the natives alone could not develop all the available arable land.[33] As late as the 1870's immigrants constituted more than half of the population of Tao-chou, in southwestern Hunan bordering Kwangsi.[34] In fact, a modern Hunan geographer gives evidence that immigration did not entirely stop even in the twentieth century.[35]

Up to the beginning of the eighteenth century much of the hills and mountains that form the Nan-ling range was practically covered with forest and relatively untapped agriculturally. Subsequently the growing pressure of population in the southeastern coastal provinces seems to have forced poor southeastern peasants to tackle the hills and mountains of inland Yangtze provinces, a process which probably went on until about the Taiping period. The earliest available evidence of the development of inland Yangtze highlands is found in the writings of Shih Yün-chang, *chin-shih* of 1649 and a famous man of letters, who described in a poem the profits of Fukien immigrants in growing jute on the mountains of west-central Kiangsi, and the clashes between immigrants and natives.[36] While jute and indigo were profitable cash crops, the immigrants relied largely on maize and sweet potatoes for sustenance, both of which had been introduced into China during the sixteenth century. As it turned out, these two relatively new food crops became the main instruments by which the Yangtze highlands were systematically developed.[37]

These pioneering peasants lived in temporarily erected shacks in mountains and were hence called *p'eng-min* or "shack people." Their numbers must have been increasing rapidly, for between 1723 and 1726 a series of imperial edicts ordered the shack people of the mountainous districts of Kiangsi, Fukien, and Chekiang to be brought under the pao-chia machinery.[38] The reddish topsoil of the hills and mountains of Kiangsi and Hunan, still covered with vegetation, was reasonably fertile when first turned into maize and sweet potato farms. While the shack people were to be found in every central and lower Yangtze province and in Fukien, they swarmed particularly to the hilly districts of Kiangsi and Hunan.

By the middle of the eighteenth century a new pattern of land utilization had emerged, if we may judge by the succinct description in the 1760 edition of the history of Yüan-chou prefecture in west-central Kiangsi bordering Hunan:

> Formerly this prefecture abounded in idle land. On account of rapid population increase more land was cultivated but it was still confined to level areas. Since the influx of immigrants from Fukien and Kwangtung, their men and womenfolk have systematically cultivated high hills and even steep mountains.[39]

"The leading crop of hills and mountains," explained another Kiangsi local history, "is maize . . . which provides half a year's food for the mountain dwellers . . . . In general, maize is grown on the sunny side of the hills, sweet potatoes on the shady side." [40]

Although the movement to cultivate virgin hills and mountains was started by peasants from Fukien and Kwangtung, not all the pioneer maize and sweet potato farmers of the inland Yangtze were immigrants from these two provinces. In parts of Kiangsi, notably some northwestern counties bordering Hupei, most of the immigrants were from the Hupei lowlands. In Wu-ning county alone there were over 10,000 households of Hupei immigrants by the middle of the eighteenth century.[41] As the economic advantages of maize and sweet potatoes became more evident, even native poor peasants trekked toward the mountains. So much of the hilly land of inland Yangtze had been taken up during the early eighteenth century that, from about the middle of the century, natives and immigrants encroached upon the area of hilly western Hunan, the home of the Miao. Local officials in general paid little heed to Miao complaints, with the result that Miao rebellions broke out in the latter half of the eighteenth century. They were put down mercilessly by government forces and the successive large-scale slaughters of the Miao lasted till the early nineteenth century.[42] Agriculturally, however, much of hilly Hunan had been gainfully utilized before 1800.[43]

Deforestation was by no means confined to Kiangsi and Hunan. Before the middle of the eighteenth century the hilly districts of northwestern Chekiang had become a prosperous agricultural colony of immigrants from food-deficient Wen-chou prefecture on the southern Chekiang coast, who had established large mountain farms for the commercial production of maize, sweet potatoes, and peanuts.[44] The hills of southern Anhwei were likewise turned into maize farms in the course of the eighteenth century.[45]

In as remote a region as Yunnan, where maize was introduced in the sixteenth century, the process of cultivating lofty mountains, chiefly with maize, went on till the middle of the nineteenth century. The virgin forests of the three southernmost prefectures of K'ai-hua, Kuang-nan, and P'u-erh, notorious for malaria, had been in large part cut down and turned into maize farms by immigrants from Hunan, Hupei, Szechwan, and Kweichow before 1850.[46] Even westernmost Yunnan bordering Burma, from which maize was first introduced into China, still received immigrant maize farmers from central Yangtze during the first half of the nineteenth century.[47] Many a mountainous locality in the heavily aboriginal area of northern Kwangsi was colonized in the course of the eighteenth century by immigrants from Hupei, Hunan, Kwangtung, and Fukien.[48] The contribution of maize to a successful province-wide utilization of mountains in Kweichow was so great that it became the object of a long eulogistic song by a famous nineteenth-century Kweichow poet.[49]

All these regional evidences indicate the geographic range within which an important revolution in land utilization was carried out during the eighteenth and early nineteenth centuries. Since much of the vast Yangtze drainage is hilly and mountainous and since the revolution in land utilization made possible by maize and sweet potatoes extended to practically the whole Yangtze drainage, the aggregate of the new agricultural land thus opened up must have been very considerable.

The development of the Yangtze highlands was not without cost, however. Migrant farmers, being tenants without a permanent interest in the land they tilled, naturally tried to get maximal returns within a short period. Without experience in cultivating mountains (the Chinese were traditionally plain and valley people), they dug deep and planted maize in straight rows. Bumper crops resulted during the first few years, but heavy rains soon washed away the topsoil. While farmers could move on from one place to another, the eroded land became so useless that it had to be permanently abandoned or could be retrieved only after major reclamation efforts. By the third quarter of the eighteenth century soil erosion had already become a serious problem in parts of Kiangsi. Yü T'eng-chiao, a native of Wu-ning county and *chin-shih* of 1745, testified:

> The shack people usually dig the mountain soil five or six inches deep. The loosened soil at first yielded ten times as large a crop [as when shallow planting was practiced]. But from time to time there were torrential rains which washed down the soil and choked rivers and streams. After consecutive planting for more than ten years none of the fertile topsoil was left and the soil was utterly exhausted. Now in such places . . . mountains are reduced to bare rocks. Unless the land is rested for several decades there is no hope of renewed cultivation.[50]

Within a few decades erosion became a serious menace in the highlands in the eastern half of China's rice belt. It silted up rivers, streams, and lakes and accounted for more frequent inundations.[51] Gentry in southern Anhwei and northern Chekiang from 1807 on petitioned the provincial authorities to prohibit intensive maize farming on mountains. A series of laws followed, to the effect that no lease should be renewed for tenant maize farmers in the lower Yangtze provinces and the mountain slopes should be used for tea planting or for the growing of shan trees of *Cunninghamias sinensis*.[52] Since then late Ch'ing editions of the local histories of the lower Yangtze provinces have generally avoided mentioning maize which had been so extensively grown on the mountains during the previous century.

The total area of hilly land permanently abandoned or reverted to tea plantation or *Cunninghamias* in the eastern half of China's rice belt after about 1820 must have been considerable. It is possible that much of the red earth of Kiangsi and Hunan now having little agricultural value because its minerals have long since been leached can be attributed to the ruthless exploitation of the mountains by the eighteenth-century maize farmers. There is also reason to believe that the cultivated areas of early nineteenth-century Kiangsi, Anhwei, and Chekiang were somewhat greater than those of the twentieth century. The possibility that these provinces grew more food during the early nineteenth century than they now produce is suggested not only by land utilization records but also by the fact that their pre-1850 recorded populations were larger than their present recorded populations.[53]

III. THE OPENING-UP OF THE HAN RIVER DRAINAGE

Geographically the Han River drainage comprises the southern tip of Kansu, the whole of southern Shensi, the northern two-thirds of the western Hupei highlands, and southwestern Honan. Although the southern third of the western Hupei highlands, which in Ch'ing times consisted of I-ch'ang and Shih-nan prefectures, lies outside the Han River drainage, it should be discussed here because of its propinquity to, and close economic and historical ties with, the Han River region. Historically this region was inseparable with northeastern Szechwan. According to local historical records, this large mountainous area whose most important range is Ch'in-ling, was sparsely populated and for the most part covered with virgin forest up to 1700, with the exception of a few historic and strategic cities which were developed early. Southwestern Hupei in particular was so backward that it was under the administration of tribal chieftains until the second quarter of the eighteenth century.

That a systematic development of this region was so long delayed was perhaps accounted for by the lack of suitable crops for lofty mountain terrain. The first steps were well described in the 1864 edition of the history of I-ch'ang prefecture: "Maize . . . had originally been grown only in Szechwan. Since our area became a prefecture [in 1735], the natives have opened up mountains and grown maize to an ever-increasing extent until now it is grown everywhere." [54] Although this statement does not rule out the possibility that maize was introduced into western Hupei before 1700, it none the less helps to date the beginning of a systematic utilization of the mountains in this area. In the northwestern Hupei highlands bordering Shensi it was said, "after several consecutive bumper crops from 1752 on, maize has become the mountain farmers' very source of sustenance and has been grown by every household." [55] The economic advantage of maize, especially in the early stage of cultivation, was indeed irresistible, for "when the virgin forest was first cut down, maize secured bumper harvests without requiring any manure." [56] Maize spread from western Hupei to the whole of southern Shensi in the first half of the eighteenth century, although this food plant had been recorded by some pre-1700 Shensi local histories.[57]

By 1800 at the latest it may be said that the Ch'in-ling range and Han River drainage had been generally opened up, with maize as the king of crops and the mountaineers' staple food and sweet potatoes as an important auxiliary. The continual influx of immigrants made it necessary to cultivate even the mountains too lofty and soil too lean for maize and sweet potatoes. The Irish potato, which was introduced into this area after 1800, answered the new needs. The 1866 edition of the history of Fang-hsien described the general pattern of land utilization:

> The Irish potato is mostly grown in the southwestern mountains. In the level area in the vicinity of the walled city rice is usually grown. In the comparatively shallow hills and mountains maize predominates. In the lofty mountains where maize cannot be successfully grown the only source of food is the Irish potato . . . . Some local people of means buy Irish potatoes, which are ground into flour, and occasionally make a fortune out of it.[58]

At the technological level then prevailing, the utilization of mountains may be said to have reached its maximum. In fact, the mountains had been so ruthlessly exploited by consecutive intensive maize planting that by the early nineteenth century many localities were already suffering from the law of diminishing returns. "While the population keeps on growing," deplored a southern Shensi local history, "the fertility of the land has steadily deteriorated, with the result that all crops yield less." [59] In some districts in western Hupei even by the beginning of the nineteenth century maize farmers could "secure almost no crop at all after a whole year's toil." [60]

But the gravest menace, as in the highlands of lower Yangtze provinces, was soil erosion and a higher incidence of floods. Lin Tse-hsü, governor general of Hupei and Hunan in 1837–38, observed:

> Formerly the bed of the Hsiang River [a tributary of the Han River which runs through southwestern Honan and northern Hupei] was several tens of feet deep. Ever since the systematic deforestation consequent upon maize growing the topsoil has been washed down by torrential rains and silted up [the Hsiang River]. Between Han-yang

[where the Han River joins the Yangtze] and Hsiang-yang [the junction of the Hsiang and Han rivers] the further upstream one goes, the shallower the river bed becomes. Small wonder that from 1821 to the present there has hardly been a single year in which the Hsiang River did not flood.[61]

This was true of practically the whole Han River area. The Han River drainage, therefore, was not opened up without cost. Perhaps it is not coincidental that during a period of increasing soil erosion and diminishing agricultural returns, this area, including parts of northern Szechwan, was the seat of the White Lotus sect rebellion of 1796–1804.

Although there are no exact figures for the immigrant population, various local histories of this region mention the main areas from which the immigrants came. In Tang-yang county, on the fringe of the western Hupei highlands, there were guildhalls established by immigrants from Hunan, Shansi, Shensi, Kiangsu, and Fukien. With the exception of the last, they were first established during the Ch'ien-lung period (1736–1795), when the onslaught on the hills and mountains was in full swing. The Fukien guildhall was established decades later, a fact which indicates the long trekking of Fukien immigrants and their continual search for new land westward. There were also two guildhalls established by immigrants from the Wu-ch'ang area during the Tao-kuang period (1821–1850), reflecting the Hupei lowlanders' quest for new land.[62] Kuang-hua county, near Honan border, had guildhalls established by immigrants from Szechwan, Honan, northeastern Kiangsi, Fukien, Shansi, and Shensi. There must have been a large number of immigrants from the Huai-ch'ing prefecture of northern Honan in order to justify a special guildhall for themselves.[63]

The 1829 edition of the history of Chu-hsi county, in the extreme northwest of Hupei, said that the overwhelming majority of the local population were immigrants, mostly from Shensi and Kiangsi and also from Honan, Chihli, and Shantung. The 1869 edition added that immigrants had since come also from Szechwan, lower Yangtze, Shansi, Kwangtung, Hunan, and the lowland districts of Hupei.[64] Since Hupei occupied a most central position in the interregional migrations, it was able to attract more immigrants than it exported. More and more people from the Hupei lowlands settled in the Hupei highlands.[65] All in all, despite emigration to

Szechwan, Hupei was one of the principal net gainers in population up to 1850. Its registered population increased from 18,556,000 in 1786 to 33,738,000 in 1850. A substantial portion of this 81.8 per cent recorded increase in two generations may have been accounted for by the full-scale development of the Hupei highlands.

A more specific account of immigration into the Ch'in-ling range of southern Shensi is given in the 1849 edition of the history of Shih-ch'üan county:

> "The locality suffered from seven visitations of war around the change of dynasty [in 1644] and the population then surviving consisted of only a little over 700 households. After the long period of recuperation under the reigning dynasty, the population began to multiply rapidly from the early Ch'ien-lung period. Consequent upon crop failures in Szechwan and Hupei in 1772–73, poor people [from these two provinces] came in ever-increasing numbers to make their living on the mountains. Since then the incessant influx of poor people from Honan, Kiangsi and Anhwei has further contributed to a manifold increase of the local population. By now they have all settled happily here and become natives. But they are scattered [in the mountains] and are full of single-member bachelor households.[66]

Another southern Shensi local history says "all such comings and goings were for the sake of maize."[67] The extremely complex geographic composition of the population of southern Shensi is illustrated by that of Hsiao-i county:

> Quantitatively it may be said that the natives constitute only about 10 per cent of the total local population, immigrants from Hupei, Anhwei, and Kiangsu 50 per cent, immigrants from Kiangsi, Shansi, Honan, Szechwan, and Kwangsi 30 per cent, and those from Chihli, Shantung, Chekiang, Fukien, Kansu, other parts of Shensi, Yunnan, and Kweichow about 10 per cent.[68]

Although these proportions may not have been typical of the rest of southern Shensi, they indicate the geographic range of the interregional migrations of the Ch'ing period. The population figures for the whole of Shensi increased from 8,390,000 in 1786 to 12,107,000 in 1850. It is true that immigration was by no means confined to the Ch'in-ling range of southern Shensi, but the opening-up of the southern Shensi mountains must have accounted for a substantial portion of this 45.4 per cent increase of the provincial recorded population in sixty-four years.

IV. POST-TAIPING REPOPULATION OF THE LOWER YANGTZE REGION

By 1850 the Chinese people, with the technological means then at their disposal, had probably approached a maximum in land utilization in China Proper. This seems corroborated by diminishing agricultural returns even in such newly developed mountainous areas as the inland Yangtze and the Han River drainage. While historians should not assume the inevitability of certain major historical events, they can nonetheless correlate to a certain extent the increasing pressure of population, the limited land and technological resources, and the series of major upheavals starting with the White Lotus rebellion in 1796 and culminating in the outbreak of the Taiping Rebellion in 1851.

The cumulative effect of fourteen years of Taiping wars was undoubtedly felt by the whole country, and particularly by the hitherto most densely populated lower Yangtze area. The combined population of Kiangsu, Chekiang, Anhwei, and Kiangsi recorded as of July 1953 was still 19,300,000 less than that of 1850, supposing the latter figure to have been accurate, in spite of the industrial development of Kiangsu and the rise of Shanghai as a metropolis of 6,000,000 people.[69] Although it is true that any decline of Kiangsi's population should probably be attributed more to the Nationalist-Communist civil war of the early 1930's than to the Taiping Rebellion, the decline of the populations of three other lower Yangtze provinces seems more easily accounted for by the events of the Taiping period.

The migration to Szechwan, the inland Yangtze mountains, and the Han River drainage seem to have been reversed by the Taiping Rebellion. With Szechwan practically filled up and the mountainous areas of the Yangtze and its tributaries suffering from diminishing returns, we may surmise that the drastically depopulated lower Yangtze region offered far superior economic opportunities to the land-hungry peasants of the central Yangtze and north China. The main trend in interregional migrations during the second half of the nineteenth century was, therefore, for tenant farmers of the mountainous and congested areas to settle in the country's best irrigated area.

This trend was explained by a prefect of Yen-chou prefecture in the valley of Fu-ch'un River in Chekiang:

> After the Taiping wars vast areas of agricultural land were laid waste and various cities and towns reduced to a shambles. Villages far and near are very sparsely populated and ridges and furrows are all covered with thorns and weeds . . . . After I arrived at my official post I discovered that the devastation and decline of population in Yen-chou are worse than in many other areas . . . . To rehabilitate this area it is vital that immigrants be attracted . . . . Peasants of Kiangsi can cultivate the idle land of Ch'ü-chou [in southwestern Chekiang]. Peasants of Ningpo and Shao-hsing can plow the once fertile land of Hang-chou. Yen-chou, lying half way [between Ch'ü-chou and Hang-chou] can only depend on Hui-chou [in southern Anhwei] for the supply of immigrants. But throughout these two prefectures abandoned fields abound and it is very unlikely that [people of Hui-chou] would care to leave their homeland for Yen-chou. The only available immigrants are the shack people, who for generations have cultivated mountains and have not mingled with the common people. After the Taiping scourge of 1860 the shack people, thanks to their location in secluded mountains, were not harmed and they are generally in better economic condition than lowland and valley farmers. Previously they were compelled to depend on mountains for a livelihood because of the shortage of irrigated paddies, but today irrigated paddies are waiting for them to till. They will be willing to come down from the mountains to till the paddies which will offer them far greater incomes.[70]

That post-Taiping migrations to the lower Yangtze region were motivated more by the desire of poor peasants to improve their lot than by sheer economic necessity is suggested by local historical evidence. In An-lu county in lowland Hupei over 10,000 peasant households left for the lower Yangtze in search of better economic opportunities during the late 1860's, with the result that some villages were entirely deserted. This was true of almost all of the prefectural area of Teh-an.[71] The Chin-hua and Lan-hsi area in the heart of Chekiang, which had suffered fairly severe losses during the Taiping period and was itself in need of immigrants,[72] witnessed the emigration of hundreds of peasant households to northern Chekiang.[73] The areas that offered the greatest attraction to immigrants were southern Kiangsu and the Hang-chou Bay delta of northern Chekiang, originally the country's best irrigated area. Even before the final pacification of the rebellion in 1864 the land tax was reduced for southern Kiangsu, and this policy was soon extended to other lower Yangtze provinces.[74] Tax reduction, together with post-rebellion government financial aid to surviving landowners, enabled southern Kiangsu landowners to offer unusually lenient terms to incoming tenants. An immigration bureau was established in Nanking in the winter of 1864[75] and similar offices were created in many other localities in the lower Yangtze area. At first the provincial governments of Kiangsu and Anhwei made every effort to facilitate the immigration from northern Kiangsu, an area which had escaped the main scourge of the Taiping wars.[76] So many northern Kiangsu peasants answered the call, especially because of the prolonged depression the area had suffered ever since the decline of the salt trade after 1800, that large tracts of land in northern Kiangsu were in turn abandoned.[77]

The acute shortage of farm labor in two-thirds of southern Kiangsu could not be met by immigrant tenant farmers from northern Kiangsu alone. Further efforts were made to attract immigrants from afar. Honan, owing to its relative immunity from war devastation and the mounting population resulting from its custom of early marriage,[78] supplied the most immigrants. During some two generations after the end of the Taiping war, Kuang-shan county of Honan alone sent over one million peasants to some sixty localities in southern Kiangsu, northern Chekiang, Anhwei, and Kiangsi.[79] The whole area of southwestern Kiangsu, with the exception of the city of Nanking, where seven-tenths of the post-Taiping population were from Anhwei and Hupei,[80] was practically an agricultural colony of Honan. Honan immigrants so dominated the area that agricultural methods, social customs, and women's apparel were all revolutionized.[81] The extent to which the greater Nanking area has been repopulated from Honan since the Taiping period is suggested by the fact that a twentieth-century scholar was able to compile an anthology of southern Honan folk songs in the vicinity of Nanking.[82]

Anhwei, because of its more central location and the unusually widespread war devastation, attracted immigrants from various provinces. Hilly southern Anhwei was resettled by immigrants from Hupei, Hunan, Honan, some counties of central and southern Chekiang, and certain northern Anhwei counties.[83] East-central Anhwei, where only between 30 and 40 per cent of the native population had survived the war, relied mainly on southeastern Honan and a few Anhwei counties for the supply of immigrants.[84] Overpopulation in Honan, together with the great drought famine of 1877–78 in north China, sent fresh streams of refugee immigrants from Honan, Shansi, and Shensi to a number of Anhwei counties north of the Huai River.[85] Peasants of a few Anhwei counties which had been less severely affected by the Taiping war were easily accommodated by the abundance of idle and once crowded agricultural land elsewhere in the provinces.[86] Despite the continual influx of immigrants, post-Taiping Anhwei as a whole had fewer than southern Kiangsu and the Hang-chou Bay delta of Chekiang. Agricultural recovery was further handicapped by frequent natural calamities, banditry, and occasional clashes between natives and immigrants.

After the Taiping war northern Chekiang as a whole, and its rich alluvium in particular, immediately attracted immigrants from overcrowded Ningpo and Shao-hsing prefectures. Land-hungry peasants from other coastal prefectures, T'ai-chou and Wen-chou, soon joined ranks with Ningpo and Shao-hsing peasants and settled first in the delta of Chia-hsing prefecture and then in the valleys and low hills of Hang-chou and Hu-chou prefectures.[87] The shortage of labor remained very acute, and immigrants were attracted from remote provinces. Table 34 shows how the 1906 edition of the history of Yü-hang county classified immigrants. Yü-hang, a relatively small and hilly county, may not be representative of the volume of immigration in the lower Yangtze region as a whole, but such divergent geographic origins for immigrants may suggest the range of post-Taiping interregional migrations.

The heavy influx of immigrants notwithstanding, economic and agricultural recovery in the lower Yangtze region was very slow and gradual. The vast destruction of life and property during the Taiping period, together with the clash of interests between natives and incomers, which by the 1880's had become a matter of grave concern to the provincial authorities, presumably delayed full rehabilitation. The conflict between native owners and immigrants was particularly serious in northern Chekiang, where native owners

Table 34. Geographic origins of immigrants in Yü-hang county, 1898.

Place of origin	Households	Immigrants
Shao-hsing	309	14,336
Ningpo	203	4,321
Wen-chou	841	2,983
T'ai-chou	330	1,324
Ch'ü-chou	4	14
Chia-hsing	4	13
Hu-chou	16	69
Ch'u-chou	21	61
Chin-hua	254	954
Other provinces		
Kiangsi	36	147
Kiangsu (S)	341	1,391
Kiangsu (N)	8	19
Honan	423	1,691
Hunan	11	23
Hupei	15	31
Fukien	5	16
Kwangtung	1	5
Szechwan	3	11
Boatsmen	40	147
Total	7,414	28,499

Source: *Yü-hang-hsien hsü-chih-kao* (1906 ed.), "Population," pp. 3b–14a.

at best tolerated only immigrants from other parts of Chekiang.[88] A sampling of late editions of lower Yangtze local histories reveals that few localities had fully recovered by the turn of the nineteenth century and many famous and populous cities, like Hang-chou, Su-chou, Chen-chiang, Yang-chou, Ch'ang-chou, Chia-hsing, and Hu-chou, had suffered permanent decline.[89]

In retrospect, the presumed reduction of lower Yangtze population during the Taiping period may be viewed as a blessing in disguise, for it gave China a breathing space in which to adjust her population to her land resources. It is a great historical irony that the greatest recipient of "surplus" population after 1864 was the very region where the pre-1850 density of population was the highest. The changing population-land relation in the lower Yangtze region during the nineteenth century is thoughtfully reviewed in a modern Chekiang local history: "It was but natural that [by the 1850's] over-population brought about disasters of the cruelest kind; but once the valley became hollow the wind was bound to blow in, hence the continual influx of immigrants [after 1864]." [90]

V. THE COLONIZATION OF MANCHURIA

Unlike these interregional migrations, the colonization of Manchuria took place in modern times and received wide attention. Despite Manchuria's long historical contacts with China, it had remained sparsely populated down to the late nineteenth century. After the internal frontiers in China Proper had been developed and post-Taiping lower Yangtze resettlement had approached its completion, Manchuria remained the only underdeveloped region which, with its fertile soil and rich mineral resources, could accommodate large numbers of surplus population from north China.

Before their conquest of China the Manchus were keenly aware of their limited manpower. Up to 1644 they are said to have captured and brought to Manchuria over one million northern Chinese during their sporadic raids on the declining Ming empire.[91] A limited number of Chinese had also taken up residence in a few Manchurian cities during Ming times. With the wholesale transfer of the Manchu Eight Banners to China Proper after 1644, there was greater need for new settlers to be stationed near a few strategic posts in Manchuria. Between 1653 and 1667 the Manchu government made repeated attempts to send convicts, many of whom were Ming loyalists and scholars, to the comparatively developed districts of Liao-yang, Kirin, and Tsitsihar. Northern Chinese peasants were particularly encouraged to settle in the Liao-yang area. Voluntary colonists were few because of Chinese suspicion of the alien conquerors and their reluctance to leave their ancestral home.[92]

In 1668 the K'ang-hsi Emperor, in a fundamental policy change, closed Manchuria to the Chinese entirely. Manchuria, the original home of the Manchus, was to be made their own economic reservoir. A further influx of Chinese immigrants would deprive native Manchus of a lucrative trade in ginseng, a medicinal root valued highly by the Chinese. Also, in order to preserve the martial tribal customs, Manchuria was to be reserved as a vast hunting ground. For these reasons Manchuria was legally made a vast vacuum area. A willow palisade was erected to isolate southern Manchuria from northern and eastern Manchuria and Inner Mongolia.

However, as the Chinese became accustomed to Manchu rule and as the population of north China increased after the late seventeenth century, a significant number of northern peasants were smuggled into Manchuria from time to time. The willow palisade (actually a boundary marker) could not prevent peasants of Shantung from landing on the Liao-tung peninsula, which was in easy reach of northern Shantung ports. Peasants from Chihli and Shansi could reach the rich Sungari River plain by going around the palisade. At times of severe drought and famine the Manchu government had to wink at northern peasant refugees' trekking toward Manchuria. The K'ang-hsi Emperor himself stated in an imperial decree of 1712 that well over 100,000 Chinese peasants had entered Manchuria from Shantung alone. The number of peasants who had entered Manchuria from Chihli, Shansi, and other northern provinces is unknown. Between 1750 and 1806 a series of edicts renewed the prohibition, but it was never strictly enforced.

Some early official figures for ting and taxed land in Manchuria are available, but they were undoubtedly much too low. The number of ting for Fengtien, modern Liaoning province, increased from 5,557 in 1658 to 26,713 in 1676, and to 46,084 in 1734. By 1734 the taxed land in Fengtien, probably converted from actually cultivated land at generous rates, amounted to 2,625,967 mou. Large numbers of Chinese settlers and large tracts of cultivated land had evidently escaped official registration entirely, since from time to time thousands of immigrant households were discovered. Moreover, since the early Manchu government's nativist policy had proved a failure and since Manchu Bannermen had become idle and impoverished, a series of laws beginning in 1741 provided that Manchu Bannermen of the metropolitan Peking area should be sent back to Manchuria as cultivating landowners. The majority of these Bannermen, having been subjected to the corrosive influence of Chinese culture for so long, had to rely on Chinese immigrant tenants for production. This further helped to bring Chinese peasants into Manchuria. By 1779 the land belonging to Manchu

Bannermen but actually cultivated by the Chinese was registered at 6,275,835 mou in Fengtien and 6,337,360 mou in Kirin. After the turn of the eighteenth century the increasing economic difficulties of Manchu Bannermen made it necessary for the government to send more and more of them back to Manchuria, which again helped to make the prohibition a dead letter. In fact, throughout the eighteenth century Manchuria was seldom tightly shut against northern Chinese peasants.

The first powerful incentive for systematic colonization in Manchuria was provided by Count Nicholas Muraviev's aggressive pursuit of his ambitions in the Amur and Ussuri regions. In 1854 the Tartar General of Heilungkiang had already memorialized that the only way to forestall Russian territorial designs was to fill up the vast northern Manchurian vacuum which the K'ang-hsi Emperor had shortsightedly created. After the cession to Russia of some 350,000 square miles of territory north of the Amur and east of the Ussuri in 1860, the colonization of Manchuria became a vital matter of policy. In 1860, therefore, the rich virgin plain of the Hu-lan River, north of Harbin, one of the most fertile areas in Manchuria, was legally thrown open to Chinese immigrants. In the next year the fertile prairie of northwestern Kirin, an unusually fertile area, was made available. By the 1870's the Hu-lan area was teeming with more than 100,000 Chinese households. By 1904 Hu-lan had attained prefectural status. The Sino-Japanese war of 1894–95 and the Russo-Japanese war of 1904–05 made the colonization of Manchuria an even more urgent national concern. Before Manchuria was made into three provinces in 1907, all legal bans to Chinese colonization had been lifted.

Manchuria entered into a new era of civil administration with the appointment of Hsü Shih-ch'ang as governor general in 1907. Provincial immigration bureaus and an increasing number of counties were created. With ample funds for police administration, a preliminary population count in the three provinces was carried out in 1907; this yielded a total of slightly less than 15,000,000, with more than one-half of the population living in the southern province of Fengtien.[93] This 1907 population count, though somewhat more respectable than those of other provinces at the same time, was necessarily incomplete. There were no figures for certain newly created counties and for the Manchu population in Fengtien, and the figures for Korean immigrants were incomplete. Sir Alexander Hosie, British consul at Newchuang, 1894–1897 and 1899–1900, estimated that in 1904 the total population of Manchuria, including the Manchus, was around 17,000,000, which is probably nearer the truth than the preliminary official enumeration of 1907.[94]

Natural conditions in Manchuria are favorable to agriculture. Although sorghum has been the leading crop, practically all other food crops can thrive. With the coming of Korean immigrants many localities in southern and eastern Manchuria have grown rice successfully, while upland rice was a famous special tribute to Peking in the Ch'ing. Millet, wheat, maize and, in the twentieth century, sweet and Irish potatoes have all been grown extensively.[95] But soybean production had the most spectacular rise to prominence; Hosie estimated production at 600,000 tons in 1900, but it increased to 2,000,000 tons in 1909. "It is only in the last three years," said a special survey of 1911 by the Chinese Maritime Customs, "that soya beans have become important in intercontinental commerce, and their rapid emergence from obscurity has, indeed, been one of the most remarkable commercial events of recent times."[96] The survey's prediction of a bright future for the soy bean trade was correct, as Manchuria had an almost complete monopoly of world exports of both soy bean seed and oil in the 1920's and early 1930's. The average production during 1931–1937 was 4,268,000 tons, as compared with an average production of 6,092,750 tons for the rest of China,[97] a factor which enhanced the attraction of Manchuria as an agricultural colony for northern Chinese peasants. The railway and industrial developments created further opportunities for employment.

In the 1920's the continual pressure of population, civil wars, heavy taxation, and famines in the low plain provinces of north China, in contrast to the abundance, relative peace, and political stability in Manchuria, were added incentives to immigration into Manchuria. Moreover, Chinese and Japanese railways and steamship companies helped to facilitate the movement by offering cut-rate fares to poor Chinese peasants. The volume of immigration is indicated by Table 35.

The main inland route to Manchuria was via the Peiping-Mukden Railways and the main ports of disembarkation were Tsingtao, Chefoo, Tientsin, and Lung-kou. A relatively small

Table 35. Volume of immigration to and emigration from Manchuria and their residuum, 1923–1929.

Year	Volume of immigration	Volume of emigration	Residuum	Ratio of residuum to immigration	Net migration 1923=100
1923	342,038	240,565	101,473	30	100
1924	376,613	200,046	176,567	47	174
1925	491,948	239,433	252,512	51	249
1926	572,648	323,566	249,082	43	245
1927	1,016,723	338,082	678,641	67	669
1928	938,472	394,247	544,225	58	536
1929	1,046,291	621,897	424,394	41	418

Source: Franklin L. Ho, "Population Movement to the Northeastern Frontier of China," *Chinese Social and Political Science Review*, vol. 15, no. 3, October 1931.

number boarded ships at Hai-chou in northeastern Kiangsu. The immigrants came chiefly from Shantung, Hopei, and Honan. At first their sentimental ties with their ancestral homes were strong and the migration assumed a seasonal character.[98] But Manchuria's economic advantages were such that more and more northern Chinese peasants took up permanent residence in Manchuria.[99] In some localities in north China in the 1920's the local educated people initiated what amounted to a voluntary propaganda campaign on behalf of Manchuria, which was strengthened by the success stories of local emigrants whose economic status had been markedly improved in their chosen home.[100] as Table 35 shows, after 1925 the migration changed in character from seasonal movements to permanent colonization. In the seven years between 1923 and 1929 alone, a total of 2,500,000 were permanently settled in Manchuria. The annual volume of immigration into Manchuria between 1860 and 1920 is not known, but it seems reasonable to assume that while the volume can not have been as large as that of the 1920's, the percentage of residuum may have been higher because of the more difficult state of communications and the lack of cheap fares to facilitate the migrants' annual return.

The extent of Manchurian colonization can be measured only roughly, for not until 1940 was there a fairly reliable census in

Manchuria and even down to the end of Japanese domination there were still few accurate land statistics. The Research Bureau of the South Manchurian Railway Company estimated the total population of the three provinces of Liaoning, Kirin, and Heilungkiang at 34,300,000 for 1930, the year before Japan launched her Manchurian conquest. Since Hosie's 1904 estimate of 17,000,000 was probably nearer actuality than Hsü Shih-ch'ang's preliminary count of 15,000,000 for 1907, and since it is generally admitted that the pre-1940 official Manchurian population figures are likely to have been too low,[101] it seems reasonable to assume that the total population of the three provinces more than doubled itself in the quarter century between 1904 and 1930. The increase was accounted for largely by immigration. In 1940 the first census yielded a total population of 44,500,000, including Jehol and the Kwantung Leased Territory. The 1953 census gave a total of nearly 47,000,000.

The future prospect for Manchuria as a region of agricultural colonization may be partly gauged from land statistics. By 1945 the cultivated area in Manchuria, including Jehol, had probably reached 42,000,000 acres. In the late 1930's the puppet Manchurian government estimated that the cultivable but not yet cultivated area amounted to 43,600,000 acres, slightly larger than the already utilized area. But an expert has pointed out that future agricultural settlement will be handicapped by poorer soil conditions, scanty rainfall, and various tenurial difficulties.[102] Since 1949 landownership and tenure may no longer be a problem, but it is doubtful whether Manchuria can continue to absorb northern Chinese peasants in such large numbers and as rapidly as in the past. However, in a land as crowded as China, Manchuria in the foreseeable future will remain the most important agricultural frontier.

VI. OTHER SIGNIFICANT MIGRATIONS AND OVERSEAS EMIGRATION

Colonization of Formosa. Following the expulsion of the Dutch in 1662 by the Ming loyalist Cheng Ch'eng-kung, commonly known to Europeans as Koxinga, Formosa (Taiwan) began to attract Chinese settlers from southern coastal Fukien and eastern coastal Kwangtung. During the wars between the Ming loyalists and the Manchus, which lasted till 1683, the inhabitants along the southeast coast of the China mainland were withdrawn inland and emigration to Formosa was strictly prohibited. After 1683 the southeast coast was thrown open and an increasing number of Chinese went to Formosa either to settle or to work as seasonal agricultural laborers on the rich alluvium of the west coast. Lan Ting-yüan (1680–1733), who put down the 1721 revolt in Formosa, said:

> The people of Ch'ao-chou and Hui-chou prefectures of Kwangtung who come to Formosa as agricultural laborers are called *k'e-tzu* (literally "guest people"). Their places of residence are called "guest" villages. They number several hundred thousands and are all without wives and children. Although they are often described as violent people, their main aim is to make a living by farm labor . . . and they generally return to Kwangtung after selling their autumn crops. With their income they buy property at home and support their families. By spring they come back to Formosa. They do this year after year.[103]

From this and other of his memorials we learn that there were two main categories of immigrants to Formosa — the seasonal workers from eastern Kwangtung and permanent colonizers from the Ch'üan-chou and Chang-chou prefectures of southern coastal Fukien.[104]

The revolt of 1721 brought about a fresh emigration ban along the southeast coast. But illegal emigration to Formosa continued throughout the eighteenth century, as officials could only patrol a few main ports of western Formosa. The growth of population forced people in Chia-ying-chou area in north-central inland Kwangtung to join the overseas emigrants as colonists in Formosa. Most of the agricultural land in Formosa cultivated by eighteenth- and early nineteenth-century settlers remained unregistered.[105] By the nineteenth century the character of immigration had undergone a basic change and most of the Kwangtung immigrants had become permanently settled, because the population count of 1811 revealed a total of 232,443 households and 1,901,833 adults and children of both sexes, excluding the Formosan aborigines.[106] In the enumeration of 1811 single seasonal agricultural workers may have been included as members of landlord households, thus inflating the size of the average household to slightly over eight persons. However the 1811 household and population figures are interpreted, there can be little doubt that the majority of the Formosan population had by then become sedentary.

Meiji Japan's territorial ambitions along the China coast, as exemplified by the Formosan and Liu-ch'iu disputes of the 1870's, brought a fundamental change in Chinese government policy toward Formosan immigration. Immigration, instead of being legally banned, was now greatly encouraged. Although Formosa still remained a prefecture under the jurisdiction of Fukien province, the famous Liu Ming-ch'uan was specially appointed as governor of Formosa in 1884. Three years later the island was made a new province. Between 1884 and his resignation in July 1891 Liu subsidized immigrants' boat fare and substantially expanded the cultivated area.[107] Under his energetic administration a population count was carried out in 1887, which yielded a total of 3,200,000, excluding the aborigines.[108] This figure seems to be reasonably accurate, for the first population enumeration made by Japan after the cession of Formosa in 1895 yielded a total of 2,697,845 "natives" of Chinese origin, and it was known that between 200,000 and 300,000 Chinese fled to the mainland in and shortly after 1895.[109] It may also be assumed that the annual inflow of Chinese seasonal farm workers, which had been declining since the late eighteenth century, had come to a temporary halt.

It was estimated that during the first years of Japanese occupation the Chinese population on Formosa consisted of more than 2,000,000 Haklos, or people from Ch'üan-chou and Chang-chou of southern Fukien, and about 500,000 Hakkas from Kwangtung, hardy pioneering farmers and camphor makers from hilly districts.[110] There were also a few Chinese from the Canton area, who were mainly urban wholesale and retail merchants, and a few northern Chinese. Although the percentages of the population increase of 1,300,000 between 1811 and 1887 attributable respectively to natural increase and immigration is not known, the 1887 return of 3,200,000 indicates that Formosa absorbed surplus population of southern Fukien, eastern coastal and north-central inland Kwangtung to a significant degree during the greater part of the Ch'ing. In comparison with the interregional migrations on the mainland, however, the migration to Formosa had only secondary importance.

Hakkas settlements in western Kwangtung and Hainan Island. The Hakkas or "guest people" were originally of northern Chinese stock. Their first southward migration took place in the fourth century A.D., when north China was invaded by the barbarians. Altogether there have been five major Hakkas migrations over the past sixteen centuries, but only the last two fall within the period being studied. After their third migration in the early Ming, the Hakkas had settled in northern Kwangtung and in parts of southern Kiangsi and southwestern Fukien. There is definite evidence that the Hakkas took part in the great interregional migrations toward the inland Yangtze mountains and particularly toward Szechwan during the first half of the Ch'ing.[111]

Our main concern is their fifth and latest migration which took place in the eighteenth and nineteenth centuries. The continual growth of their population and the limited means of sustenance offered by the Nan-ling range forced the Hakkas to start moving downhill and to settle in central and lowland Kwangtung after the first quarter of the eighteenth century. They gradually moved down to the Kwangtung coast and migrated to Kwangsi and Formosa. Hardy and frugal people, they were able to expand their property steadily and hence clashed seriously with the Kwangtung natives. The conflict reached a climax in 1856 in a number of counties in the Pearl River estuary southwest of Canton and resulted in considerable bloodshed. Although the historic clashes were eventually settled in 1867, thanks to the governor's personal intervention and arbitration, the Hakkas found it wise to move out of the densely settled lowland districts. They were thus instrumental in the development of the hitherto underpopulated western Kwangtung region, including the Lei-chou peninsula and the western half of Hainan island. While it is impossible to know how many of the aggregate estimated Hakkas population of 16,000,000 were involved in this last Hakkas migration, it is safe to say that the migration, like the colonization of Formosa, was of regional significance.

Famines, wars, and migrations. In addition to major interregional migrations, from time to time there were migrations caused by serious regional famines and wars which involved large numbers of people. The great drought-famine of 1877–78, which affected all the northwestern provinces and particularly central and northern Shensi, started new waves of migration. The drastic reduction of population in many Shensi counties[112] led the surviving landowners and local and provincial authorities to try to attract immigrants first from neighboring provinces like Hupei and Szechwan and then from such remote provinces as Hunan, Anhwei, Kiangsi, and Shantung.[113] This interprovincial migration to central and northern Shensi seems to have involved millions of people. History probably repeated itself after the appalling drought-famine of 1928, but detailed local evidence has not yet come to light.

The Moslem rebellions of the 1860's and 1870's, which are said to have substantially reduced the populations of Shensi and Kansu, likewise created many local vacuums which were gradually filled by immigrants from other provinces.[114] Migrations due to regional famines and wars have been too numerous to be listed and traced in full. Although less important than the major interregional migrations, famines and wars have played a significant role in adjusting population and land resources in modern times.

Overseas emigration. Even though overseas emigration seems to have had almost no effect on the rates of population growth in China, it has been the subject of several monographic studies.[115] Although the Chinese of the southern Fukien coast went out to trade and colonize southeastern Asian islands as early as the fifteenth and sixteenth centuries, systematic emigration overseas dates back only to the opening of China in the 1840's. In many localities emigration from Kwangtung and Fukien began to assume notable proportions only relatively late in the nineteenth century.[116] The background of the emigrating communities and the stories of Chinese emigrants in Latin American and Caribbean countries, North America, Malaya, Dutch East Indies, Siam, Philippines, and other southeastern Asian islands and countries have all been told, but up to the present there are no accurate figures for the whole overseas Chinese population. H. F. MacNair's 1921 estimate of 8,600,000 was widely accepted and the current estimates range between 10,000,000 and 13,000,000. Without trying to separate the aggregate net volume of emigration from the natural increases of the comparatively old established Chinese communities abroad, these estimated total figures still suggest the significance of emigration in alleviating pressure of population in Kwangtung and Fukien.

While emigration has had demographic significance only to the southeast coastal region, its economic significance has been far greater. C. F. Remer has estimated that the average annual total of remittances from overseas Chinese for the period 1914–1930 was in the neighborhood of 200,000,000 Chinese dollars.[117] Later and more detailed studies show that his estimate was reasonably cautious. In the year 1937 the total overseas remittances received by Kwangtung alone amounted to $180,000,000.[118] The aggregate overseas remittances for the five-year period 1931–35 amounted to 1,300,000,000 Chinese dollars.[119] Until the 1930's, when China's imports increased by leaps and bounds, overseas remittances almost always helped to balance China's international payments. The economic importance of emigration has therefore far exceeded its demographic importance.

The changing historical population-land relationship can be traced in broad terms from records on interregional migrations. It may be hypothesized that up to about 1800 the supply of land was probably not a serious problem. The migrations to Szechwan and the opening-up of the Yangtze highlands and the Han River drainage were presumably in response to the growing population of the late seventeenth and the entire eighteenth centuries. By the first half of the nineteenth century the pressure of population on land resources seems to have been mounting everywhere in China proper, but most of all in the Yangtze region. Only a drastic reduction in the population of the lower Yangtze area could give China a respite in which to adjust her population to land resources. From the late nineteenth century Manchuria began to absorb millions of northern peasants, and southeast Asia and the New World afforded fresh outlets for the excess population of the southeast coast. While all these migrations have in the long run helped China to accommodate her increasing population, the population-land ratio seems to have become every decade more unfavorable, except in the post-Taiping lower Yangtze region. With an actual stoppage of emigration and with Manchuria more than half filled, China's growing population will now have to be sustained largely by industrialization and by raising the technological level of her agriculture.

CHAPTER VIII

Land Utilization and Food Production

One factor beneficial to population growth has been the continual improvement in land utilization in Ming, Ch'ing, and modern times. The acreage in cultivated land has been expanding and land has been yielding more, partly as a result of increasingly labor-intensive farming and even more of the steady improvement in the cropping system. It is partially true that for centuries there has been no technological revolution in Chinese agriculture, evidenced by the fact that the same kind of agricultural implements have been used by Chinese peasants of certain areas for centuries.[1] Yet such a generalization requires qualification. During the Ming period there were significant improvements in agricultural implements, particularly in various kinds of water-pumps. The increase of draft animals, especially oxen and water buffalo, and the construction of irrigation works in the southeast both deserve further study.[2] In any event, not every important change in the technique of production is necessarily associated with the introduction of new implements and machinery. The first major revolution in the iron and steel industry of the modern West, to name an outstanding example, was caused by the use of a new fuel rather than by machinery. The core of Chinese agriculture has been its cropping system. In the absence of major technological inventions the nature of the crops has done more than anything else to push the agricultural frontier further away from the low plains, basins, and valleys to the more arid hilly and mountainous regions and has accounted for an enormous increase in national food production.

I

The first long-range revolution in land utilization and food production in China during the past millennium was brought about by the early-ripening rice which began to be disseminated extensively at the opening of the eleventh century. Agricultural and etymological evidence indicates that although ancient and medieval China had a limited number of indigenous early-ripening rice varieties, they were far less important than the heavy yielding late- and medium-ripening varieties which usually require well-watered lowlands and rich clayey soil. *Ch'i-min yao-shu*, the earliest systematic Chinese agricultural treatise extant, compiled in the first half of the sixth century, states as a general rule that the rice plant required 150 days to mature after it was transplanted from the nursery bed to the paddy field.[3] Since rice seeds are grown in nursery bed for four to six weeks, the total time required by ancient and medieval rice was evidently at least 180 days. Such a long period made it difficult for the same piece of land to be sown to a second crop after the harvest. Furthermore, it is common geographic knowledge that the well-watered lowlands in China suitable for the cultivation of late- and medium-ripening rice are of rather limited extent. For these reasons the area of rice culture up to the end of the first millennium of the Christian era had been practically confined to the deltas, basins, and valleys of the Yangtze region.[4]

The increase in the population of the rice area throughout the T'ang (618–907) and Five Dynasties (908–960) brought about a shift of the economic and demographic center of gravity from the northwest to the southeast and made it necessary for the early Sung emperors to expand the frontier of rice culture. The Sung Emperor Chen-tsung (998–1022), who was deeply concerned with the food supply, introduced to Fukien an early-ripening and relatively drought-resistant rice from Champa, a state in central Indochina. In 1012, when the lower Yangtze and Huai River regions suffered from drought, 30,000 bushels of Champa seeds were shipped from Fukien to be distributed in the drought areas. To familiarize peasants of the lower Yangtze and Huai areas with this valuable strain the government circulated pamphlets in which the methods of cultivating it were explained. The publicity given to the Champa rice in 1012 not only accounted for its cultivation in the southeastern provinces but called the attention of Chinese rice farmers to the importance of developing further early-ripening varieties with which to use land of poorer quality.

Sung local histories of Chekiang and southern Kiangsu recorded that the original Champa rice matured one hundred days after transplantation. It was suited to relatively well-watered hills on which indigenous late- and medium-ripening varieties could not be successfully grown. By the twelfth century ingenious Chinese farmers had developed new strains which matured in but sixty days after transplantation. The effect of the introduction of Champa rice and the further development of early-ripening varieties was manifold. First, these early-ripening varieties helped greatly to ensure the success of the double-cropping system, for which Chinese agriculture, particularly that of the rice area, is world famous. Since rice in the Yangtze area is usually a summer crop, a shorter period of growth made it possible for the same piece of land to be sown to wheat, rape, or other winter crops after the rice was harvested. The average late-ripening rice of ancient and medieval China would have made such a rotation of crops precarious if not entirely impossible in many places. Secondly, for centuries peasants had worried about the precarious food supply during the long interval between the harvest of the winter crop and the harvest of rice. Early-ripening rice now became a valuable fill-in crop, much superior in taste and food value to other auxiliary crops. Thirdly, the early-ripening rice, which requires much less water than the other varieties, made possible the cultivation of higher land and hilly slopes that could be fed by spring or rain water. The first long-range revolution in land utilization and food production, based on the conquest of relatively well-watered hills, thus got under way.

The dissemination of the early-ripening rice, like the propagation of other food crops, was necessarily a slow process. As far as can be ascertained, up to the end of the Southern Sung (1127–1270) the early-ripening varieties seem to have been disseminated to a significant extent only in Chekiang, southern Kiangsu, Fukien, and Kiangsi. The lowlands of southern Anhwei and large sections of Hupei and Hunan, which are the main rice baskets of modern China, were markedly lacking in early-ripening varieties and therefore agriculturally underdeveloped. The Huai River plain, which serves as an area of demarcation between the rice-producing south and the wheat-producing north and into which Champa rice was introduced by the imperial order of 1012, suffered a long-term economic decline after the downfall of the Northern Sung dynasty in 1126. After that it became a major battleground between the Chinese and the Juchens, who had conquered north China, and throughout the twelfth and thirteenth centuries remained a vast desolate area. Any agricultural gains that this area may have made during the eleventh century must have been nullified by these wars and extensive southward migrations by the Chinese. Hupei suffered for similar reasons.

Although the dissemination of Champa rice during the Sung was confined mainly to three or four provinces, its effect on food production and population growth seems to have been already felt within one century of its introduction, for in 1102 and 1110 the number of officially registered households for the entire country exceeded 20,000,000. Since it has been shown by some leading Japanese historians that the official Sung household returns, in sharp contrast to the notoriously under-registered "total" population, were comparatively accurate,[5] it seems reasonable to believe that by the opening of the twelfth century China's actual population exceeded, for the first time in her history, the 100,000,000 mark.

The dissemination of Champa varieties must have continued throughout the subsequent periods. From available Ming records and local histories it appears that the sixty-day variety, which had been confined essentially to the lower Yangtze area during the Sung, had reached as far as Kwangtung and become fairly common throughout the rice area. The "white-Champa," "red-Champa," and other early-ripening varieties had struck permanent root in Kwangsi before the compilation of the 1531 edition of its provincial history. The 1574 edition of the history of Yunnan province places early-ripening rice as the foremost crop of Meng-hua prefecture. While giving no detailed listing of rice varieties, the 1621 edition of the history of Ch'eng-tu prefecture, in the heart of the Szechwan basin, particularly points out that the "white-Champa" had been instrumental in conquering dry fields. Even in Honan during the sixteenth century the early-ripening rice varieties were very common.

Most of the vulgar names for early-ripening varieties are too fanciful to be of real use for our purpose, but the continual dissemination of early-ripening rice may be conveniently indicated by the names of those varieties which reveal their geographic origins. In the lowland districts of southern Anhwei, an area that had been agriculturally underdeveloped during the Sung, there were in the fifteenth-century varieties such as the "white sixty-day" and "red sixty-day," and also varieties bearing names like the "Kiangsi early,"

"Hu-kuang (Hupei and Hunan) early," and "Kwangtung early." These lowland districts around Wu-hu have since become the leading producing centers of early-ripening rice. By the early seventeenth century, possibly much earlier, in as southerly a place as Ch'üan-chou in southern coastal Fukien, there were early-ripening varieties called the "Su-chou red" and "Honan early." In the eighteenth century another valuable northern variety was added which bore the name "Shantung seed." Hupei and Hunan, two provinces that during the Ming forged ahead as leading rice-exporting areas, must have introduced and developed many new rice varieties in the course of time. Unfortunately, neither the handful of extant Ming local histories of these provinces nor the 1684 edition of the combined history of Hupei and Hunan yield any systematic information. However, when Hunan compiled its first independent provincial history in 1820, there were not only certain extremely early-ripening varieties called "fifty-day Champa" and "forty-day Champa" but over a dozen varieties that, judging by their names, had originated in Su-chou, Nanking, Liang-shan in southwestern Szechwan, Kuei-yang and Ssu-nan in Kweichow, Kiangsi, Kwangtung, Yunnan, and Indochina.

For centuries after the introduction of Champa rice the quickest ripening variety required 60 days after transplantation. The availability of a fairly large number of quickly ripening varieties ranging from 120 to 60 days had unquestionably helped to solve the problem of crop rotation in the rice area. But there were marginal rice areas which called for varieties that would ripen even more rapidly or for special strains that could survive unusually menacing natural conditions. The marshy flats of Kiangsu north of the Yangtze, visited by the annual midsummer flood and consequently submerged for a greater part of the year, made up one such blighted area. For this reason Kao-yu and T'ai-chou, both in the heart of the Kiangsu flats, became virtually the experimental farm for extremely early-ripening varieties. To beat the annual midsummer flood, peasants of Kao-yu developed the fifty-day in the sixteenth century. Some inland districts in southwestern Chekiang and lakeside Kiangsi may also have independently developed this fifty-day variety. It was the fifty-day that saved Kao-yu's peasants from total crop failure during the terrible flood of 1720–21. In the eighteenth century the forty-day was developed, probably independently, in Kao-yu and in Heng-chou in southern Hunan. When Kiangsu was particularly hard hit by the flood of 1834–35, the thirty-day variety, which was said to have been developed in Hupei, was rushed in and distributed to peasants of the Kiangsu flats. The development and dissemination of these extremely early-ripening varieties by the peasants of marginal rice areas indicate that rice culture in China Proper was about to reach its saturation point. The thirty-day variety is probably the quickest ripening rice ever recorded in world history.

There was also a need in certain localities for a rice that could be planted late and would also ripen quickly. This need was created by the limited local supply of good land and the local rotation system. By Ming times at the latest certain Champa varieties had been developed that could withstand cold so well that they were planted in midsummer after the field was entirely cleared of spring crops or early rice. Since they were harvested in late fall or early winter, they were called "cold Champa" or "winter Champa," names that are recorded in a number of early modern local histories of Chekiang, Fukien, Kiangsi, and Hunan. In addition to facilitating a multiple cropping system, they gave peasants of lakeside Kiangsi a weapon with which to combat the annual flood. A particular reddish rice was usually planted in the lake districts of Kiangsi in the sixth lunar month, when the flood began to recede. It became so famous that when in 1608 a northern Chekiang county suffered from an unusually severe flood and was on the point of losing all its rice crop, this reddish variety was introduced by the magistrate to be planted after the retreat of the flood waters. In spite of its late planting, it ripened quickly enough to ensure the peasants a fair harvest. T'ai-chou in the Kiangsu flats was equally famous for this late-planting but quick-ripening reddish rice. Since then the reddish rice of Chiu-chiang, the main port of Kiangsi, and T'ai-chou has been recorded in a number of local histories. By the early eighteenth century the practice of growing quick-ripening and late-planting rice after the retreat of flood waters had become increasingly common in the rice area.

It is true that in the history of so slowly moving an industry as agriculture, let alone the self-sustaining agriculture of China, few events can really be called revolutionary. Yet the introduction of Champa rice and the subsequent dissemination of all early-ripening varieties, native and imported, ultimately produced results beyond the fondest dream of the early Sung emperors. Within two centuries of the introduction of Champa rice the landscape of the eastern half of China's rice area had already been substantially changed. By the thirteenth century much of the hilly land of the lower Yangtze region and Fukien, where water resources, climatic and soil conditions were sufficiently suitable for the cultivation of early-ripening rice, had been turned into terraced paddies. The early-ripening rice not only ensured the success of the double-crop system but also prolonged the economic hegemony of the Yangtze area. One certain indication of such hegemony was that the population of the rice area was increasing much more rapidly than the population of north China throughout the Sung, Yüan and Ming periods. During the Yüan and Ming the cultivation of early-ripening rice became common in the southwestern provinces and in Hupei and Hunan, the two provinces that have since become China's rice bowl. By the time of Matteo Ricci (1553–1610) double, and sometimes triple, rice cropping was common.[6]

In the 1830's, when early-ripening rice culture was approaching its saturation in the main rice area, it was estimated by Lin Tse-hsü (of Opium War fame and an authority on early-ripening rice) that the total national output of early-ripening rice approximately equaled that of the late- and medium-ripening rice.[7] There can be little doubt that, with the extension of irrigated areas and some not insignificant improvements in water pumps and other agricultural implements during the early modern centuries, the national production of late- and medium-ripening rice must have increased somewhat since the eleventh century; hence it seems reasonable to conclude that China's total rice output had greatly increased, perhaps even doubled between 1000 and 1850. As the per-acre yield of early-ripening rice is considerably less than that of late- and medium-ripening rice, it seems possible that the total area under early-ripening rice may have exceeded the area under the latter. In other words, China's rice area, too, may have more than doubled itself in the course of some eight and a half centuries. Since there is a good deal of qualitative evidence showing that the Chinese agricultural system was becoming increasingly labor-intensive, owing to an almost continual growth of population during Ming and early Ch'ing times, and since in most places the early-ripening rice indirectly enhanced local food production by promoting a busier and better rotation system, the effect of early-ripening rice on the overall food supply of early-modern China must have been even greater than Lin's statement indicates.

In retrospect, it may be suggested that during the greater parts of the past millennium the food situation was in all probability better in China, where the early-ripening rice brought about a major long-range revolution in land utilization and food production, than in Europe, where important changes in agriculture did not occur until the eighteenth century. If this is correct, it seems worth further notice by historians of world population because, in terms of food supply, China's population could probably increase, and apparently did begin to increase substantially, from 1000 onward. If it did not assume a continuous or more or less linear growth until after the founding of the Ming dynasty, this was due to political and institutional factors rather than to the most basic of economic restraints imposed by a shortage of food.

II

While the effect of early-ripening rice in increasing the food supply is rather obvious, the dissemination in the rice area of various dryland crops, originally from north China, is difficult to trace because systematic records are unavailable. Yet the ultimate aggregate effect of the propagation of northern crops in central and south China is not insignificant, for much of the history of Chinese agriculture can be written in terms of the conquest of inferior land by suitable crops. The need to utilize new land became more pressing with the increase of population in early modern and modern times. Despite the scattered and fragmentary nature of available sources, an attempt should be made to trace the main stages in the utilization of marginal land in China, particularly in the rice area. In the absence of major technological inventions the Chinese peasant's main weapon in his struggle with new land was crops; hence our discussion will center on crops.

Sorghum began to be disseminated extensively after the Mongol conquest; [8] maize, sweet potatoes, Irish potatoes, and peanuts were introduced into China in post-Columbian times; the other dryland crops, like wheat, barley, and millet, had been grown since ancient times.[9] Available ancient sources contain at best only very general references to the staple crops of each region. It is often difficult to know whether such dryland crops were grown to any significant extent in central and south China in early times,[10] but it is fairly certain that the Yangtze region as a whole was agriculturally primitive, with rice as the main cereal crop. The southwest, particularly Szechwan, with its subtropical summer, long growing season, and fertile soil, seems to have had a much wider range of crops than any other southern agricultural region.[11] Since crop migration cannot be treated systematically without a large number of local histories, which are unavailable except for relatively modern periods, the main stages in the dissemination of dryland northern crops in the rice area can be traced for the most part only from official exhortations.

By far the earliest known systematic government effort at encouraging wheat and barley cultivation in the rice area was made in A.D. 318 when north China had fallen to barbarians and a Chin prince had just declared himself emperor at Nanking. In order to ensure an adequate food supply, now even more vital than before, an imperial decree read: "In the two provinces Hsü (Huai River region in northern Kiangsu and northern Anhwei) and Yang (greater part of Kiangsu and southern Anhwei) the land is suited to the planting of three kinds of wheat and barley. The people should be ordered to plant these in dry land at the approach of autumn. The crop ripened by summer can be used to fill the gap between the old and the new grain." [12] It may be assumed that some of the estimated one million Chinese who migrated to the south during the fourth century brought with them a knowledge of wheat cultivation.[13] From time to time the exhortations of conscientious local officials also helped in the dissemination of wheat and other dryland crops in the south.[14] These scattered references to wheat, though indicating wheat cultivation in certain Yangtze districts during early medieval times, are far from suggesting that it had acquired a significant place in the agricultural system of the rice area.

A major step toward extensive wheat cultivation in the rice area began with the Sung period (960–1270). A decree of the late tenth century said: "While people north of the Yangtze grow various cereals, people south of the river rely solely on *keng* rice (nonglutinous and late-ripening rice). Although there may be climatic and topographical reasons for this, it is in accordance with ancient practice to grow a wider range of crops so as to stave off flood and drought." All southern provinces, except those of the extreme southwest, were ordered to grow wheat, millet, legumes, and other dryland crops. If seeds were lacking, they were to be supplied by local officials of the northern Huai region. Peasants north of the Huai River were exhorted to experiment with the cultivation of *keng* rice. Further to encourage the growing of these dryland crops, it was ordered that southern tenant farmers should not be required to pay additional rent for the harvest of dryland crops.[15] A series of similar decrees was proclaimed after north China was lost to the Juchen in 1126.[16]

The repeated exhortations of the Sung emperors, who may in a sense be regarded as physiocrats, seem by and large to have secured permanent results. Local histories, particularly those of the lower Yangtze region, which became more numerous from the Sung, usually contain brief references to wheat. Ch'iu Chün, in his famous new commentaries on the *Great Learning*, which were presented to the emperor in 1487, praised Sung T'ai-tsung and Sung Chen-tsung highly: "In our time people south of the Yangtze all grow various cereals [besides rice] and people north of the river also cultivate rice. Formerly the keng rice secured only one crop in autumn; now there is the early-ripening rice as well. The graces of these two emperors have since reached far into the populace." [17] Further evidence of the fairly extensive cultivation of wheat and other dryland winter crops in the lower Yangtze area from Sung times is the fact that the exemption of the wheat harvest from rent has since become customary in most lower Yangtze districts, where the tenants pay rent only for the autumn rice harvest.[18] So great had been the advantage of wheat growing that in the first half of the nineteenth century tenant farmers in Kiangsu were reluctant to grow early-ripening rice because "wheat still occupies the land," although the increasing flood hazards made the extremely early-ripening rice a more suitable crop.[19]

The repeated Sung exhortations, however, could not prevail over climatic and topographical factors; neither could they force the majority of rice farmers to adopt a more labor-intensive system of double-cropping in a period when land was still comparatively plentiful in the inland Yangtze region. As far as historical tenurial records go, the official Sung efforts probably secured permanent results mainly in Kiangsu and Chekiang. In Hunan, for example, a memorial of 1179 testified that wheat was grown in only two or three prefectures, despite imperial exhortations.[20] A warm and humid climate, together with competition from other cereals, seriously limited wheat cultivation in Hunan as late as the early eighteenth century.[21] Even in Hupei, today one of the leading wheat-producing provinces in the Yangtze region, wheat was not extensively grown on the lowlands until after the 1730's.[22] By Ming times wheat, millet, barley, and sorghum were grown in many parts of Fukien, but none of them was a major crop.[23] A prefect of Canton testified at the end of the ninth century that wheat often failed to ripen because of the tropical climate.[24] In modern times, however, all these dryland crops are fairly common even in the most southerly province of Kwangtung, although the extent of their cultivation is limited by natural conditions.

Although a detailed treatment of the southward migration of northern dryland crops is impossible, there are valuable references to their propagation in marginal or submarginal agricultural districts during the Ming and Ch'ing. The famous Christian prime minister Hsü Kuang-ch'i (1562–1633) called the attention of officials and farmers to the advantage of growing wheat on land that was submerged for greater parts of the year. He explained in his famous agricultural encyclopedia:

> In the north the worst category of land is the submerged. On it natives usually grow sorghum and can secure, on the average, only one crop in several years. For this reason they are poverty-stricken. I have taught them to grow wheat, which may not be affected by [annual flood]. For the flood usually comes in late summer or early fall and can do no harm to wheat. In places where drainage is possible after the recession of the flood the land dries up in autumn and is therefore suitable for autumn wheat. In localities where drainage is not feasible the land dries up in winter and is suited to spring wheat . . . . This method can practically assure nine crops in every ten years.[25]

This was no mere theorizing; it was the result of experience gained in his own experiments near Tientsin on several occasions when he took sick leave.[26]

Hsü's fame and his sound common sense were such that his

method was adopted by some provincial officials in the eighteenth century, when the population was evidently increasing very rapidly. Mai Chu, governor general of Hupei and Hunan, 1727–1733, stated in a memorial that wheat was not grown extensively in Hupei, excepting in two northern prefectures. But after the serious flood of 1727 the peasants in lowland Hupei, under the persuasion of the provincial government, began to grow wheat widely. The wheat harvest preceded the annual flood; hence the advantage of wheat farming was increasingly appreciated.[27] Hupei has since become one of the five principal wheat-producing provinces in modern China.[28] Fang Kuan-ch'eng, who served as the governor general of Chihli from 1749–1768 save for a brief interval, took special pains to drain the marshes of the Chihli-Shantung border area and along the Tzu-ya River in central Hopei, and farmers were exhorted to grow wheat on the partially drained land. His project was so successful that farmers in these swampy districts gladly agreed to pay taxes on what had originally been tax-free wastes.[29] Through the same principle some marshes in northern Kiangsu were partially utilized, thanks to the exhortations of Ch'en Hung-mou (1696–1771), another famous eighteenth-century provincial official.[30] The knowledge that wheat was an effective weapon against annual floods had gained so much currency that by the early eighteenth century farmers in certain swampy districts of Yunnan were relying mainly on wheat and buckwheat for sustenance.[31] In some lowlands where post-flood drainage was slow and difficult a late-planting but early-ripening wheat was needed. It was for this reason that from 1740 spring wheat was introduced into Ch'en-chou prefecture in east-central Honan, which is otherwise the heart of the winter wheat area.[32] In some of the worst swamp districts in Honan only one crop could be grown each year and that crop was almost invariably wheat.[33]

Barley, in certain areas, served the same purpose as wheat in the struggle against annual flooding. In Chen-chiang in southern Kiangsu there was a special variety of barley which could be planted any time between the lunar tenth and first months. After the late seventeenth century, if not earlier, it contributed greatly to the utilization of riverside lowlands. Its cultivation became more widespread during the first half of the nineteenth century, when the rising river bed caused more frequent floods and made cultivation of the riverside lowlands impossible until the first lunar month. This particular kind of barley gave the needy farmers bumper crops year after year and has since made south-central Kiangsu one of China's leading barley-producing areas.[34]

Although wheat was the most common crop for lowland districts subject to periodic inundation, the northern Hunan flats around the Tung-t'ing Lake had accidentally become an important millet-growing area since the early seventh century. Prince Li Yüan-tse, brother of T'ang T'ai-tsung (627–649), while a prefect of Li Chou in northern Hunan, insisted that peasants pay part of their taxes in millet stalks to provide sufficient fodder for his war horses. Since millet was a northern crop the peasants had to go to northern Hupei for the stalks; and to save the cost and labor of transportation they began to grow millet locally. The plant's unusual adaptability, and the fact that it was harvested in the first ten days of the fifth lunar month before the lake waters rose, made millet one of the staples of the area, which has since become nationally famous for its production.[35] In the course of time northern Hunan farmers developed certain special varieties of millet which could be grown in the fall after the retreat of the flood waters and which quickly ripened in early winter.[36] It may reasonably be assumed that millet was brought into the northern Kiangsi flats around the Po-yang Lake from northern Hunan, because millet is rather extensively recorded in the local histories of northern Kiangsi.[37]

The country-wide dissemination of sorghum during Mongol and Ming times is even more difficult to trace than the southward migration of ancient dryland crops. Although an expert on Chinese botany and sinology, in a learned and methodical article, concluded that sorghum was introduced into China around the time of the Mongol conquest from western Asia, his qualification that it might have been brought into southwestern China, particularly Szechwan, before the late thirteenth century appears more reasonable.[38] The argument in favor of an earlier introduction is based on various reasons. First, *Po-wu chih*, one of the early works on natural history attributed to Chang Hua, a famous scholar of the late third century, clearly mentions sorghum, which is called *Shu-shu*, or literally "the millet of Shu (Szechwan basin)." It says: "The land on which Shu-shu is grown for three consecutive years is likely to be full of snakes in the next seven years." It has been held that the work may not be by Chang, but to attribute a book to a famous scholar was fairly common in traditional China and this does not necessarily mean that the contents of the book are not authentic. That the earliest, and still very common, name for sorghum should be associated with Szechwan is highly significant. In addition, the statement on sorghum accords well with its later method of planting, for Ming local histories and Hsü Kuang-ch'i's agricultural encyclopedia testify to the plant's remarkable adaptability to wet soil, although some modern strains are also known for their ability to resist drought. Secondly, *Po-wu chih*, while not necessarily Chang's own work, was nonetheless cited in the important agricultural treatise of the sixth century, *Ch'i-min yao-shu*, in which the passage on sorghum is given in the chapter on plants of exotic origin.[39] The *Hua-yang kuo-chih*, the earliest systematic geographic treatise on Szechwan, completed in the late third century, states that Szechwan had a much wider range of food crops than most other regions and describes Szechwan's contacts with the aborigines farther to the southwest. The fact that Chang Ch'ien, the famous envoy to Bactria, found in central Asia in the late second century B.C. certain types of cloth and bamboo that were said to have come from Szechwan via India suggests that there must have been considerable contact between Szechwan and the nations beyond China. Before recorded history, rice, which is indigenous to India and parts of southeast Asia,[40] had already been introduced into China from the landward side. American food plants were definitely brought into China by both maritime routes and the overland India-Burma-Yunnan route.[41] Why then, was not sorghum, which is native to Ethiopia, introduced into China until the Mongol conquest? [42] Finally, the 1175 edition of the history of Hui-chou in southern hilly Anhwei, written by the famous natural historian Lo Yüan, clearly describes the reddish and black sorghum, both glutinous and non-glutinous.[43] This indicates beyond much doubt that sorghum had been introduced into China long before the Mongol conquest.

Partly because of the publicity given it in Mongol times, sorghum began to be disseminated extensively. It is not easy to trace the main stages of its geographic propagation, for Ming local histories of various provinces, northern as well as southern, usually contain a brief reference to it. The 1574 edition of the history of Yunnan province gives the impression that sorghum, though not a major crop, had a province-wide distribution, an indication of its early appearance in the southwest. Sorghum was grown at first mostly on flats or even on partly submerged land in north China but it was adapted to highlands in certain southern localities. It is worth mentioning that before American food plants were introduced into China, sorghum was a significant and relatively new crop with which Chinese farmers sowed land that was unsuited to other dryland crops of ancient origin.

In brief review, northern dryland crops appear to have been grown in the lower Yangtze provinces and the region south of the Huai River after the repeated exhortations of Sung emperors. Their dissemination in the rest of the rice area went on apace throughout the subsequent periods. Since the sixteenth century they have been subjected to competition from maize, sweet potatoes, and Irish potatoes. These northern dryland crops have always been of secondary importance in the rice area; nevertheless their continual dissemination throughout the centuries has ultimately contributed in no small degree to better land utilization and an increase in total national food production. Apart from routine references to wheat and barley in local histories, practically all the special accounts on wheat and barley during the late Ming and early Ch'ing periods were connected with the utilization of marginal land, which reflected the fact that by then China's population had grown far beyond the Chinese chroniclers' high watermark of 60,000,000.

III

I have argued that rice culture in China Proper seems to have reached its saturation point by about 1850. Yet three centuries before this long-range revolution in land utilization and food production had run its course another similar revolution had already begun and is still continuing. The main agents of this second agricultural revolution have been American food plants, such as maize, sweet potatoes, Irish potatoes, and peanuts.[44] Within fifty years of Columbus's discovery of the New World peanuts were being grown in the sandy loams south of the Yangtze near Shanghai. Before the middle of the century maize was grown fairly extensively in southwestern China and even in one Honan district. In the 1560's the sweet potato had already been welcomed by the people of

Yunnan, and toward the end of the century it was becoming a poor man's staple food in Fukien. Available evidence indicates that the peanut was introduced into China by sea, and that maize and the sweet potato were introduced by both maritime routes and the overland India-Burma-Yunnan route.

Early-ripening rice aided the conquest of relatively well-watered hills. American food plants have enabled the Chinese, historically a plain and valley folk, to use dry hills and mountains and sandy loams too light for rice and other native cereal crops. There is evidence that the dry hills and mountains of the Yangtze region and north China were still largely virgin about 1700. Since then they have gradually been turned into maize and sweet potato farms. In fact, during the last two centuries, when rice culture was gradually approaching its limit and encountering the law of diminishing returns, the various dryland food crops introduced from America have contributed the most to the increase in national food production and have made possible a continual growth of population.

The dissemination of American food plants in a country as large and varied as China was necessarily a slower process than some previous writers would have us believe. Thanks to the unique body of successive editions of Chinese local histories, of which more than three thousand are available in the eastern United States, we can trace the main steps in the geographic propagation of these new food plants rather accurately.

So far as can be ascertained from written records, the peanut was the first American food plant introduced into China, probably by the Portuguese, who arrived in the Canton area in 1516 and subsequently traded at southern Fukien ports and Ningpo, which is within a day's voyage from Shanghai. By the 1530's peanuts were being grown in some lower Yangtze localities and they attracted the attention of some gentry-scholars. Nevertheless, it took more than one and a half centuries for peanuts to be disseminated extensively in the sandy loams north and south of the lower Yangtze and in southeastern coastal provinces. Although before 1700 not a few of the coastal localities had specialized in extensive peanut and peanut-oil production, sometimes for export to the rest of China, peanuts were not yet a common and cheap food in the southeast; they were still regarded as a delicacy and were served at formal banquets.

In the eighteenth and early nineteenth centuries peanuts made systematic inroads into hitherto underdeveloped western Kwangtung, including the Lei-chou peninsula and the north coast of Hainan island, which became an important exporting area. Many previously backward districts in Kwangsi and Yunnan had been transformed by this valuable new crop into prosperous areas of specialized farming.[45] On the sandy bars of the innumerable rivers and streams in Szechwan, peanuts were particularly extensively grown, though they were not famous for their quality. They were also to be found in several localities in central Yangtze provinces, such as Hunan and Kiangsi. Nan-k'ang prefecture, traditionally a poor and mountainous area at the southwestern corner of Kiangsi, owed its prosperity to specialized peanut farming.

In north China, however, peanuts remained a comparative rarity down to the late eighteenth century, although the plant is mentioned in a few local histories compiled before 1750.[46] Hao I-hsing, scholar and native of Shantung, was impressed during his visit to Peking in 1787 by the fact that "longevity nuts," one of the common vulgar names for peanuts, were a "must" at any formal banquet in the nation's capital.[47] Today peanuts are a very common food in north China, even for the poor. Various local histories of Hopei, the second largest producer of peanuts in twentieth-century China, take pains to explain that peanuts were grown extensively for commercial purposes beginning during the latter half of the nineteenth century. It was not until the beginning of the twentieth century that the T'ai-an area, at the foot of the sacred mountain T'ai-shan, and the localities in the lower Yellow River alluvium of northwestern Shantung and northeastern Honan, became the leading peanut-producing area in the country.[48]

Throughout the last three centuries peanuts have gradually brought about a revolution in the utilization of sandy soils along the lower Yangtze, the lower Yellow River, the southeast coast, particularly Fukien and Kwangtung, and numerous inland rivers and streams. Even within the crowded cropping system of some rice districts, they usually have a place in the rotation because farmers, without knowing the function of the nitrogen-fixing nodules at the roots of the peanut plant, have learned empirically that it helps to preserve soil fertility. Peanuts, unlike rice and wheat, are necessarily a secondary food crop in so large a country as China, but China, excluding Manchuria, with an estimated annual output of 2,800,000 metric tons during 1931–1937, ranks with India as one of the world's leading peanut-producing countries.

The sweet potato is first recorded in some local histories of Yunnan in the 1560's and 1570's, which suggests an overland introduction from India and Burma. But it was also independently introduced into coastal Fukien two or three decades before it was officially blessed by the governor in the famine year 1594. It then made rapid headway in the southeastern coastal provinces. Ho Ch'iao-yüan, a scholar from southern coastal Fukien and the compiler of the famous 1629 edition of the history of Fukien province, and the well-known Christian prime minister and agriculturalist Hsü Kuang-ch'i were very enthusiastic about this new plant. Its unusually high per-acre yield, its nutritiousness (in terms of calories next only to rice), its pleasant taste, preservability, and value as an auxiliary food, its relative immunity to locusts, its greater resistance to drought as compared with native Chinese yams, and the fact that it can easily adapt itself to poorer soils and hence does not compete with other food crops for good land, are among the many advantages systematically pointed out by these two scholars.[49]

Apparently farmers of the southeast coast did not accept the sweet potato without field experiments. A twentieth-century edition of the history of Foochow has preserved a late Ming peasant song in which the battle between the new intruder and native tubers was vividly described.[50] The battle, however, was of short duration because the sweet potato soon established its superiority. Even in T'ai-ts'ang county, north of the Yangtze not far from Shanghai, an area nationally famous for its delicate native Chinese yams, farmers were forced, though not without reluctance, to replace the yam with the sweet potato in the early seventeenth century because of its heavy yield and greater assurance against famine.[51] In sampling available local histories we find that only Huai-ch'ing prefecture in Honan, which traditionally brought Chinese yams of distinctive quality to Peking as tribute every year, clung to its old staple, probably for reasons other than economic.

In the southeastern coastal provinces, which were always deficient in rice and where the people were long accustomed to Chinese yams and taros as secondary food, the sweet potato suited the dietary habit of the maritimers and was promptly welcomed. It soon became the poor man's staple. In the red rescripts of the Yung-cheng emperor, officials in Fukien and eastern Kwangtung annually estimated the degree of regional food sufficiency and the quantities of food that had to be imported in terms of rice and sweet-potato harvests. In the eighteenth century the sweet potato gradually spread to all inland Yangtze provinces, among which Szechwan was a leading producer.

Since China's population was increasing rapidly after 1700, a series of imperial edicts and provincial circulars exhorted the northern peasants to grow sweet potatoes on a large scale, in order to stave off famine. By about 1800 sweet potatoes, in the southeast as well as in the north, had become the poor man's staple. Along the rocky Shantung coast sweet potatoes often accounted for nearly half a year's food for the poor. The selling of roasted and boiled sweet potatoes in shops and by peddlers became a familiar practice in many large northern cities, particularly Peking, a phenomenon which deeply impressed a Korean official who was instrumental in stimulating sweet-potato cultivation in Korea.[52] In 1931–1937 China, excluding Manchuria, with an average annual output of 18,500,000 metric tons, was easily the world's largest producer of sweet potatoes. Next to rice and wheat, the sweet potato is now the most important source of food for the Chinese.

Like the sweet potato, maize was introduced into China through both the overland India-Burma and the maritime routes before the middle of the sixteenth century. The overland introduction probably came slightly before the maritime. Owing to mountainous terrain and relatively backward economic and agricultural conditions, maize scored an early success in Yunnan. The 1574 edition of the history of Yunnan province lists maize as a product of as many as six prefectures and two department counties. From Yunnan maize gradually spread to Kweichow and Szechwan. In the late eighteenth and early nineteenth centuries — perhaps much earlier had early Ch'ing editions of southwestern local histories contained more detailed listings of local crops — [53] many mountainous districts of the southwest depended on maize as a primary food crop. An exhaustive examination of nineteenth-century Szechwan local histories reveals that it was grown in practically every county except the lofty mountains of the northwestern corner, with heavy concentration on the peripheries of the Red Basin. Except in certain special localities, maize was not so extensively grown in Kwangsi as it was in three other southwestern provinces.

Despite its early appearance in coastal Fukien and Chekiang, maize remained relatively neglected; the people preferred rice and sweet potatoes and, even more important, maize competed with native cereal plants for good land. Up to 1700, therefore, it was grown mostly in the southwest and in a few scattered districts on the southeast coast. In the eighteenth century, when the Yangtze lowlands had been entirely filled up, a few million migrants from the overcongested southeast found in maize the key crop with which to utilize the hills and mountains of the inland Yangtze provinces. One stream of these migrants went as far as Szechwan and Yunnan and another populated the whole drainage of the Han River, an area that comprises southern Shensi, western Hupei, and southwestern Honan. The hills and mountains along other tributaries of the Yangtze were likewise turned into maize fields. In these newly developed mountainous districts maize and sweet potatoes were sometimes complementary and sometimes competitive, depending on the dietary habits of the local people and on the locality's demand for grain cereals. In the whole Han River drainage, which was one of the leading maize-producing areas in China, although sweet potatoes were also grown, maize reigned supreme. As the population of these areas grew rapidly, the Irish potato, which appeared in northern Fukien before 1700, was belatedly introduced and made possible the utilization of mountains that were too lofty and soils too poor for maize and sweet potatoes. By the middle of the nineteenth century Wang Shih-to, an important observer of the population problem, testified that "all the deep ravines and secluded mountains have been developed into thoroughfares." [54] The ruthless destruction of virgin forests and consecutive intensive maize farming, often lacking in foresight and sown in straight rows, resulted in serious soil erosion which in turn led to the silting of river and lake beds and more frequent inundation of the Yangtze.

Maize was known early to a few scattered districts in north China. It was mentioned in the 1555 edition of the history of Kunghsien in Honan, where the low plain and the hills meet. Although introduced into southern Manchuria in the seventeenth century,[55] maize was cultivated in north China to only a limited extent, judging by the scant references to it in local histories.[56] While it is not unlikely that many northern local histories of the eighteenth century might have continued to overlook this comparatively new crop, the fact remains that maize was not widely grown on the low plain of north China until relatively late in the nineteenth century. In view of the large amount of maize produced by Hopei, Shantung, and Honan in the twentieth century, it is possible that during the last hundred years it has been slowly and steadily gaining at the expense of some old cereal crops.

The Irish potato was slower in winning wide acceptance by the people. From northern Fukien it slowly spread to a limited number of localities in the Yangtze interior in the eighteenth century. During the first half of the nineteenth century it became a significant crop in some high mountainous areas, particularly the peripheries of the Szechwan basin and the Han River drainage. Unlike the sweet potato, it seldom competed with maize because it could be adapted to climatic and soil conditions unsuitable for either maize or the sweet potato. Whether served whole or dried and ground into flour, the Irish potato, as had the sweet potato along the southeast coast, became the poor mountaineer's staple. During the late nineteenth and early twentieth centuries it was grown extensively in poorer districts of the loess high plain, Kansu, Inner Mongolia, and Manchuria.[57]

This revolution in land utilization and food production made possible by the propagation of American food plants is, in a sense, continuing at the present. Maize has been gaining steadily at the expense of old dryland crops like barley, millet, and sorghum. It was estimated by Professor J. L. Buck that between 1904 and 1933 the acreage under maize in China Proper increased from 11 to 17 per cent of the total area sown to grain cereals, while barley, millet, and sorghum declined by 5, 5, and 10 per cent respectively.[58] Manchurian agriculture showed some signs of stagnation or even decline under Japanese occupation after 1931, but maize production increased at the average annual rate of 10.3 per cent in 1936–1940 and 7.3 per cent in 1940–1944.[59] Buck's survey shows that peanut production enjoyed similar expansion during the first three decades of the present century. All this seems to confirm the view that in modern times these American food plants have continually demonstrated their superiority over native dryland crops.

The Sino-Japanese war of 1937–1945 gave further impetus to the extension of maize, sweet potatoes, and Irish potatoes. The loss of much of the eastern plains forced many southwestern provinces to try to increase food production. There was comparatively little room for rice and wheat expansion; maize and sweet and Irish potatoes, all of which give better per-acre yield than the old dryland crops, answered the urgent needs of the time. In Kwangsi from 1938 the provincial government exhorted the people to plant maize in alternate rows with t'ung trees. It was estimated that between 1938 and 1942 the area under sweet potatoes increased by more than half a million mou and the area under Irish potatoes, which were practically unknown to the province a few years earlier, expanded by over 70,000 mou.[60] This trend seems to have continued after the establishment of the People's Republic in October 1949. Many of Peking's recent utterances on agricultural policy have touched upon the propagation of these crops, particularly in connection with the increase of food production in Anhwei. In 1955 Anhwei, whose agricultural systems are based at once on northern dryland crops and on the southern staple rice, carried out important agricultural experiments. One method of enhancing food production was to increase the area sown to summer crops, among which maize and sweet potatoes were prominent. These were planted in alternate rows with other cereals and the newly expanded area under these two crops amounted to 1,700,000 mou. Furthermore, the province in 1954 secured new and better varieties of Irish potatoes, the area under which expanded by 1,280,000 mou in 1955.[61] Judging from present national needs and from the pattern of land utilization it seems probable that the new government's reliance on these crops will be increased rather than reduced.

Sung Ying-hsing, a foremost authority in his day on traditional Chinese technology, estimated in 1637 that rice accounted for approximately 70 per cent of China's total cereal production, with wheat as a poor second crop.[62] Some modern students may think that Sung, as a native of rice-rich Kiangsi, may have overstressed the importance of rice, especially since his estimate does not accord with our modern knowledge. When modern students examine the historical circumstances peculiar to the beginnings of modern China — the permanent economic decline of the northwest after the late eighth century, the frequency of wars and disturbances in north China, as contrasted to the prolonged peace and commercial development of the south, the steady southward shift of economic, cultural, and demographic centers of gravity, the relative scarcity of crops suitable to northern drylands that could not grow wheat, the crude agricultural system of the north as compared with the intensive rice cultivation of the south, and the almost incessant construction of irrigation works and the expansion of the frontier of rice culture in southern China since the opening of the eleventh century — Sung's estimate may not seem to have been a great exaggeration.

Sung's statement was made at a time when the second revolutionary change in China's cropping system and land utilization had just got under way. Although it was not until two centuries after Sung's lifetime that rice culture approached its saturation in China Proper, the new dryland crops introduced from America had already begun to enable the Chinese to use dry hills, lofty mountains, and sandy loams. The opening-up of millions of acres of highlands in modern centuries must have helped to redress the old agricultural balance so badly upset by the dissemination of early-ripening rice. As the south had been for centuries much more densely populated, it was the northern provinces that probably witnessed a more rapid growth of population under the benevolent despotism of the early Manchu rulers. In 1393 the aggregate population of the area equivalent to six modern provinces of north China was recorded in round numbers as only 15,500,000 (as compared to a total population of 35,000,000 for six modern southern provinces); by 1787 the same areas were given a total population of 103,000,000 and 136,000,000 respectively.[63] This apparent change in the geographic distribution of population, which took place mainly after Sung's death, had of course, an important bearing on the percentages of various food crops in the total national food output. Also, the rise of a more variegated national economy, brought about in part by the continual influx of European and Japanese silver since the early sixteenth century,[64] had created new demands for industrial and cash crops, like cotton, vegetable oils, indigo, sugar cane, fruits, and tobacco. Not all these cash crops gained at the expense of rice, but, as the south was economically and commercially more advanced, there was some shrinkage of the area under rice because of the rise of specialized farming.

The approximate percentage of rice in the total national food output may possibly have declined by almost one-half, from perhaps 70 in the early seventeenth century to about 36 in 1931–1937. During the past three hundred years various dryland crops, old and new, have increased to about 64 per cent, and American food plants alone to approximately 20 per cent, of the total national food production.[65] However, China's cropping system is seldom stationary. The enormous recent effort to expand the country's irrigated area will again increase the area of rice culture. This important recent development, coupled with the present government's emphasis on high-yielding miscellaneous crops like maize and sweet and Irish potatoes, will bring a further percentage decline of areas under various old dryland crops.

The problem of food self-sufficiency in a country as populous and varied as China is so complex that not even highly trained experts can arrive at any definitive conclusions. Little beyond a few general remarks based on historical studies and on present-day observations can be attempted here. The normal state of food supply may be partly and conveniently shown by the long series of statistics on food imports compiled by the Chinese Maritime Customs which is shown in Appendix III. From this series it can readily be seen that modern China's food imports have never reached an alarming scale. It is true that in 1923, 1927, and 1932 rice imports exceeded 20,000,000 piculs, but they amounted to less than 4 per cent of the estimated total national rice output. Moreover, even imports on such a proportionately small scale were by no means certain indications of national food deficiency. The circumstances in which such an "unusual" quantity of foreign rice was bought are authoritatively explained in the Chinese Maritime Customs reports. The 1932 rice imports cost China 119,200,000 taels, hence the Customs took particular pains to analyze the problem:

> [In 1932] the quantity of rice bought from abroad was 22,500,000 piculs as against 10,700,000 piculs in the previous year, and 19,900,000 piculs in 1930. While the cost of this foodstuff was very great in comparison with that of other imports, the quantity involved was a mere bagatelle compared with the figures for the total home production of rice, which was placed at 873,000,000 piculs for the year. Why the quantity imported should vary so much from year to year is not always

apparent. As was shown in reviewing the possible causes of the decrease in the import position in 1931 (the year of great floods), good first crops were harvested in some cases before the inundation, there was a large carry-over from the previous year in the great Wuhu rice center, good hill-crops were reaped in some otherwise flooded areas. In the year now under discussion there is no doubt that crops were excellent throughout the whole rice belt, and the doubled imports of foreign grain cannot, therefore, have been due entirely to an unusual shortage of home supplies. The importations appear to have been made for the following reasons: (a) fear of a shortage from home crops, owing to the difficulties foreseen in farming the desolated regions affected by the flood, led to large speculative purchases; (b) fear of a shortage for local consumption in rice areas led to embargoes on exports from these districts until the excellence of the crops was assured; (c) before these embargoes were lifted in the autumn, other parts of the country were forced to cover their requirements by the purchase of foreign rice, and the markets were already stocked before Chinese supplies were available; (d) foreign rice was cheap, and the absence of quick overland transport from the purchasing districts retarded the movement and increased the price of the native article, which was already high owing to local taxation.[66]

From an analysis of Customs figures and from detailed geographic studies a modern Chinese geographer has come to the conclusion that under normal circumstances China can achieve self-sufficiency.[67]

It should be further pointed out that for a great many people rice and wheat remain a comparative luxury. We may recall that the British working class could not afford to eat white wheaten bread till the Industrial Revolution had nearly run its initial course, and that right down to the middle of the last century there lingered in France a long series of proverbs reminding the peasants that wheaten bread had been for ages a treat rather than a daily staple. If the Chinese nation, which is just beginning to industrialize, relies exclusively on rice and wheat, the two aristocrats among grain cereals, there will of course be a deficiency. But the fact is that the Chinese, particularly the peasants of northern China and Manchuria, are accustomed to a preindustrial low standard of living and to various coarser foods. If self-sufficiency is defined, not according to the exceptional standards of the more fortunate parts of the modern West but in terms of bare human sustenance, then there is little evidence of a national food deficiency in China as a whole. The one persistent fact is that in modern times China has always exported a large quantity of auxiliary foods.

The most authoritative, and in general the most rational and modest, official Chinese statement so far on the problem of food supply was made in June 1956 by Chang Nai-ch'i, Minister of Food. With the establishment of an omnipotent monolithic state since 1949 China's new statistics are no doubt superior to those of the previous period.[68] China's aggregate 1955 grain output, which by definition includes legumes, tubers (converted to grain at the ratio of four catties to one) but not oil-bearing crops like peanuts, rapeseeds, and sesame,[69] was reported as 386,000,000,000 catties or 184,000,000 metric tons. It thus averages 610 catties, or roughly 670 pounds, per capita.[70] This per-capita figure is higher than the average rate of cereal consumption in industrialized nations which consume a much larger amount of animal food. Since the majority of the Chinese rely very largely on cereals and since this per-capita figure includes the amount of grain intended for seed, fodder, brewing, and various miscellaneous uses, "the margin of surplus is by no means a wide one."

Apparently the margin will continue to be narrow. As Liu Shao-ch'i, Vice-chairman of the Republic, recently explained, the present means for increasing agricultural production for the double purpose of feeding the nation and enabling the country to carry out full-scale industrialization are: agricultural cooperation, construction of more irrigation works, selection of better seeds, more effective pest control, encouragement of close-planting, increase in the double-cropping and multiple-cropping area, collection of more compost, and introduction of better agricultural implements—rather than an over-all effort to open up new land. In fact, comparatively speaking, so little emphasis has been placed on agriculture that even the completion of the Second Five-Year Plan in 1962 will not create an increase of more than 10 per cent in the total cultivated area, although in the northwest and in northern Manchuria the amount of land available for future development is considerable. Such high priority has been given to heavy industries that the state is planning to supply chemical fertilizers only to the amount of three catties to every cultivated mou by 1962.[71]

In conclusion, our historical survey and brief analysis of recent trend show that China has so far been able to achieve self-sufficiency in food. In case of a greater emphasis on agriculture she can be expected to increase her food production substantially because there is possibility of raising her average per-acre yield which at present is considerably lower than that of Japan. It seems possible for China even to increase food production at a rate higher than that of her population growth for a limited number of years if advanced agricultural technology is widely introduced. But the long-range prospect is bound to be quite different. In the first place, her present population is already very large and even a moderate sustained growth will be a great strain on her agriculture. Secondly, more labor-intensive cultivation and introduction of advanced agricultural technology cannot in the long run prevent agriculture from reaching the point of diminishing returns.

CHAPTER IX

Other Economic and Institutional Factors

Unlike the preceding three chapters, which are based on an extensive examination of local evidence, this chapter will necessarily be in the nature of a brief historical review, because each of the topics covered requires further detailed study. It will deal with the general character of the early-modern and modern Chinese economy and certain institutional factors, such as the fiscal system and land tenure.

I

Not only were increased means of livelihood provided by a vastly expanding and more intensive agriculture; the employment opportunities offered by an immense domestic trade, by a highly lucrative if somewhat limited foreign commerce, and by some newly rising industries and crafts throughout the later Ming and early Ch'ing were also considerable. Ever since the latter half of the eighth century the influence of money had been increasingly felt, at least in the Yangtze regions, which, thanks to an incomparable network of rivers, lakes, and canals, constituted a vast single trading area.[1] The economic development of the Yangtze area was further stimulated by the continual influx of silver from the Europeans and the Japanese after the early sixteenth century.[2] True, the Yangtze area and the southeast coast were not representative of the whole country. But when the southeast coast was brought into the sphere of a worldwide commercial revolution, the effects reached far into inland China. The commutation of labor services, which by 1600 had become nationwide, is one of the eloquent testimonials to the increasing influence of money. Although the majority of the people were engaged in subsistence farming, as they still are today, there were relatively few localities that did not depend to some extent on the supply of goods and products of neighboring or distant regions.

Whatever the institutional and ethical checks on the growth of capital, the late Ming period witnessed the rise of great merchants. The unusually observant Hsieh Chao-che, *chin-shih* of 1602, later governor of Kwangsi and author of the famous cyclopedia *Wu-tsa-tsu*, gives the following account:

> The rich men of the empire in the regions south of the Yangtze are from Hui-chou (southern Anhwei), in the regions north of the river from Shansi. The great merchants of Hui-chou take fisheries and salt as their occupation and have amassed fortunes amounting to one million taels of silver. Others with a fortune of two or three hundred thousands can only rank as middle merchants. The Shansi merchants are engaged in salt, or silk, or reselling, or grain. Their wealth even exceeds that of the former.[3]

In fact, many regions in later Ming times boasted resourceful long-distance merchants. People of the congested islands in the Tung-t'ing Lake in the heart of the lower Yangtze delta, for example, were driven by economic necessity to trade in practically every part of the country and for a time vied with the Hui-chou merchants in wealth. Merchants of the central Shensi area, while active in trading almost everywhere, specialized in transporting and selling grains to garrisons along the Great Wall, in the salt trade in the Huai River region, in the cotton cloth trade in southern Kiangsu, and in the tea trade with various vassal peoples along the thousand-mile western frontier stretching from Kokonor to the Szechwan-Tibet border.[4] The southern Fukien ports, Ch'üan-chou and Chang-chou, which handled the bulk of the Sino-Portuguese trade in the sixteenth century, probably produced some of the largest individual fortunes.[5]

As interregional merchants became more numerous, they gradually established guildhalls in commercial centers. In the early Ch'ing period there were guildhalls in Peking established by money-lenders from Shao-hsing in Chekiang, wholesale dye merchants from P'ing-yao in Shansi, large tobacco dealers from Chi-shan, Chiang-hsien, and Wen-hsi in Shansi, grain and vegetable-oil merchants from Lin-hsiang and Lin-fen in Shansi, silk merchants from Nanking, and Cantonese merchants who specialized in various exotic and subtropical products.[6] From the late seventeenth century the accounts in local histories of guildhalls established by distant merchants became more and more common, which indicated the continual development of the interregional trade.

The dimensions of individual and aggregate merchant fortunes were growing along with the volume of interregional trade. It has been estimated that some of the Hui-chou salt merchants of the eighteenth century had individual fortunes exceeding 10,000,000 taels and that the aggregate profit reaped by some three hundred salt merchant families of the Yang-chou area in the period 1750–1800 was in the neighborhood of 250,000,000 taels.[7] It was known to the Western merchant community in Canton during the early nineteenth century that the Wu family, under the leadership and management of the famous Howqua, had built up through foreign trade a fortune of 26,000,000 Mexican dollars.[8] Commercial capital had made giant strides since China's first contacts with the Europeans.

A sampling of the biographies in the histories of Hui-chou prefecture reveals that the Hui-chou merchants, though their headquarters were in the cities along the lower Yangtze, carried on trade with various parts of north and central China, Yunnan, Kweichow, Szechwan, and even the remote aboriginal districts and Indochina.[9] In the national capital alone there were 187 tea stores in 1789–1791 and 200 in 1801 which were owned and operated by merchants of She-hsien, the capital city of Hui-chou prefecture.[10] So ubiquitous were the Hui-chou merchants that there was a common saying: "No market is without people of Hui-chou." [11] The radius of the trading activities of these and other comparable merchant bodies is one indication of the increasingly mobile character of the national economy. The fact that it was trade as well as agriculture that sustained the local population and made its multiplication possible is well attested by various local histories, particularly those of the active trading areas, such as Hui-chou, a number of counties in Shansi, Shensi and Kansu, the lower Yangtze counties, the Ningpo and Shao-hsing areas in Chekiang, Chang-chou and Ch'üan-chou in southern Fukien, and the Canton area. Even people of the poor and backward western Hupei highlands depended to a substantial degree on trading with Szechwan as a means of livelihood.[12]

The interregional and local trade consisted of an exchange of a few staple commodities, like grains, salt, fish, drugs, timber, hardwares, potteries, and cloths, and of a number of luxury and artistic goods of quality for the consumption of the ruling classes. The quantity of internal trade in late Ming and early Ch'ing China, although not unusual according to modern Western standards, certainly left a profound impression upon the Jesuits of the seventeenth and eighteenth centuries. In fact, few modern scholars are in a better position to compare the dimensions of the domestic trade of early Ch'ing China with that of early modern Europe than were the Jesuits, who, knowing both about equally well, measured the Chinese economy with the standards of pre-industrial Europe.

Du Halde, whose famous description of China may well be regarded as the synthesis of seventeenth- and early eighteenth-century Jesuit works on China, said of Chinese commerce:

> The riches peculiar to each province, and the facility of conveying merchandise, by means of rivers and canals, have rendered the domestic trade of the empire always very flourishing . . . . The inland trade of China is so great that the commerce of all Europe is not to be compared therewith; the provinces being like so many kingdoms, which communicate to each other their respective productions. This tends to unite the several inhabitants among themselves, and makes plenty reign in all cities.[13]

This generalization probably referred only to the vast Yangtze area, but it can nevertheless be applied to many other parts of China. The trade of mountainous Fukien during the late sixteenth century was described by the educational commissioner Wang Shih-mao:

> There is not a single day that the silk fabrics of Fu-chou, the gauze of Chang-chou, the indigo of Ch'üan-chou, the ironwares of Fu-chou and Yen-p'ing, the oranges of Fu-chou and Chang-chou, the lichee nuts of Fu-chou and Hsing-hua, the cane sugar of Ch'üan-chou and Chang-chou, and the paper products of Shun-ch'ang are not shipped along the watershed of P'u-ch'eng and Hsiao-kuan to Kiangsu and Chekiang like running water. The quantity of these things shipped by seafaring junks is still harder to reckon.[14]

Wang's description of the large quantities of commodities shipped along the difficult mountain pass of northern Fukien is borne out by the later Jesuit testimony that in the watershed at P'u-ch'eng there were "eight or ten thousand porters attending to the barks, who get their livelihood by going continually backwards and forwards across these mountains." [15] Wang's comment on the large coastal trade between Fukien ports and the lower Yangtze area is also corroborated by other sources. The demand of remote markets for Fukien sugar was so great that by the late sixteenth century a considerable percentage of the rice paddies in the Ch'üan-chou area had been turned into sugar-cane fields.[16] Throughout the late Ming and early Ch'ing annually "hundreds and thousands of junks" discharged sugar in Shanghai and went back to southern Fukien ports with full loads of raw cotton which were made into cotton cloth locally.[17]

Even in landlocked north China the interregional trade was very lively. Despite the lack of cheap water transportation in many northern areas, daily necessities as well as luxury goods from distant regions were carried by wheelbarrows, carts, mules, and asses. "The prodigious multitudes of people" and "astonishing multitudes of asses and mules" engaged in the shipping of commodities in north China never failed to impress those Jesuits commissioned by the K'ang-hsi Emperor as imperial cartographers.[18] Silk and cotton fabrics of various kinds and luxury goods from the lower Yangtze region and Chekiang were to be found in practically every northern provincial town, including the late Ming military posts along the Great Wall. Generally speaking, it was the technologically advanced southeast that supplied the inland Yangtze and northern provinces with finished products, for which the recipients paid in rice, cotton and other raw materials.[19] Even in westernmost Yunnan bordering Burma, trade in precious and common metals, ivory, precious stones and jades, silk and cotton fabrics was constantly going on during the late Ming.[20] In fact, so great was the volume of China's interregional trade that for centuries it consistently impressed the Europeans.

This growing internal trade stimulated industries and crafts and made possible regional specialization in commercial crops. In the late Ming and early Ch'ing, rural industries and crafts of regional importance were so numerous that it is possible here to mention only a few outstanding ones. The pottery or porcelain industry of Ching-te-chen in northern Kiangsi expanded greatly during the sixteenth century, thanks to increasing government demand for high-quality porcelains and the investment of the Hui-chou merchants in privately owned kilns.[21] By the K'ang-hsi period (1662–1722), when Chinese porcelain "had materially altered" the artistic tastes of the English aristocracy,[22] the Ching-te borough had about five hundred porcelain furnaces working day and night to meet the national and foreign demand. At night, with its flame and smoke, this township, which stretched one and a half leagues along a river, looked like "a great city all on fire, or a vast furnace with a great many vent-holes." Since all the provisions and fuel had to be supplied by the surrounding districts, the cost of living in this industrial town was high. Yet, in the words of a contemporary Jesuit and longtime resident, "it is the refuge of an infinite number of poor families, who . . . find employment here for youths and weakly persons; there are none, even to the lame and blind, but get their living here by grinding colours." [23]

Another outstanding industry was cotton textiles, in the Sung-chiang area, of which Shanghai was a rising city. Thanks to an early start and to its moist climate, Sung-chiang was the Lancashire of early modern China. Although an enormous quantity of cotton was grown locally, Sung-chiang in the seventeenth and eighteenth centuries depended on remote northern provinces like Honan and western Shantung for the supply of raw cotton. The Jesuits reckoned that in the late seventeenth century there were in the Shanghai area alone "200,000 weavers of calicoes." [24] Since at least three spinners were needed to supply the yarn for one weaver,[25] the total number of spinners must have been several times larger. Cloth of many grades and designs was made to meet the varied demands of the people of Shansi, Shensi, the Peking area, Hupei, Hunan, Kiangsi, Kwangtung, and Kwangsi.[26] Contemporaries remarked that Sung-chiang clothed and capped the whole nation. The Su-chou area was also an important textile center, supplying much of western Shantung with its finished products.

The area around Nanking, from which the name of the famous cotton fabric nankeen was derived, produced cloth of high-quality which was exported to the West from Canton. Exports increased constantly until over one million pieces were being exported annually to Great Britain and the United States during the early nineteenth century. H. B. Morse, a New Englander and the famous historian of the Chinese Customs, said:

Cotton manufactures in 1905 constituted 44 per cent of the value (excluding opium) of all [China's] foreign imports, but in this industry the West could compete with cheap Asiatic labour only after the development springing from the inventions of Richard Arkwright and Eli Whitney, and in the eighteenth and early nineteenth centuries the movement of cotton cloth was from China to the West, in the shape of nankeens to provide small-clothes for our grandfathers.[27]

From the late sixteenth century Sung-chiang was subject to increasing competition from the rising cotton textile centers in north China. The low plain area of north China could produce cotton at lower cost and in larger quantity than the densely populated lower Yangtze region. This increased production of raw cotton in turn stimulated spinning and weaving, which were becoming very important rural industries in north China. The rapid development of the cotton industry in southern Pei-chihli, or modern Hopei, greatly impressed the Christian prime minister Hsü Kuang-ch'i (1562–1633), a native of Shanghai, who estimated that the cotton cloth produced by Su-ning county alone amounted to one-tenth of the cloth produced by the entire Sung-chiang prefecture.[28] In the course of the seventeenth century many northern districts became regionally famous for their finished cotton products, although few could vie with Sung-chiang in skill and quality. Toward the end of the seventeenth century the Hankow area had already deprived Sung-chiang of much of its old market in the northwest and the southwest.[29]

Cotton cultivation, which had been extensive in the Ming period, further expanded under the repeated exhortations of the early Ch'ing emperors. Many counties in southern and western Chihli, western Shantung, Honan, the Wei River valley in Shensi, the Fen River valley in Shansi, the Hupei lowlands, and central Szechwan derived a major portion of their incomes from cotton.[30] Cotton spinning and weaving became a common rural industry even in Yunnan and Kweichow. A great many people must have made their living partly or entirely on the growing of cotton or cotton spinning and weaving.

The cotton growing was by no means the only example of farming for profit. Thanks to the expanding economy and the nationwide grain trade, many areas became specialized in one or a number of commercial crops which found ready markets elsewhere. The economy of the mountainous southwestern corner of Kiangsi, the middle Fu River valley in Szechwan, the Lei-chou peninsula and the north coastal region of Hainan Island, the Hsün and Yü river valleys in Kwangsi, the Mi-lo area in Yunnan, and many more scattered and relatively backward areas, had been revolutionized by a single crop, the peanut, in the course of the eighteenth century.[31] Since the late Ming sugar cane and indigo had helped to transform the economy of many southern regions, particularly Fukien, Kwangtung, and southern Kiangsi. Many counties in Szechwan also benefited from extensive sugar cane cultivation. In 1727 the governor of Kwangsi, in a memorial protesting the extensive purchase of Kwangsi rice by the people of Kwangtung, attributed the rice shortage in Kwangtung to the fact that a considerable portion of its good farmland had been devoted to such commercial crops as fruits, sugar cane, tobacco and indigo.[32]

Tobacco, one of the most profitable crops introduced into China during the late Ming period, had brought wealth to many localities, particularly Lan-chou in Kansu, P'u-ch'eng and Lung-yen in Fukien, northeastern and southeastern Kiangsi, Chi-ning in Shantung, and northeastern Chihli near the strategic pass of Shan-hai-kuan.[33] The extent to which rice paddies and good farmland were converted to tobacco farms so alarmed some provincial authorities that they brought about several imperial decrees prohibiting the cultivation of tobacco in the eighteenth century. But as an early nineteenth-century Manchu poet pointed out, repeated government prohibitions could not in the long run compete with a 200 per cent profit, which accounted for a continual conversion of cereal-growing land into tobacco farms in the Lan-chou area.[34] It was estimated by an early nineteenth-century official in Fukien that in some Fukien counties tobacco occupied some 60 or 70 per cent of the farmland.[35] In 1829 an able scholar and economic expert testified that in Chi-ning county in Shantung tobacco was monopolized by six local families who employed more than four thousand workers and annually grossed about two million taels.[36]

Many areas, such as the Ch'eng-tu plain in Szechwan, Heng-yang in Hunan, and Han-chung in southern Shensi, were regionally if not nationally famous for their tobacco production.[37] The people of Jui-chin, a hilly district in southeastern Kiangsi, said that, despite the loss of good farmland to tobacco, every spring the sight of the green tobacco leaves in the fields gave them assurance, because tobacco meant ready cash with which rice and other necessities could easily be bought from northern Kiangsi. This county furnishes one of the best examples of the extent to which self-sustaining agriculture had given way to specialized commercial farming in many areas. By the nineteenth century, perhaps much earlier, none of the cereal crops was grown there on any significant scale and the new staples were all commercial crops — tobacco, tea oil, an indispensable ingredient for curing tobacco, peanuts, and ginger.[38]

A well-traveled European during the 1840's commented on the general state of commerce:

> One excellent reason why the Chinese care little about foreign commerce is that their internal trade is so extensive . . . . This trade consists principally in the exchange of grain, salt, metal, and other natural and artificial production of various provinces . . . . China is a country so vast, so rich, so varied, that its internal trade alone would suffice abundantly to occupy that part of the nation which can be devoted to mercantile operations. There are in all great towns important commercial establishments, into which, as into reservoirs, the merchandise of all the provinces discharges itself. To these vast storehouses people flock from all parts of the Empire, and there is a constant bustle going on about them — a feverish activity that would scarcely be seen in the most important cities of Europe.[39]

From this and earlier Jesuit comments it becomes clear that the early Ch'ing economy, if somewhat less variegated than that of Europe, was reasonably complex and able to meet both the basic and the more sophisticated demands of the nation.

However, even during the period of steady economic growth there were inherent weaknesses in the traditional Chinese economy. It was capable of small gains but incapable of innovations in either the institutional or the technological sense. Institutionally, despite the availability of commercial capital on a gigantic scale (witness the Yang-chou salt merchants and the Canton Hong merchants), the traditional Chinese economy failed to develop a genuine capitalistic system such as characterized the Europe of the seventeenth and eighteenth centuries. The reasons were many and varied. In the first place, by far the easiest and surest way to acquire wealth was to buy the privilege of selling a few staples with universal demand, like salt and tea, which were under government monopoly. The activities of the Hong merchants, and of other powerful merchant groups, also partook of the nature of tax-farming rather than genuine private enterprise.

Secondly, the profit and wealth accruing to these merchant princes was not reinvested in new commercial or industrial enterprises but was diverted to various noneconomic uses. Ordinary commercial and industrial investments were less profitable than money lending and tax-farming in the broad sense. Furthermore, the cultural and social values peculiar to the traditional Chinese society fostered this economic pattern. In a society where the primary standard of prestige was not money but scholarly attainment, official position, or literary achievement, rich merchants preferred to buy official ranks and titles for themselves, encourage their sons to become degree-holders and officials, patronize artists and men of letters, cultivate the expensive hobbies of the élite, or simply consume or squander their wealth in conspicuous ways. Consequently, up to a certain point wealth not only failed to beget more wealth; it could hardly remain concentrated in the same family for more than two or three generations.[40]

Thirdly, the lack of primogeniture and the working of the clan system proved to be great leveling factors in the Chinese economy. The virtue of sharing one's wealth with one's immediate and remote kinsmen had been so highly extolled since the rise of Neo-Confucianism in the eleventh and twelfth centuries that few wealthy men in traditional China could escape the influence of this teaching.[41] Business management, in the last analysis, was an extension of familism and was filled with nepotism, inefficiencies, and irrationalities. These immensely rich individuals not only failed to develop a capitalistic system; they seldom if ever acquired that acquisitive and competitive spirit which is the very soul of the capitalistic system.

Fourthly, the Confucian cultural and political system rewarded only the learned and studious. Technological inventions were viewed as minor contrivances unworthy of the dignity of scholars. Despite the budding scientific spirit in Chu Hsi's philosophy, China failed to develop a system of experimental science; moral philosophy always reigned supreme. Major technological inventions are seldom accidental and are necessarily based on scientific knowledge; hence traditional China could not produce a major technological revolution, which depends as much on the application of scientific knowledge to practical industrial problems as on a coordination of various economic and institutional factors. By the last quarter of the eighteenth century there was every indication that the Chinese economy, at its prevailing technological level, could no longer gainfully sustain an ever-increasing population without overstraining itself. The economy during the first half of the nineteenth century became so strained and the standard of living for the majority of the nation deteriorated so rapidly that a series of uprisings occurred, culminating in the Taiping Rebellion.

Finally, throughout the Ch'ing by far the most powerful control over the economy was exerted by the state, through the bureaucracy. Such key enterprises as the salt trade and foreign commerce were jointly undertaken by the bureaucracy and a few individuals who were resourceful enough to assume the financial responsibility demanded by the state. Even in the late Ch'ing and early Republican periods the few new industrial enterprises launched by the Chinese were almost invariably financed by bureaucratic capitalists. In the cotton textile industry, for example, out of a total of twenty-six mills established between 1890 and 1913, nine were established by active and retired high officials, ten by mixed groups of officials and individuals with official titles, and seven by the new breed of treaty-port compradores, practically all of whom had official connections.[42] It is common knowledge that after the founding of the Nationalist government in 1927 a few top-ranking bureaucrats who enjoyed Chiang Kai-shek's confidence exerted ever more powerful control over the modern sector of the national economy through the incomparably superior apparatus of four major modern banks. Genuine capitalism based on private enterprise never had a chance of success in modern China, which could only choose between bureaucratic capitalism and bureaucratic collectivism.

In the midst of the storms and stresses of the middle nineteenth century, the Chinese economy came into full contact with the Western economy. Western technology, which could be the answer to China's economic ills, was adopted piecemeal only after prolonged resistance from the scholar-official class who had a vested interest in the Confucian system. In a nation so accustomed to state paternalism, the post–1850 government was prostrate and unable to take the much needed leadership. The net results of the economic innovations of the period 1860–1894 were insignificant except that they gradually broke down the dogged resistance to the Western system and helped to establish more firmly the concession that Western technology was superior.

The treaties of the 1840's opened a number of sea ports to Western commerce. The treaties of 1858–1860 further opened up inland river ports. The Sino-Japanese treaty of 1895 threw the entire country open to foreign investments. During the second half of the nineteenth century, therefore, there gradually came into being a kind of economic dichotomy, with the treaty ports along the coast and the main river partially Westernized and with the vast hinterland practically unaffected. By the time Communist control was established in 1949, the industrial sector of the Chinese economy comprised only the treaty ports and Manchuria. For all the evils of imperialism, the treaty ports turned out to be the harbinger of China's industrialization and a training ground for a new type of entrepreneur. In addition, up to 1930 foreigners had invested in China over three billion United States dollars.[43] To this must be added Japan's investments in Manchuria between 1931 and 1945.

All this does not mean that imperialism did not impede China's economic progress. On the contrary, eighty-eight years of a treaty tariff, many items in which were at 5 per cent ad valorem, put both old and new Chinese industries in an unfavorable competitive position with regard to foreign industries and foreign-owned industries in China. As an illustration, between 1860 and 1890 Lancashire drove native cotton cloth entirely out of urban markets. Except during World War I, newly established Chinese cotton textile mills could never compete with the better equipped and better managed foreign-owned mills.[44]

China's comparatively meager industrial foundation and the extent of foreign domination are summed up statistically by an economist:

> In the early 1930's, there were about 3,450 modern manufacturing factories in China with an estimated annual production value of about US $547 million. The foreign factories in China, being larger per unit, were only 283 in number, but they yielded 31.8 per cent of the total production value. Topping the list of foreign industrial enterprises in China were: cotton textiles; electric, water, and gas supplies in metro-

politan cities; tobacco and food processing industries. In coal mining, the foreign-owned mines produced 56.3 per cent of China's total production. In the field of communication, out of a total shipping bottoms of 1,433,000 tons in China, 49.8 per cent was under foreign flags. Foreign capital solely or jointly owned 21.3 per cent of Chinese railways which had a total of 16,972 kilometers, while the remaining mileage, with 45 per cent of the investment representing loans from foreign sources, was therefore heavily influenced.[45]

After the outbreak of the Sino-Japanese war in July 1937 the modern sector of the Chinese economy in particular suffered from war losses, blockade, and inflation. Only Manchuria, under complete Japanese domination, quickened its pace of industrialization. All in all, although China's economy during the past century has become more variegated, the tempo of industrialization has not kept pace with the increasing population. Agriculture, by and large, has been little affected by the incipient industrialization. Interregional migrations and the colonization of Manchuria have helped to accommodate more people, but at a bare subsistence level, and the standard of living for the overwhelming majority of the nation may have been further lowered.

If we hypothesize a change from a comparatively stable and slowly expanding economy in the early Ch'ing to the stagnant economy insufficiently stimulated by industrialization which has characterized the last hundred years, we may see it reflected in the changing rates of population growth.

II

Favorable as general economic conditions were during early Ch'ing, they probably did not contribute as much to the phenomenal growth of population in the two centuries following 1650 as did political conditions. Many of the institutional barriers that for ages had hampered population growth were removed by the early Manchu government. There seems to have been some truth in the belief current among some seventeenth- and eighteenth-century Jesuits and particularly some French philosophers that China represented benevolent despotism par excellence.

The improvement in the welfare of the nation brought about by the first several Manchu rulers can be demonstrated in contrast to the miseries the people had suffered during the late Ming. Although the first half of the fifteenth century was a period of government retrenchment during which the national landtax quota was somewhat reduced, the labor-service burden of the people was soon increased. Because of the unusually poor pay for officials and the limited funds for local administration, local governments gradually made it a practice to charge various local services to the original corvée quotas. After 1500 the burden of the expanded labor services, many of which were commutable and designed to benefit the local-government personnel, became oppressive in a large number of localities. The gentry and the local-government underlings together contrived to make the new burden fall almost invariably on the poor and illiterate, with the result that an increasing number of peasants were forced to desert their home villages. It was by no means uncommon for peasant desertion to reach the point of local depopulation. In certain areas the benefit of a reduced land tax was more than offset by extortionate labor-service charges. It was this increasingly heavy and unjust corvée burden that led many conscience-stricken local and provincial officials to the fiscal reforms known as the single-whip system.

Throughout the late fourteenth and fifteenth centuries both the central and the local governments lived largely on payments in kind, while the net cash receipts of the imperial treasury remained only slightly above 2,000,000 taels of silver. Increasing court expenditures led in 1514 to the first imposition of an additional 1,000,000 taels on the land tax in the southeastern provinces, an imposition that was intended to be temporary. After 1550 the increasingly extravagant court and the wars against the Tartars and the Japanese pirates made the surtaxes a permanent feature in the fiscal system. The amounts of the surtaxes in the 1550's ranged between 3,000,000 and 6,000,000 taels.[46] Toward the close of the sixteenth century the Korean expeditions, campaigns against the rebellious southwestern tribal chieftains, and the strengthening of defenses in the northeast against the rising Manchus brought about tremendous increases in the surtaxes. By 1639 the surtaxes reached almost 20,000,000 taels, or ten times the original annual cash receipts of the central government.[47] It has been commonly held that it was the crushing burden of the land tax rather than the Manchus, that caused the downfall of the Ming dynasty.

The heavy surtax was but one of the nation's miseries. During the latter half of the long reign of Ming Shen-tsung, commonly known as the Wan-li period (1573–1619), various means were devised for increasing the cash revenue of the state; the emperor himself was perhaps the most silver-conscious ruler in Chinese history. His trusted eunuchs were appointed administrators of various old and newly created customs posts and officials in charge of mineral resources. In their frenzy for silver they resorted to the most shameless forms of extortion. Internal duties, tolls, and forced levies on mines real and imaginery became so unbearable that revolts broke out in various provinces. They were mostly ill-organized and on a local scale, but they were expressions of a universal despair.[48] These uprisings heralded the large peasant rebellions led by Chang Hsien-chung, who carried out systematic slaughter in several provinces, particularly in Szechwan, and Li Tzu-ch'eng, who sacked Peking and brought about the extinction of the Ming dynasty in 1644.

To win the good will of the Chinese people, the conquering Manchus abolished all the surtaxes of the late Ming, although their highhanded measures against Ming loyalists incurred the wrath of the Chinese nation. In the fixing of the quotas for land and ting taxes during the late 1640's and early 1650's large exemptions were made for wartorn areas. In areas unaffected by the peasant wars and the Manchu wars of conquest the quotas were never higher than those of the 1570's. There were usually local and regional tax remissions during bad years. After 1683, when the empire began to enjoy universal peace, tax remissions reached a gigantic scale. The K'ang-hsi Emperor took pride in pointing out that between 1662, the year of his accession, and 1701 more than 90,000,000 taels of taxes had been remitted.[49] By 1711 the total amount of taxes remitted had exceeded 100,000,000 taels.[50]

Inspired by the personal example of the emperor, there were provincial and local officials who were not only incorruptible but actively compassionate toward the people. Yü Ch'eng-lung (1617–1684), whose self-denial and sense of justice earned from the grateful people the appellation "Yü of the Clear Sky," T'ang Pin (1627–1687), whose personal austerity and kindliness won the deep affection of the people of Kiangsu, Lu Lung-ch'i (1630–1693), whose wife paid the household kitchen bill partly with cotton cloth which she wove herself — are but a few examples. In fact, the prevalent ethic among late seventeenth- and early eighteenth-century local officials was that they should try their best to prevent an enlargement of local tax quotas.

The general contentment of the nation during the late seventeenth century was vividly described by the famous writer P'u Sung-ling (1640–1715):

> Houses cluster together like fish-scales and people are as numerous as ants. Since local administration is simple, the district is often quiet. While everywhere on the fields mulberry, hemp, and various cereal crops are grown, the streams abound with carp and other fish. Despite the occasional comings and goings of local-government underlings, no household is disturbed. White-haired people, exempt from labor services, listen at leisure to the wind sighing in the pine trees, or birds' speech, or schoolboys' chants. The rest of the adults are all occupied with the plow and the loom.[51]

A greater blessing to the nation was yet to come. In 1712 the emperor froze the total amount of the ting payment permanently on the basis of the 1711 ting returns, irrespective of future increase in the population. "This," wrote a usually critical nineteenth-century scholar, "marked the end of two thousand years of government oppression." [52] "The people outside the ting quota," commented a Hupei local history, "have thus been benefited without end." [53] Then during the reign of the succeeding Yung-cheng Emperor (1723–1735) the ting quota was merged, on an almost national scale, with the land tax, a reform which "has benefited the people even more." [54] Thenceforth the poor and landless bore little or none of the ting payment.

Further to ensure an honest administration, the Yung-cheng Emperor introduced a premodern form of a government auditing system, supervised by his able and conscientious half-brother Prince I. The administration so improved that this auditing system was eventually discontinued. He also realized the error of paying government officials poorly, which had accounted for a great deal of peculation during the Ming. The pay system throughout officialdom was revised. The *huo-hao* or *hao-hsien*, charges on "wastage and loss," which always accompanied the collection of the main taxes were legalized.[55] Previously these charges had often been excessive; now a ceiling was put on them. Portions of the revenue from these charges were now earmarked for public construction work and for subsidies to increase officials' salaries. The increased portions of salaries were known as *yang-lien* because they were intended to "nourish incorruptibility." The salary scale ranged from one or two thousand taels a year for a county magistrate to ten or twenty thousand a year for a governor or governor general. Sun Chia-kan (1683–1753), a high and upright official who had wide experience in both central and provincial administration, testified in 1742 that it was difficult to say which officials during the Yung-cheng reign were particularly noted for their incorruptibility, because every official was adequately paid and could afford to be honest.[56]

During the early Ch'ing perhaps the greatest blessing to the nation was that the people did not have to perform labor services personally. The compulsory labor and military services were usually irksome, even during the much-praised reigns of Han Wen-ti (179–155 B.C.) and T'ang T'ai-tsung (627–649). Since the ting service had begun to be commuted in the sixteenth century, during the Ch'ing public works as a rule were accomplished through government-hired laborers. Memories of the sufferings attending forced labor and military services grew dim, so dim that people sang old folk songs like "The Maiden Meng-chiang Weeping at the Great Wall" without realizing that the sorrowful theme had been a reality to millions throughout the centuries.

When in the former Han period a poll tax was levied on male and female children under the age of three, the poor people often responded by killing their newly born infants. Consequently, the age subject to poll tax was raised to seven.[57] The kindly Sung Emperor Chen-tsung (998–1022), who introduced Champa rice into China, substantially reduced the poll tax for the Fukien and Chekiang areas in the hope that the common practice of abandoning infants might be halted.[58] When in the Ming period the burden of land tax and labor services became too heavy or was shifted from the rich to the poor, the hapless peasants fled their homes, sometimes completely abandoning a village or a locality. Compared with their ancestors, the common people of early Ch'ing times seem to have had a happier fate.

The end of the campaigns against the three southern feudatories and against the Ming loyalists in Formosa in 1683 ushered in a prolonged period of peace and prosperity seldom paralleled in China's long history. To add further to the good fortune of the nation, effective means were taken to ensure an adequate storage of public grain to be rushed to the scene in case of famine. Readers of the Yung-cheng Emperor's red rescripts cannot fail to be impressed by the diligence with which most of the provincial officials made their regular weather reports, crop forecasts, and preparations for the eventuality of local or regional famine. The amounts of grain stored at public granaries were annually reported to the throne. In fact, so much attention was given to public granaries that an able eighteenth-century governor specified large purchases by various provinces as a factor contributing to the steady rise in the prices of rice.[59] The price of rice in Yüeh-chou, the important rice-exporting port in lakeside Hunan, had risen from one-sixth of a tael to one-half of a tael a bushel in 1748,[60] largely because of increased government purchases.

China's material conditions in the late seventeenth and eighteenth centuries may not have been as favorable as those of contemporary England, a country that was blessed with almost unlimited economic opportunities. But it is doubtful whether the lot of the poor in China could have been much worse than that of the poor in pre-Speenhamland England. The average peasant in early Ch'ing China in all probability was a happier person than the average peasant in France in the age of Louis XIV and Louis XVI; and he certainly may have been much more fortunate than his Prussian counterpart, who even in the early nineteenth century was "a hapless missing link between a beast of burden and a man." [61]

Material and political conditions in early Ch'ing China, contrary to the impression of some modern observers who are prone to judge the past by the standards of twentieth-century China, seem to have been more favorable to the people's livelihood than those in Tokugawa Japan, where, despite the rise of commercial capitalism, the peasants were commonly said to be taxed to the extent that they could neither live nor die.[61] In any case, Chinese chronicles in the eighteenth century were by no means ungrateful to their government. The 1779 edition of the history of Honan prefecture, of which Lo-yang was the capital city, was typical of dozens of eighteenth-century local histories which, though singing the praises of the reigning dynasty out of courtesy, claimed to express the true feeling of the people:

> After recuperating and growing stronger for a century and several decades under our reigning dynasty the people have not only achieved self-sufficiency but have attained such a high level of wealth and prosperity as is entirely unparalleled in history. In addition, there have been repeated remissions of taxes. Recently the *Fu-i ch'üan-shu* has been further revised so that anything that inconveniences the people has been eliminated. How fortunate people are to be living in this age! [63]

We may assume that the chances for an early Ch'ing Chinese child to grow to manhood once he had survived the usually hazardous infancy, were not worse than those of an English child during the age of Queen Anne, none of whose thirteen children survived her, or even during the age of Dr. Johnson and Edward Gibbon. "So feeble was my constitution," recollected Gibbon in 1792, "so precarious my life, that in the baptism of each of my brothers, my father's prudence repeated my Christian name of Edward, that in the case of the departure of the oldest son, this patronymic appellation might still be perpetuated in the family." [64]

In China longevity was reported to be no longer a rare phenomenon. In 1686, when an era of peace and prosperity had just dawned upon the nation, the provinces reported 169,830 people over eighty years of age, 9,996 over ninety, and 21 over one hundred. Septuagenarians were so common that the provinces did not trouble to report them to the throne.[65] The joyful scene of elders being invited to the K'ang-hsi Emperor's banquet was vividly described by the gifted poet Singde, son of a Manchu grand secretary.[66] By 1726 the number of people between seventy and a hundred years of age or even older reached 1,421,652. "This," remarked a nineteenth-century scholar, "indeed marked the highest peak of peace and prosperity in history." [67] There is reason to believe that these reports on elderly people were far from complete. It is unfruitful to check these figures against the tables of elderly people in local histories, which as a rule preserve only the lists of names of those officially reported and honored with official ranks. However, the unusually observant diarist Wang K'ai-yün (1833–1916), in editing the 1868 edition of the history of Kuei-yang county in southeastern Hunan, stated that most of the four-hundred-odd natives of the locality, all of the Ch'ing period, who lived to the advanced age of eighty or more, had not been reported to the imperial government.[68]

Even these incomplete figures on the aged suggest the feeling of peace and plenty of the early Ch'ing period. (In the 1953 census, out of a total population of 582,600,000 — which is possibly four times as large as the total population in the early eighteenth century — people aged eighty or more numbered only 1,854,696.) The prevalence of healthy old people in early Ch'ing China was attested to by many contemporary writers.[69] In the second half of the eighteenth century the imperial banquet in honor of a "thousand elders" became almost an institution. A scholar of Su-chou consoled himself for his repeated failures in civil service examinations by composing a long poem of one hundred rhymes singing of his good fortune in having lived in a unique era of tranquility and contentment.[70]

The benevolent administration could not be maintained forever, for its success depended primarily on the ability and character of the ruler and not every ruler could be of the caliber of the K'ang-hsi and Yung-cheng emperors. Whatever the authoritarian character of the government, these two emperors apparently had the interest of the people at heart and worked very hard to further it. Readers of Yung-cheng's rescripts cannot fail to be impressed by his diligence, judicious choice of officials, and profound knowledge of state administration. Somewhat reminiscent of Han Hsüan-ti (73–49 B.C.), during whose reign the administrative efficiency of the Former Han government reached its height, Yung-cheng relied more on the Legalist doctrine of a strict observance of the law than on the Confucian principle of moral influence. He realized that wide discrepancy between law and practice would give rise to official corruption and official corruption would victimize the segments of the population who could least protect themselves from the bureaucracy and powerful local interests. Through close surveillance of officialdom he achieved a level of administrative honesty and efficiency unsurpassed by any other Manchu ruler.

Many of his ideas and measures were naturally disliked by the majority of the officials of the Confucian school. Almost immediately after his death in 1735 a movement was set afoot among the conservative officials to bring about a reversal of his policies.[71]

His successor, the Ch'ien-lung Emperor, young and eager to court popularity, tacitly reverted to orthodox Confucian *laissez faire* and the discrepancy between law and practice gradually widened.

In the midst of the peace and long prosperity that prevailed, except on the borders, the ruling classes and the commoners alike subscribed more and more to easy and more luxurious living. The increasing desire for material enjoyment attested by contemporary writings and local histories had a great deal to do with the gradual revival of official peculation. After the 1760's several serious scandals were discovered involving a number of provincial and salt officials. Peculation reached its height during 1780–1799, when the shrewd Manchu courtier Ho-shen monopolized the favor and confidence of the aging Ch'ien-lung Emperor.[72] Ho-shen was ordered by the new Chia-ch'ing Emperor (1796–1820) to commit suicide, and his property, confiscated by the government, was estimated to be worth several hundred million taels.[73]

Since the quotas of the main taxes were more or less inflexible, the source of official peculation had to be in increasing old surtaxes or creating new perquisites to accompany the main taxes. This additional fiscal burden invariably fell on the people, particularly on those who lacked the means and knowledge to cultivate intimate relationships with local officials and subofficials. An indication that official peculation must have been serious and the people's fiscal burden unbearable in many parts of the country is a major slogan used by the rebels when White Lotus rebellion broke out in the Hupei-Shensi-Szechwan border area: "We of the people are forced to revolt by the officials."

The Ch'ien-lung Emperor's extravagance, his campaigns in China's peripheries, and the prolonged military action in putting down the rebellion of the White Lotus sect between 1796 and 1802 rapidly drained the imperial treasury. By 1850 the silver reserve in the treasury of the Board of Revenue reached a perilously low level of eight million taels, as compared with sixty million taels during the Yung-cheng period.[74]

The outbreak of the Taiping Rebellion in 1851 forced the government to raise funds irrespective of means. The most important new revenue was the *likin*, duties on commodities in transit. Likin, which was not abolished until 1930, not only had far-reaching evil effects on China's newly established industries but proved to be one of the most irksome taxes to the nation as a whole. Even in the post-Taiping period there was no substantial enlargement of the main tax quotas, but surcharges and perquisites were mounting until they were probably two or three times the taxes themselves.[75] All this increase was absorbed by the peasantry which had been further victimized by the changing ratio between silver and copper coin since early in the nineteenth century.

If the fiscal burden of the peasantry was heavy during the late Ch'ing period, it was probably worse in the early years of the Republic. However corrupt the late Ch'ing fiscal administration may have been, it was by and large based on time-honored quotas and on the idea that the state should live within the bounds of its revenue. The early Republican warlords had not even this moral scruple and were shameless extortioners. It is common knowledge that they increased excise duties and created new ones almost at will and often collected land tax in advance. Even down to the early 1930's the land tax was collected in some Szechwan counties decades in advance. The Nationalist government's fiscal innovations in the late 1920's and early 1930's, though not insignificant in the technical sense, had very little ameliorating effect on the peasantry. After 1927 the Nationalist party seldom had the interest of the peasantry at heart and its main backers were Yangtze landlords and Shanghai financiers.

III

Another major institutional factor that has an indirect, contextual bearing on population growth is land tenure. The early Ch'ing government, despite its various kindly acts toward the poor, believed in the sanctity of private property and refused to intervene in the ownership of land. In the absence of detailed study on the changes in land tenure during the Ming and early Ch'ing times, it is impossible to draw generalizations for the whole country. While landlordism was not a new thing to certain parts of China, particularly the densely populated southeast, some local and regional evidence seems to suggest that the combined economic and political conditions of the early Ch'ing favored the growth of a landowning gentry.

In the lower Yangtze area land ownership seems always to have been relatively concentrated. One of the reasons Ming T'ai-tsu adopted unusually harsh measures against the landlords of southern Kiangsu and northern Chekiang was his uneasiness over their economic power and social prestige. Forced migration to Nanking, the early Ming capital, to Feng-yang in northern Anhwei, birthplace of the founder of the Ming dynasty, and even to outlying Yunnan,[76] may have reduced landlordism in the lower Yangtze but it was at best only temporarily curbed. The 1506 edition of the history of Su-chou prefecture testified to the prevalence of tenants and agricultural laborers and the 1556 edition of the history of Wu-chiang county described in detail how the tenants were further victimized by the gentry-usurers. In fact, throughout the centuries it had generally been believed that most of the farmers of the Su-chou area were tenants.[77]

The effect of changes in the tax burden and in agricultural prices on the growth of landlordism in the Sung-chiang area was described by a Shanghai scholar of the late seventeenth century. During the late Ming the biggest landlords in the prefecture owned but a few thousand mou. The sharp rise in the price of rice during the 1640's, which was caused by the wars of Manchu conquest, created an unprecedented demand for land. The gentry and the medium and small landowners vied with one another in seizing any land that was offered in the market. Then in the 1660's the labor-service payment was greatly, though temporarily, increased by the government in order to pay and appease the southern feudatories. Almost all of those medium and small landowners who had expanded their property in the 1640's became bankrupt. One way in which they could escape this crushing fiscal burden was to mortgage or surrender their land to the resourceful gentry, at a greatly deflated price. When taxes were drastically reduced and prosperity returned after 1683, the prefecture was full of great landlords whose estates ranged from 10,000 to well over 50,000 mou.[78]

Northern Chekiang also is known to have been a stronghold of landlordism. The population was large and the supply of land limited; consequently, tenants were forced to accept terms dictated by landlords or their agents.[79] In Fukien, where limited agricultural land and a dense population always kept the price of rice at a high level, rich merchants and retired officials found in land a most attractive investment. Throughout the Ming and Ch'ing, Fukien province in general, and the congested southern coastal prefectures in particular, was considered a land of landlordism.[80] Landlord-tenant relations were often strained. Tenant uprisings in 1448 and 1640 spread through a number of counties and were put down only after considerable bloodshed. Violence was barely avoided in 1746, when tenants in southwestern Fukien agitated for rent reduction in the wake of regional tax remissions.

In the Kwangtung lowlands also the ownership of land was highly concentrated. In Shun-teh county, for instance, it was said that the majority of the people "cultivated rich men's land."[81] Generally speaking, the gentry of Kwangtung, thanks to their intimate alliance with the venal local-government underlings, could encroach upon the poor without fear of government reprisal.[82] They were so ruthless that the poor were afraid to rent their land. Some conscientious provincial officials repeatedly memorialized during the Yung-cheng period that unless security of tenure was guaranteed by the government there was no hope of attracting the surplus population of eastern Kwangtung to develop the fertile land that was plentifully available in the west-central section of the province.[83] In fact, one of the most eloquent proofs of the existence of landlordism in Kwangtung was said to be its unusually large amount of agricultural land that had successfully evaded the land tax throughout the Ming and early Ch'ing.[84]

In some inland Yangtze provinces landed property was also unevenly distributed. The classic example is the frequently cited 1872 edition of the history of Pa-ling, modern Yüeh-yang, the important northern Hunan port:

> If the land [of the locality] is divided into ten portions, mountains and streams take up seven. If the population is divided into ten portions, scholars and merchants occupy four. If the agricultural population is divided into ten portions, tenants constitute six. If the agricultural labor is divided into ten portions, hired labor accounts for five. If the local income is divided into ten portions, daily necessities eat up one-half [of the total].[85]

With some slight changes in percentages, this description may hold for Chiang-chin county in Szechwan during the nineteenth century.[86] In the Ch'ang-sha area in Hunan during the early eighteenth century small landowners were outnumbered by landless tenants, among whom were those "who toiled year long but were unable to support their parents." [87] In Hunan a 50 per cent rental in crops was common, with the result that tenants could hardly improve their status.[88]

Whatever the value of these scattered impressions, there can be no doubt that there had been maladjustment between population and landed property in the country. Ku-tsung, for example, a high-ranking Manchu and economic expert, whose uprightness earned for him the nickname of "Iron Bull," [89] suggested in a memorial in 1743 that the land owned by a single household should be limited to 3,000 mou.[90] This proposal, which was rejected by the government, indicates that in certain parts of China, especially in Kiangsu, where Ku-tsung had served for a number of years as the chief commissioner in charge of tribute grain, great landlords with estates larger than 3,000 mou appear to have been fairly common. Yang Hsi-fu, a native of Kiangsi and longtime governor of Hunan, in his famous 1748 memorial analyzing the causes of rising rice prices, stated, "lately some 50 or 60 per cent of the land has passed into the hands of landlords, and small landowners have sunk to the status of tenants." [91] Although he did not say specifically whether this trend occurred mainly in certain Yangtze provinces or in the country as a whole, he probably was speaking only of the areas of which he had firsthand knowledge. Hung Liang-chi (1746–1809), a native of southern Kiangsu and a geographer and historian who has been called "the Chinese Malthus," was likewise of the opinion that the uneven distribution of landed property aggravated the economic difficulties of the poor.[92]

Since much of the available information is fragmentary and all of it refers to the Yangtze region and the southeastern coastal provinces, it is dangerous to generalize, as have some modern students, that during the eighteenth and early nineteenth centuries China as a whole was becoming a country of landlordism.[93] So far there has been little information on the unchanged aspects of land tenure, and a general quantitative statement for the entire country is impossible. The relative scarcity of references to land tenure in the local histories of the northern provinces tends to corroborate our information, based on modern surveys, that north China in general has been a land of small landowning peasants, although extremely powerful individual landlords are known to have existed in the eighteenth and early nineteenth centuries.[94] In the light of the impressions of Ch'ing writers, the best one can say is that land ownership seems to have become somewhat more concentrated in the rice region during the first two hundred years of the Ch'ing; and the increasing concentration of ownership and a continually increasing population probably aggravated the economic difficulties of the tenant farmers, who joined the Taiping forces by the million after they reached middle and lower Yangtze.[95]

The writings on tenurial problems by modern Chinese students often display a certain amount of moral indignation which leads them to exaggerate the extent of landlordism in modern China. Few have taken the trouble to study the countermovement in land tenure in favor of the small men. By far the most important change in land tenure during the past century took place in the most densely populated lower Yangtze provinces, where the concentrated land ownership was substantially broken up not by the primitive communistic land policies of the Taipings, which were not actually carried out on any significant scale,[96] but by the results of the rebellion and the economic forces it engendered.

Many large landowning families in the severely wartorn Yangtze provinces died out during the fourteen years of the Taiping war. The population of southern Kiangsu except the Shanghai area, northern Chekiang, practically the whole province of Anhwei, and parts of Kiangsi and Hupei were so drastically reduced that every effort was made by the provincial governments and surviving landowners to attract immigrants from afar at terms unprecedentedly favorable to the latter. The acute labor shortage and the burden of taxation led many surviving landlords to offer their property for sale at a fraction of the pre-Taiping price. Concerned chiefly with the quickest possible recovery of the revenues from the land tax, provincial governments stipulated that all land unclaimed by the original owners within a certain time limit was to be regarded as the property of immigrant tenants. Consequently, to these rice-rich but partly emptied areas, millions of peasants immigrated from Honan, the more congested parts of Hupei and Hunan, northern Kiangsu, and the Ningpo and Shao-hsing areas of Chekiang. In sharp contrast to the pre-1850 descriptions of the evils of a continually rising landlordism, the tenurial accounts in the lower

Yangtze local histories of the late nineteenth and early twentieth centuries reveal a most unusual fact, that native owners could not endure the bullying and encroachments of immigrants, with the result that immigration had to be regulated by the provincial governments or stopped by law. That the Taiping Rebellion indirectly helped to level off land ownership in the traditional region of great landlords seems undeniable.

In view of the fact that small landowning peasants historically predominated in north China and that concentration of ownership in the lower Yangtze was partially broken up by post-Taiping economic forces, there is some reason to believe that there is truth in the common impression of the 1920's and early 1930's that about 50 per cent of the peasants were occupying owners, about 30 per cent were tenants, and about 20 per cent owned part of their land while renting the remainder.[97] The tenurial problem seems to have been very serious. The differences in proportions among the three categories of peasants in different provinces were quite striking. The proportions of tenants in some central Yangtze provinces, southernmost China, and the outlying regions of Inner Mongolia and Manchuria appeared to be very high. The distribution of landed property was probably likewise uneven. Two Nationalist-endorsed surveys in the 1930's revealed that 10 per cent of the farm population owned 53 per cent of the cultivated land and that the average landlord holding was 128 times as large as that of the peasant.[98]

With the sole important exception of the post-Taiping lower Yangtze area, it may be hypothesized that in historical as well as modern China institutional and economic forces have generally been against the small men. The rural triple alliance between the gentry, village usurers (who more often than not were themselves members of the gentry), and local government underlings, which exploited the peasantry, is as old as the institution of private landed property. At the time of K'ang-hsi and Yung-cheng the small man was reasonably protected by the government and was victimized by the rural scion only through his own folly or the latter's exceptional cunning; in the modern age of political anarchy, however, he was a helpless prey to the evil gentry, the tax collector, and any man with a gun — soldier or bandit. The words of an eminent economic historian who studied China's rural problem in the late 1920's were prophetic:

The revolution of 1911 was a bourgeois affair. The revolution of the peasants has still to come. If their rulers continue to exploit them, or to permit them to be exploited, as remorsely as hitherto, it is likely to be unpleasant. It will not, perhaps, be undeserved.[99]

To add a historical footnote to his prophecy it may be pointed out that the Nationalist revolution of 1927 was also a bourgeois affair, so bourgeois, in fact, that it spurned outright one of the goals it set out to achieve — equalization of land ownership, so dear to the founder of the Nationalist party, Dr. Sun Yat-sen. A theoretical justification for the rejection of Dr. Sun's land program was eventually offered by Ch'en Kuo-fu, the leader of the CC clique, shortly before his death on the island of Formosa:

[During the 1930's] we . . . did not attempt to put a special check on great landlords. A more truthful way of putting it is that fundamentally we did not pay any attention to this problem. This was because we believed that in the Chinese ethical society the landlord and his tenants lived together like members of the same family. Moreover, our social institution was based on the custom of equal inheritance by all sons. However large a landlord might be, after two generations his estate was bound to be divided into a number of small households.[100]

There were, of course, a series of laws fixing a legal rent, but they were all halfhearted. The Nationalist government allowed the peasantry to be exploited as ruthlessly as before. Small wonder that, according to a survey carried out in the early 1930's by the Central Agricultural Experimental Station, 56 per cent of the farm households had to resort annually to cash loans and 48 per cent had to borrow food every year to be able to subsist.[101] After the outbreak of the Sino-Japanese war in July 1937, the Japanese sea blockade, the loss of industries on the eastern seaboard, the greater dependence of the government and the people in the unoccupied southwest on local food supply, the increase of rent, and the ever-worsening inflation all made land a favorite investment outlet. But only the most resourceful were in a position to invest, for the professional and middle classes were soon reduced to an economic nonentity by the continual inflation. Although increasing concentration of land ownership in the Yangtze and southwestern provinces since 1937 cannot be fully established until documents on the Communist land reform between 1949 and 1952 are published,[102] a 1947 Nationalist government survey shows that landowning peasants constituted only between 40 and 45 per cent of the total agricultural population.[103]

It is difficult to assess the effect of land tenure on population growth during the long period covered in this study. However, it can at least be said that its effect varied from time to time and with the changing ratio between population and land supply. Unequal distribution of landed property and exorbitant rent must inevitably affect the tenant's standard of living; it could even postpone his age of marriage or restrict the size of his family. But it is important to bear in mind that the Chinese, among the world's large nations, are as capable as any of adjusting themselves to a lowering living standard over a considerable period, since for ages they have heard Confucian moralists preaching the virtue of living simply and frugally.[104]

Up to a certain point, the Chinese poor were almost always able to eke out a meager living somewhere and somehow, despite the uneven local distribution of landed property. In seventeenth-century Su-chou, for example, traditionally an area of great estates, the poor made a living not only by seasonal farm and orchard work, but also by carpentry, masonry, spinning and weaving, hemp, silk, and cotton, ropemaking, the manufacture of straw mats and carpets, sericulture, embroidery, boating, and local and itinerant peddling.[105] In nineteenth-century Pa-ling county in lakeside Hunan, where land ownership was most concentrated, many thousands of the local poor lived on seasonal employment as agricultural laborers, cloth dyers, masons, and brewery workers in southern Hupei counties.[106] In fact, the economic history of early nineteenth-century China can probably be summarized by a progressive lowering of the customary standards of living for the whole nation and a continual growth of population.

In the long run, however, the adverse effect of irrational land tenure on population growth must depend on the basic population-land ratio. Throughout the late Ming and early Ch'ing, when the total national population was comparatively smaller and the supply of agricultural land more plentiful, local and regional congestion could be solved partly by emigration. Although in some areas the greater portion of land was owned by a relative few, the land was in no case idle or turned to nonagricultural use. There was never any agricultural or agrarian revolution in late Ming and early Ch'ing times comparable to the extensive conversion of arable land into pastures in sixteenth-century England, which accounted for considerable rural depopulation. Nor was there any thing comparable to the conversion of millions of acres in the Scottish Highlands into deer forests, which deprived thousands of crofters of their livelihood in the latter half of the nineteenth century. The growth of regional landlordism had a comparatively slight effect on regional employment in an age when farming was becoming more labor-intensive. It was estimated by various seventeenth- and eighteenth-century writers and officials that in the rice area a male adult could till between 10 and 15 mou of land, and that in the low plain of north China the utmost that a peasant could cultivate intensively was 30 mou.[107]

The suggestion that landlordism had a comparatively slight effect on employment when the total population was relatively small is supported by the 1593 edition of the history of Shang-yüan county, a part of modern Nanking:

> The immigrants [who have become great landlords] are all rich men, on whom the livelihood of the local poor depends. The reason is that prior to the equalization of the land tax and labor services by the "single-whip" system, people frequently went bankrupt on account of the heavy burden of *li-ch'ai* [i.e., those labor services that in theory had to be performed personally]. Landed property was therefore regarded as a liability . . . . Since then the profit of farming and the value of agricultural land have been duly understood, and considerable land has been bought by the rich. The development of the wasteland, the stoppage of rural exodus, and the steady rise in land values are all accounted for by the large-scale purchase of land by the rich. Previously poor owners could not afford the plowing ox, the fertilizers, the digging of pools and ponds, and the construction of dikes and dams. When the locality was subjected to flood or drought, they could only cry to heaven. Nowadays the rich have anticipated all this and there is no crop failure except in years of excessive flood and drought. What is more, the rich cannot themselves cultivate the land, which must be let to the poor. Although the poor have lost their titles to the land, they actually share every benefit with the rich. . . . In addition, the ox and seeds are all provided by the rich, who alone bear the burden of the land tax. In times of natural calamity the rich help the poor with loans. To share the crops, the poor only have to contribute labor. This is why

the immigrant rich men are indispensable to our locality, and are sources of poor men's sustenance.[108]

With due allowance for its landlord bias, this passage clearly suggests why regional or local concentration of ownership was not likely to be a major impediment to population growth at a time when the total population could be sustained by land and other resources.

One may assume with Malthus that the growth of a population will be affected by the long-range prospect of demand for labor more than by the distribution of landed property. But, after the Chinese population had reached a certain point, for example, suppose it was 300,000,000 in 1800, the over-all opportunities for gainful employment in the nation began to be drastically reduced amidst continual population increase and technological stagnation. At the point where the margin above bare subsistence became so much smaller than the traditional or customary living standard, the effect of irrational land tenure on the marginal segments of the population presumably became disproportionately greater. Whereas landlordism in the Yangtze region in the early Ch'ing had prevented the tenants from attaining their maximum possible material welfare, landlordism by the middle of the nineteenth century seems to have compelled millions of landless tenants to swell the ranks of the Taiping rebels. Had landed property been more equally distributed and tenurial terms more reasonable in modern China, there would have been much less of the social misery so familiar to students of twentieth-century China, although the national standard of living would still have steadily deteriorated. Land tenure must therefore be regarded as a significant, though not basic, factor relating to population change.

CHAPTER X

Catastrophic Deterrents

I

China is known as a land of famine. Various factors — natural, economic, political and social — which contribute to frequent natural calamities are so well known and well expounded as to require little discussion.[1] Since 1878 some Western and Chinese writers have approached the problem statistically. These statistics, though indicating the historical and geographic distribution of droughts and floods, are misleading. In the first place, they are based on a few general and officially compiled works whose data on natural calamities are necessarily very brief and incomplete. Secondly, there are both exaggerations and omissions in the information provided in these general works, with omissions in most cases the more serious. One of probable reasons for recording a local or regional calamity in official works was connected with tax remission. Fewer droughts, for example, were recorded during the Ch'ing period for Shensi and Kansu, two of the loess provinces where rainfall is scanty and precarious, than for Kiangsu and Chekiang, two of the best irrigated provinces in monsoon China with a usually abundant rainfall. This statistical eccentricity was best explained by the Honan-Shensi-Kansu Famine Relief Commission in 1928: "It may be said that since the Revolution of 1911 there has not been a single year in which Shensi was not a scene of civil wars and natural calamities. Such calamities have seldom received publicity because of poor communications and because so few natives have become prominent in government service."[2] What is true for the twentieth century must also have been true for the Ch'ing. On the other hand, throughout modern centuries Kiangsu and Chekiang have produced more holders of higher degrees and officials than any other province and thus were able to get more "droughts" recognized by the imperial government than the arid and culturally backward loess provinces. Official calamity data are further distorted by the changing character of the central government. Under the benevolent despotism of K'ang-hsi and Yung-cheng the imperial government practiced financial retrenchment and was often eager to grant tax remission upon partial local crop failure. The prevalence of "model" and compassionate local and provincial officials also helped to secure tax remission and to exaggerate the incidence of famine. It is thus shown in a monograph on calamities that the total number of officially recorded famines in the eighteenth century far exceeds that of the nineteenth century, a statistical feature that is against our historical knowledge. But by far the strongest objection to tabulating official calamity records is that no one is able honestly to classify them. Due to necessary economy of expression, the information on calamities given in such overcrowded general official works often fails to indicate the scope and severity of a famine. A partial crop failure in a rich agricultural district, officially recognized as a famine, cannot be treated on an equal footing with a famine covering a larger number of localities or even several provinces.

So far the best sources from which to compile comprehensive historical calamity records are local histories. The better ones as a rule give more specific information as to the intensity and dimension of a natural calamity. A fairly large number are of mediocre and poor quality and, though they yield a more detailed listing of calamities in the locality, give little quantitative information. Moreover, as local histories were compiled at various dates it is often difficult to collect data for all provinces for the same long period. A better quality general illustration, shown in Appendix IV, is provided by Hupei, a centrally located province whose climatic, topographical, and agricultural conditions are more varied than many provinces and whose historical calamity records give more specific information.

After 1880 the calamity records are far from complete because most of the late Hupei local histories were compiled within twenty years of the pacification of the Taiping Rebellion in 1864. For the last four decades of the Ch'ing the editors of the Hupei provincial history, which was belatedly printed in 1921, had to rely upon the so-called local surveys, and these were carried out at best only sporadically in times of political disorganization. Nevertheless, its calamity records are far more comprehensive and specific than those of a well-known modern study based on the *Tung-hua lu*. The statistical evidence of Appendix IV shows that droughts occurred in 92 years, and floods in 190, during the entire 267 years of the Ch'ing, as compared with the *Tung-hua-lu's* record of 42 occurrences of drought and 75 of flood.[3] In addition, the Hupei records indicate in most cases the total number of counties affected by natural calamities, which is 1898 for the entire Ch'ing period. The average is thus 7 counties a year, or almost exactly one-tenth of the total of 71 counties in the province. Since the records after 1880 are incomplete and since the figures on Hupei do not include frequently recorded disasters such as earthquakes and early frosts, it would appear that on the average slightly more than one-tenth the area of the province was annually struck by one or another type of calamity.

As to the frequency and intensity of natural calamities, it may be observed that, after the province had recuperated from the disastrous late-Ming peasant rebellions and the Manchu wars of conquest there was a prolonged period of peace during which the people and government could cope with natural menaces fairly effectively. Despite the large number of counties affected by droughts and particularly floods, the late seventeenth and the entire eighteenth centuries as a whole had remarkably few serious famines. The largest number of natural calamities are found in the 1830's, 1840's, and 1860's, the three decades that preceded and followed the Taiping Rebellion.

Hupei statistics do not permit a definite conclusion of increasing incidence of natural calamities in the nineteenth century, owing to possible slight exaggerations for early-Ch'ing records and more serious omissions for late nineteenth-century records, but an independent set of data (Table 36) concerning the inundations of the Han River seems to suggest that the incidence was increasing during the nineteenth century.

All local histories of southern Shensi and western and northern Hupei in the nineteenth century testified to the ruthless attacks on virgin forests, the intensive and consecutive maize planting in straight rows on mountain slopes, and the increasingly serious soil erosion in the whole Han River drainage. This was observed by the famous official Lin Tze-hsü (1785–1850), the capable historian

Table 36. Frequency of Han River inundation, 1796 1911.

Period	Number of inundations	Average interval between inundations
1796–1820	6	4.17
1821–1850	16	1.88
1851–1861	7	1.43
1862–1874	8	1.50
1875–1908	14	2.43
1909–1911	2	1.50

Source: *Hu-pei-sheng nien-chien* (1937), pp. 90–98.

and geographer Wei Yüan (1794–1856), and the keen commentator on the population problem Wang Shih-to (1802–1889).[4] Wei Yüan, also known as an expert on river conservation, took special pains to explain the relation between the growing population, interregional migrations, the opening-up of mountains and hills in the inland Yangtze area, soil erosion, and increasingly frequent floods in the Yangtze region in general and in the Han River drainage in particular.

Whether Hupei calamity records are representative of the entire country can only be surmised. The province's flood incidence may have been considerably higher than that of most provinces except Hopei, Shantung, Honan, and Kiangsu, but it definitely suffered from fewer droughts than most of the northern provinces. In view of the fact that Hupei is centrally located, commercially developed, easily accessible by river, streams, and lakes, and is normally a food-surplus area, there is reason to believe that it has been somewhat more fortunate than the average, if indeed any province can be called average. At any rate, from the detailed Hupei data it is apparent that natural calamities must be regarded as a normal rather than exceptional phenomenon in a country as large and as dependent upon the forces of nature as China.

II·

Although a full statistical study of natural calamities and human disasters like wars and rebellions as demographic factors is impossible, an examination of a few major cases of destruction may help us to understand that along with basic economic factors favoring population growth there have been various forces tending to hold the population in check. While dealing with this subject descriptively rather than statistically, available materials will none-theless suggest the extent to which population growth was held back by natural calamities and human disasters.

Great drought famines. One of the worst drought famines in modern times was that of 1877–78, which struck four northern provinces, particularly Shensi and Shansi. It was reported that practically no rain fell in Shensi, Shansi, Honan, Hopei, and parts of Shantung between 1876 and 1879. Communications were so poor that it was not until relatively late in 1877 that the news widespread famine reached the capital. Yen Ching-ming, a native of Shansi and an upright official appointed as special imperial commissioner for famine relief in Shansi, reported the conditions in Shansi in the winter of 1877:

> The famine area in Shansi already comprises over eighty counties. The number of people in need of relief is estimated at between five and six million. From my inspectional tours which have covered some 3,000 li all that my eyes could see were those thin and emaciated human figures and all that my ears could hear were the howls of males and screams of females . . . . Sometimes my cart had to detour in order to avoid rolling over human skeletons which piled up on the highways. Many of those still alive fell flat on the ground after calling for help.[5]

In Shensi the worst happened. Cannibalism occurred again and again. Provincial officials ordered local magistrates to connive at evasion of the laws prohibiting the sale of children, so as to enable parents to buy a few days' food. The fertile Wei River valley in central Shensi, which consists of over forty counties, was particularly hard hit. The 1934 edition of the Shensi provincial history, based on local histories, summarizes the loss of population in this area:

> The population [of central Shensi] reached its peak by about 1850, after having rested and multiplied for over a hundred years . . . . In the early 1860's the outbreak of the Moslem rebellion killed almost half a million people. During the severe droughts of 1877–78 human skeletons lay along roads. On the average a large county lost between

100,000 and 200,000 lives and even a small one lost some 50,000 or 60,000.[6]

In some counties in central Shensi the only possible way to dispose of dead bodies was to dig huge holes, which today are still called "ten-thousand-men holes"; dead children were thrown in water wells.[7]

The acuteness of the famine in the landlocked northwest was enhanced by poor transport facilities. Despite earnest government and private efforts, large amounts of grain could not be quickly shipped inland from coastal areas. The chairman of the Foreign Relief Committee in Tientsin said in his report:

> In November 1877, the aspect of affairs was simply terrible. The autumn crops over the whole of Shansi and the greater part of Chihli, Honan, and Shensi had failed . . . . Tientsin was inundated with supplies from every available port. The Bund was piled mountain high with grain, the Government storehouses were full, all the boats were impressed for the conveyance of supplies toward Shansi and the Hochien districts of Chihli. Carts and wagons were all taken up and the cumbersome machinery of the Chinese Government was strained to the utmost to meet the enormous peril stared it in the face. During the winter and spring of 1877–78, the most frightful disorder reigned supreme along the route to Shansi. Hwailu-hsien, the starting point, was filled with officials and traders, all intent on getting their convoys over the pass. Fugitives, beggars, and thieves absolutely swarmed. The officials were powerless of creating any sort of order among the mountains. The track was completely worn out, and until a new one was made a deadlock ensued. Camels, oxen, mules, and donkeys were hurried along in the wildest confusion, and so many were killed by the desperate people in the hills, for the sake of their flesh, that the transit could only be carried on by the banded vigilance of the interested owners of grain, assisted by the train bands, or militia . . . . Night traveling was out of the question. The way was marked by the carcasses or skeletons of men and beasts, and the wolves, dogs, and foxes soon put an end to the sufferings of any wretch who lay down to recover from or die of his sickness in those terrible defiles.[8]

The Committee's estimate that from nine to thirteen million humans perished from hunger, disease, or violence during the famine of 1877–78 certainly does not appear to be an exaggeration; scores of late northwestern local histories testify to the permanent decimation of population brought about by this calamity. Shansi, for example, a province which no longer has any great prospect for agricultural expansion, had a registered population of 15,131,000 in 1850 but only 14,314,485 by as late as July 1953.[9] So greatly reduced was the population of Shensi that in the last decades of the Ch'ing it became one of the leading recipients of immigrants.[10]

The loess high plain provinces suffered from other drought famines, notably in 1892–94, 1900, 1920–21, and 1928. The severity of these famines cannot be accurately assessed because of lack of reliable statistics. W. W. Rockhill, the American diplomat and sinologue, who was in close touch with foreign famine relief personnel, estimated the loss of population during the drought of 1892–94 at approximately one million.[11] Thanks to the construction of certain railway trunk lines in north China, the famine of 1920–21, which at its peak rendered some 20,000,000 northern peasants destitute, was more effectively relieved by concerted government and private philanthropic efforts, Chinese and international. The total death toll was estimated in the neighborhood of half a million.[12] The great famine of 1928, which affected seventy-five counties in Shensi and dozens of counties in western Honan and Kansu, was comparable in extent and severity to that of 1877–78. By the summer of 1930 it was known that in Shensi alone more than 3,000,000 people had died of starvation and disease. Because of the exposure of corpses epidemics were rampant which were likely to cause more deaths than famine itself.[13] How many of the estimated 20,000,000 destitute were wiped out in all famine-stricken areas is unanswerable.[14] It seems fairly certain, however, that with the coming of railways to the land-locked provinces the total death toll in 1928 was considerably smaller than in the famine of 1877-78. The famine of 1877–78 leads to the conjecture that in pre-railway China severe droughts, especially in landlocked provinces, may have had greater effect on human life than floods, although the good administration and efficient public granary system of the early Ch'ing may have mitigated the famine situation.

Great floods. Many fairly detailed accounts of great floods during the Ch'ing are available, all of them descriptive rather than statistical. For the three great modern floods, the Yangtze in 1931 and 1935 and the Yellow River between 1938 and 1946, there are special committee reports or monographs. While none of the statistics are too accurate, they should indicate better than nonquantifiable materials the scope and severity of such calamities. The Flood Relief Committee of the Nationalist government gave estimates, shown in Table 37, for the losses caused by the Yangtze flood of

Table 37. Losses caused by the Yangtze flood of 1931.

Area	Total loss in Chinese currency	Percentage of loss suffered by each province	Number of families affected
Northern Hunan	248,300,000	19.3	424,200
Hupei	530,800,000	41.2	1,022,700
Northern Kiangsi	114,700,000	8.9	243,300
Southern Anhwei	314,500,000	24.4	613,200
Southern Kiangsu	80,300,000	6.2	224,300
Total	1,288,600,000	100.0	2,527,700

Source: Chung Hsin, *Yang-tzu-chiang shui-li k'ao* (CP, 1936), pp. 39–42.

1931, one of the worst in modern times. On the average each peasant family suffered a loss of $509.80. The number of people affected by the flood, if estimated at five persons to each family, would exceed 12,000,000. Four years later the same committee estimated that during the Yangtze flood of 1935 some 73,000,000 mou of agricultural land were submerged and 14,000,000 people dislodged from their homes. There were no detailed reports on the actual loss of life, but Han-ch'uan county in Hupei was caught entirely unprepared and saw 220,000 of its entire population of 290,000 carried away by the waves.[15]

For strategic reasons the Nationalist forces broke the dike of the Yellow River in July 1938 at two places in Honan, in order to halt the advance of the Japanese army. The Yellow River soon changed its course for the seventh time in recorded history and forced its exit to the sea through the lower Huai River. It was not until March 1947, nearly nine years later, that it was diverted to its original course. Shortly after the end of World War II the Institute of Social Sciences of the Academia Sinica and the China National Rehabilitation and Relief Agency jointly carried out a detailed survey of the extent of the damage. While necessarily imperfect, this is by far the best study of its kind and its findings are summarized in Table 38.

The areas of agricultural land submerged and the figures for people who were displaced by floodwaters or actually died are based

Table 38. Losses caused by the Yellow River inundation, 1938–1947.

Province	Number of counties flooded	Loss of agricultural land: Total agricultural land (1936) (mou)	Agricultural land flooded (mou)	Per cent of flooded land to total agricultural land
Honan	20	23,227,000	7,388,000	32
Anhwei	18	21,997,000	10,819,000	49
Kiangsu	6	12,411,000	1,777,000	14
Total	44	57,635,000	19,934,000	35

Loss of food crops (percentages)

Province	Summer crops	Winter crops
Honan		
1938	100	90
1939–45 (average)	90	90
1946	90	90
Anhwei		
1938	100	50
1939–46 (average)	90	50
Kiangsu		
1938	100	50
1939–45 (average)	90	50
1946	90	50

Loss of human life

Province	Total population as of 1936	Population displaced		Loss of lives	
		Number	Per cent	Number	Per cent
Honan	6,789,098	1,172,639	17.3	325,589	4.8
Anhwei	9,055,857	2,536,315	28.0	407,514	4.5
Kiangsu	3,581,238	202,400	5.7	160,200	4.5
Total	19,426,193	3,911,354	20.1	893,303	4.6

Source: Han Ch'i-t'ung and Nan Chung-wan, *Huang-fan-ch'ü ti sun-hai yü shan-hou chiu-chi* (Shanghai, 1948).

on postwar field surveys which, though they may be incomplete, should not be too far from fact. But the total cultivated land and the total populations of the flooded areas are based on the 1936 figures of the Directorate of Statistics, which are generally regarded as too low. The percentages of agricultural land submerged and of populations displaced and eliminated are therefore slightly too high. Since this artificially created flood was of unusual duration and its damage more severe than peacetime major inundations, this case study serves as a useful reminder that figures on the destruction of human life by floods in Chinese history are likely to have been exaggerated. At the very least, this monograph confirms the opinion of experts that drought famine was the most menacing of natural calamities, especially in pre-railway China.

III

The effect of man-made disasters like rebellions, civil and international wars on population must also be studied. The widespread peasant rebellions in the second quarter of the seventeenth century are said to have drastically reduced the populations of many localities in north China and nearly exterminated the population of the Red Basin of Szechwan. Official accounts may have exaggerated the bloodthirstiness of such rebel leaders as Li Tzu-ch'eng and Chang Hsien-chung. The conquering Manchus were by no means less inhuman.[16] Two basic historical facts stand out as a monument to the great destruction of human life in the areas of civil strife. First, the area of officially registered land shrank from over 700,000,000 mou for the greater part of the Ming period to a mere 405,000,000 mou by 1645.[17] Secondly, for two full centuries after the Manchu conquest Szechwan was unquestionably the largest and most important recipient of immigrants.[18]

The Taiping and Nien rebellions. With the outbreak of the insurrection of the White Lotus sect in the Hupei-Szechwan-Shensi border area in 1796 China was ushered into an era of great civil disturbances which culminated in the Taiping Rebellion of 1851–1864. Overlapping in time with the Taiping Rebellion was the prolonged guerrilla warfare waged by the Nien rebels in the Huai River area, southern Chihli, and western Shantung. The abundance of local histories for the wartorn areas compiled during the 1860's and 1870's and the voluminous recent publications on these wars by Mainland Chinese historians make it possible roughly to assess their effect on population.

Before making regional assessments it is necessary to explain briefly some of the characteristics of these civil wars. These wars were marked by slaughter of both military and civilian populations on a scale difficult to imagine by modern war ethics. Recent revelations have made it known that the government forces not infrequently were more cruel than the rebels who sympathized to a degree with the poor and were at times even tolerant of the rich. One thing that marred the career and reputation of Tseng Kuo-fan (the most outstanding of the anti-Taiping leaders and a conservative general and statesman) and indeed of practially all of his associates was his belief in the full-scale slaughter of rebels. As the commander-in-chief he pursued a policy of exterminating rebel forces with such singleness of purpose that he actually prolonged the dogged resistance of the rebels. Li Hsiu-ch'eng, "Loyal Prince" and mainstay of the Taipings in the late phase of the struggle, recalled in his autobiography, written in captivity, that the Taipings would have disintegrated of their own accord had Tseng and his generals not persisted in invariably executing, rather than accepting the surrender of, rebels who spoke the dialect of Kwangsi, where the rebellion had originated.[19]

After the Taipings had captured Nanking in 1853 the war became one of attrition. Provisions and basic supplies were the chief concerns for both sides rather than the capture of cities and towns. After 1853 the Taipings depended for sustenance on provisions from upriver, particularly Hupei and Kiangsi, which were shipped by captured private and government boats.[20] As the war dragged on, government forces adopted the scorched-earth policy to starve the rebels. This made it necessary for the Taipings, especially during the late phase, repeatedly to break through government encirclement and to "feed' in Anhwei province. In the winter of 1861 and again in 1863, over 100,000 rebel troops were reported starved to death.[21] When Nanking was recaptured by government forces in 1864, all the remaining Taiping faithful, numbering over 100,000, were exterminated through combat, fire, and particularly starvation.

The scorched-earth policy was also applied in the Huai River area and north China. In the Yangtze region this was mainly the government policy; in the Huai River area it was adopted by both the government and the Nien rebels. A modern monograph states:

> After 1856 the Nien pursued a scorched earth policy within a radius of 200 li outside their central area. By 1857 the Nien had completely stripped the Anhwei-Honan borderland. Between Po-chou and east Honan extended an area 300 li wide and 200 li long [about 6,000 square miles] which contained no trace of human habitation.[22]

Manpower was also a decisive factor in these wars. While government forces could rely on Hunan and the Ho-fei area of Anhwei for a supply of recruits, the Taipings had to resort to the forced enlistment of peasants from all over the central and lower Yangtze region. Many Yangtze local histories testified to the capture by Taipings of tens of thousands of adult males and children. On the other hand, as the combined effects of war, devastation, famine, and epidemic began to be felt, even larger numbers of poor peasants voluntarily swarmed into the Taiping ranks. Millions of otherwise peaceful peasants were thus drawn into this prolonged struggle against one another and against starvation.

In scope, duration, intensity and barbarity, these nineteenth-centuries civil wars in China must not be compared with major modern wars fought by and large under established codes of conduct. The Taiping Rebellion is deservedly called the greatest civil war in world history. In sheer brutality and destruction it has few peers in the annals of history.

Geographically, the areas most severely devastated by the Taiping and Nien wars were the entire province of Anhwei, southern Kiangsu except Sung-chiang prefecture, which received adequate defense due to the rapid rise of Shanghai as the leading treaty port, northern Chekiang, northern Kiangsi, and parts of the Hupei lowlands. North China, with the exception of southeastern Honan and parts of western Shantung, escaped large-scale disaster. Anhwei, a most contested area, received the deepest wounds.

The catastrophe in Kuang-teh departmental county in southern Anhwei was described by local scholars who had survived the prolonged ordeal:

> From the second [lunar] month of 1860, when the rebels broke into the county, they were highly elusive. The inhabitants were either

killed, committed suicide, or were captured, or otherwise sickened and starved. The death toll [during the first half of 1860] exceeded one-half [of the total local population.] Those who survived had no safe place to go and generally moved to the Bamboo Grove Fortress of the South Hill. People there, confident of their numbers and of their strategic advantages, became arrogant after having repeatedly repulsed the rebels' attacks. But finally the Fortress was stormed and captured by the band of the rebel chief Hung Yung-hai, which carried out a full-scale slaughter. Few were able to escape. From 1860 to 1864, for five years, people could not grow food. [Toward the end] all roots and herbs in the hills were exhausted and cannibalism occurred. Consequently epidemics struck. At that time corpses and human skeletons piled up and thorns and weeds choked the roads. Within a radius of several tens of li there was no vestige of humanity. The county's original population was over 300,000. By the time the rebels were cleared only a little over 6,000 survived. This was a catastrophe unique for the locality since the beginnings of the human race.[23]

Critical students are prone to doubt the accuracy of traditional Chinese quantitative statements, for not infrequently scholars used vaguely worded pseudo-numbers and twisted facts in the interests of a better literary style. Here, fortunately, as in many other lower Yangtze areas, a check on the authenticity of local accounts is possible.

For one thing, after such destruction of life and property a local population count became the foremost task of the county government, since population and land were the very foundations of the fiscal structure. Although the nationwide pao-chia registration system began to fall into desuetude after the outbreak of the Taiping Rebellion in 1851, the population counts of wartorn areas were carried out by local officials and specially appointed personnel. For Anhwei Tseng Kuo-fan ordered the task to be undertaken immediately after the recapture of Nanking in the spring of 1864.[24] The survivors, too, were eager to return home, especially with the assurance of tax remission and even government financial assistance. From the point of view of both government and the people nothing was more urgent than getting resettled and raising enough food. Since government rehabilitation aid was usually given in proportion to the size and needs of the household, there was less reason for the householder to under-report the number of survivors. The detailed population figures for Kuang-teh county are given in Table 39. The immigration from 1865 on and its rapid increase not only reflects the extent of local depopulation during the Taiping period but serves as an excellent summary history of the repopulation of this area by people from Hunan, Hupei, Honan, southern Chekiang, and certain districts of northern Anhwei. Various types of data and material, such as population figures, local government figures and documents concerning immigrants, the amount of land abandoned and later reclaimed, and occasionally even the accounts of distant emigrating localities all tally and confirm the drastic reduction of population in Kuang-teh during the Taiping wars.

Table 39. Changes in the population of Kuang-teh county, 1850–1880.

Year	Native		Immigrant		Ratio of immigrant population to native population as 100
	Household	Population	Household	Population	
1850	58,971	309,008	....	....	....
1855	59,106	310,994	....	....	....
1865	2,629	5,078	381	1,250	24.6
1869	3,222	14,720	4,995	17,993	122.2
1880	5,298	19,981	23,560	109,567	548.3

Source: *Kuang-teh CC* (1880 ed.), 16.14a–16b.

Kuang-teh was representative of a much larger area in southern Anhwei. She-hsien, the capital city of Hui-chou prefecture, saw its population reduced to one-half, from 617,111 in pre-Taiping times to 309,604 in 1869.[25] That this drastic reduction of population was not unusual for the whole prefectural area is proven by the autobiography of Hu Ch'uan (1841–1895), a native of Chi-hsi county in Hui-chou and the father of Dr. Hu Shih, who was elected by his kinsmen in 1865 to enumerate the survivors within the clan, in order to impose a head tax for the reconstruction of the common ancestral temple. After months of thorough investigation by various branches of the clan it was discovered by the time of the winter solstice that out of a pre-rebellion total of over 6,000 members, only 1,200, or one-fifth, had survived. As he commented, the whereabouts of a few clan members might still be unknown so shortly after the restoration of peace, but the task of rebuilding the ancestral temple and of recompiling a clan genealogy was so important that the count must be regarded as fairly complete.[26] In Nan-ling county, about 200 li north of She-hsien, a scholar appointed by Tseng Kuo-fan to take charge of local rehabilitation reported that only one-fourth of the members of his clan had survived, a ratio which was fairly representative of his locality and many neighboring counties.[27]

Although many post-Taiping local histories of the lower Yangtze give late nineteenth-century population figures and some quantitative statements, dozens of them fail to give pre-Taiping population figures, owing to the destruction of local archives. Table 40 illus-

Table 40. Population loss during the Taiping wars.

Locality	Province	Pre-Taiping population	Post-Taiping population	Percentage of population loss
Kuang-teh	Anhwei (S)	309,008 (1850)	5,078 (1865)	83.3
She-hsien	Anhwei (S)	617,111 (1827)	309,604 (1869)	50.0
Shu-ch'eng	Anhwei (C)	396,334 (1802)	107,196 (1869)	73.0
Shou-chou	Anhwei (NW)	765,757 (1828)	379,663 (1888)	48.1
Ch'ing-yang	Anhwei (SC)	432,049 (1776)	51,032 (1889)	88.2
Ying-shang	Anhwei (NW)	271,886 (1825)	162,679 (1867)	40.2
Ssu-hsien	Anhwei (NE)	588,112 (1777)	148,291 (1886)	74.8
Hang-chou [a]	Chekiang (N)	2,075,211 (1784)	621,453 (1883)	70.0
Chia-hsing [a]	Chekiang (NE)	2,933,764 (1838)	950,053 (1873)	67.7
Liu-ho	Kiangsu (WC)	318,683 (1781)	115,155 (1882)	63.9
Li-shui	Kangsu (SW)	230,618 (1775)	37,188 (1874)	83.9

Source: *Kuang-teh CC* (1880 ed.), 16.14a–16b; *She-hsien chih* (1937 ed.), 3.3b–4a; *Shu-ch'eng HC* (1907 ed.), 12.2b–3a; *Shou-chou chih* (1890 ed.), 8.12a; *Ch'ing-yang HC* (1891 ed.), 10.3b; *Ying-shang HC* (1878 ed.), 3.8b; *Ssu-hung ho-chih* (1888 ed.), 5.2b–3a; *Hang-chou FC* (1923 ed.), 57, *passim*; *Chia-hsing FC* (1877 ed.), 30, *passim*; *Liu-ho HC* (1883 ed.), 2.1b–2a; *Li-shui HC* (1883 ed.), 6.8b–10a.

[a] Prefecture.

trates the extent to which populations of severely war-damaged areas were reported to have been reduced. The actual percentage losses for several localities should be greater than shown in this table, because their pre-Taiping figures are those of the late eighteenth century or early decades of the nineteenth century, which are smaller than those of 1850, and their post-Taiping figures include immigrants.

It is regrettable that many Kiangsu local histories followed the old practice of listing only ting numbers. But even from changes in ting numbers which are of more fiscal than demographic significance, the extent of population loss can still be roughly gauged. The ting quotas of populous Su-chou prefecture and Chin-k'uei county in Ch'ang-chou prefecture were reduced from 3,412,494 and 258,934 respectively in 1830 to 1,288,145 and 138,008 in 1865.[28] The ting quota of Kao-ch'un county in southwestern Kiangsu was reduced from 188,930 in 1837 to 55,159 in 1869.[29] Chin-t'an county, about fifty miles southeast of Nanking, with a pre-Taiping population of over 700,000, had less than 30,000 natives in the countryside and less than 3,000 in the walled city by 1864.[30] All these figures are corroborated by the memorials of provincial officials.[31] The once rich and densely settled southern Kiangsu area was so depopulated that whole villages of peasants from some Hupei counties rushed in as settlers in the hope that they might "cultivate ownerless fertile land and occupy ownerless nice houses." [32]

Unlike Western residents in the treaty ports, Baron von Richthofen, the famous geologist and traveler, was a good foreign observer. His reports on post-Taiping Chekiang and southern Anhwei are invaluable:

> The valleys, notwithstanding the fertility of their soil, are a complete wilderness. In approaching the groups of stately whitewashed houses that lurk at some distance from underneath a grove of trees, you get aware that they are ruins. Eloquent witnesses of the wealth of which this valley was formerly the seat, they are now desolation itself. Here and there a house is barely fitted up, and serves as a lodging to some wretched people, the poverty of whom is in striking contrast with the rich land on which they live. The cities which I have mentioned, Tung-lu [T'ung-lu], Chang-hwa [Ch'ang-hua], Yü-tsien [Yü-ch'ien], Ning-kwo-hsien [Ning-kuo], are extensive heaps of ruins, about a dozen houses being inhabited in each of them. Each is the devastation wrought by the Taiping rebels thirteen years ago. The roads connecting the district cities are now narrow paths, completely overgrown in many places with grasses fifteen feet high, or with shrubs through which it is difficult to penetrate. Formerly the valley teemed

with population. The great number and size of the villages is evidence thereof, while the fine style of the houses, all of which were built of cut stone and brick and had two storeys, gave proof of the more than usual comfort and wealth that reigned here. The fields in the valley, as well as the terraced rice ground on the hillsides, are covered with a wild growth of grass, no other plants being apparently able to thrive on the exhausted soil. Plantations of old mulberry trees, half of them decayed from want of care, tell of one of the chief industries of the former inhabitants; in other places the ground is covered with perfect forests of old chestnut trees. . . .

It is difficult to conceive of a more horrid destruction of life and property than has been perpetrated in these districts, and yet they are only a very small proportion of the great area of country that has shared a similar fate. One must have seen places such as these to value at their full extent the ravages which the races of eastern Asia are capable of performing when full sway is left to their excited passions. There can be little doubt that the destruction of life, of which the province of Chekiang was repeatedly the theatre during its history, was not less fearful than it has been in the last instance. I used to enquire in different places into the percentage of population that has escaped death of the Taiping rebels. It was generally rated at three in every hundred. Of four hundred monks who lived before in the temple of Si-Tienmu-shan [West T'ien-mu mountain], only thirty survived after the rebellion; but the ratio is less in the villages and cities. Most people died of starvation, in the recesses of the mountains to which they fled, but still the numbers of men, women and children killed by the hand of the rebels is excessively great.

He also made a forecast for northwestern Chekiang and southern Anhwei:

There is reason to expect that these regions will revive. The course of immigration has set in. In the Fansui [Fen-shui] valley I found quite a number of new settlers, mostly from Ningpo and Shauhing [Shao-hsing] in Chekiang, but also a few from other provinces. They are less numerous than in Nganhwei [Anhwei], but the influx of people will probably increase. It is an interesting subject of speculation for the national economist, to trace the causes of the exceedingly slow rate at which the country is recovering its productive power. While there was formerly overpopulation, a few individuals are now masters of the soil, and new comers can purchase, at 1,000 cash (80 cents) a *mow*, as much as they like of the same ground which was worth, formerly, 40,000 cash a *mow* . . . yet the area put under cultivation is increasing at an incredibly slow rate. It appears, indeed, that a Chinese is capable of cultivating only a certain number of square yards of ground to every head of the population, and cannot overstep that limit with impunity.[33]

Richthofen was right: the repopulation of many lower Yangtze localities was indeed a very slow and gradual process. The inability of the average farmer to cultivate more than two or three acres of rice paddies was one reason, but there were other factors too. After a long period of neglect, the land required major reclamation, for which the surviving and impoverished owners had little ready capital.[34] In the years immediately following the Taiping wars the acute shortage of labor in the southeast made immigrants from Hupei, Hunan, Honan, and northern Kiangsu most welcome,[35] but the continual influx of immigrants soon created serious economic and social problems. The immigrants, aware of their bargaining power, often dictated terms and sometimes resorted to bullying. Disbanded soldiers, immigrating as tenant farmers, were an even greater menace to rural peace and order. Shen Pao-chen, governor general of Kiangsu, Anhwei, and Kiangsi, 1875–1879, repeatedly memorialized from 1877 on that it would be prudent to stop immigration even at the cost of delayed economic recovery.[36] In 1883 the accumulated grievances of the natives of two northern Chekiang counties resulted in a massacre of hundreds of immigrants.[37] To remedy the acute shortage of labor, a scholar and member of the gentry in southern Kiangsu advocated the purchase of Western farm machines and tractors.[38] The agricultural rehabilitation was so slow that even by the latter half of the 1890's the wisdom and necessity of immigration to the lower Yangtze area were still debated.[39] Liu K'un-i, governor general of Kiangsu and Anhwei, 1891–1902, stated in a memorial in the late 1890's that there were still some 100,000 mou of formerly taxed land in Hsin-yang county of Su-chou prefecture which remained untilled.[40] This unusually slow recovery in the area that was once the richest and most highly developed is another eloquent indication of the exceedingly heavy population losses in the lower Yangtze region.

The other Yangtze provinces affected by the Taiping wars were Kiangsi and Hupei, whose post-1850 population returns are completely useless. Since these provinces compiled their population figures arbitrarily after 1851, comments on the extent of population reduction for certain prefectures contradict the official figures. Hupei was an area much fought over in the early phase of the Taiping wars and many battles took place in its leading cities. But after 1856 the province was stabilized by the able governor Hu Lin-i and made into an important supply base for government forces. During the first five years of the struggle and occasionally during later years Hupei's population loss was evidently considerable. Many local histories recorded the capture by Taipings of tens of thousands of adult males and even females; in 1853 these were said to amount to between 300,000 and 500,000. In certain localities war ravages must have been serious to drive a reported 300,000 poor land-hungry peasants voluntarily to join the Taipings in 1860–61.[41] After the war some counties testified to the abundance of land and shortage of farm hands and some received immigrants.[42] The fact that after 1864 Hupei as a whole became an emigrating province must not be attributed to its relative immunity from heavy war tolls but to the irresistible economic temptations offered by many lower Yangtze localities.

Population losses in Kiangsi seem to have been heavier than in Hupei. Northern lakeside Kiangsi was one of the major theaters of war and there were surprise attacks by the Taipings in many other Kiangsi localities. Although the total provincial loss of population is impossible to ascertain, out of fourteen prefectures and one large departmental county only two were altogether immune to war ravages.[43] I-ning county in the northwestern corner of the province, for example, carried out in 1855 a pao-chia registration of the population within the walled city and the refugees from the nearby suburbs, which totaled 100,600. When the city fell after twenty-one days of fierce battle, less than 10,000 survived. It was said that for some 100 li the Hsiu River was red and navigation was blocked by corpses. Subsequently, dead bodies had to be cremated and buried in a huge tomb which has since been called the "one-hundred-thousand-men tomb." [44] From various counties poor peasants voluntarily joined the Taipings until by 1861 the total number, including peasants from Hupei, reached a million.[45] As the war of attrition went on, the Taipings had to raid Kiangsi repeatedly for provisions and manpower. The evidence in local histories suggests that the number of adult males and sometimes even of elderly people, women, and children who were carried away by the rebels must have been larger than in Hupei. In many localities in the west central part of the province the Taiping raid of 1861 practically depopulated whole villages. After 1864 only about 30 or 40 per cent of the Taiping captives were eventually repatriated.[46]

For reasons explained in Chapter V there are no reliable provincial total population figures after 1851 except possibly those of the 1953 census, which are so far removed in time as to have mainly suggestive interest. However, a comparison of the 1850 figures and the 1953 provincial totals for the lower Yangtze region (Table 41) will suggest the permanent decimating effect on population

Table 41. Populations of Kiangsu, Chekiang, Anhwei, and Kiangsi in 1850 and 1953.

Province	1850	1953	Percentage of change
Kiangsu	44,155,000	47,456,609[a]	7.5
Chekiang	30,027,000	22,865,747	−23.8
Anhwei	37,611,000	30,343,637	−19.3
Kiangsi	24,515,000	16,772,865	−31.4
Total	136,308,000	117,138,441	−14.0

Source: The 1850 figures are from Yen Chung-p'ing, *et al.*, eds., *Chung-kuo chin-tai ching-chi-shih tzu-liao hsüan-chi* (Peking, 1955), Appendix.
[a] This includes the population of Shanghai.

of the Taiping wars. The aggregate population of these four provinces as of July 1953 is still 19,200,000, or 14 per cent, less than that of 1850. Only in Kiangsu, because of the rise of Shanghai as the nation's largest metropolis and also of various economic opportunities not found in other parts of the country except Manchuria, does the 1953 population figure slightly exceed the 1850 level. Although twentieth-century civil and international wars must also have affected the populations of these provinces, the above figures may reflect the permanent wounds that the populations of the lower Yangtze provinces received in the great upheaval of the middle of the nineteenth century.

Some nineteenth-century Western observers estimated the total population loss during the Taiping period at 20,000,000 to 30,-

000,000. Their estimates, however intelligent, were the guesswork of treaty port residents. They do not accord with Richthofen's keen observations, with detailed evidence from local histories, or with table 41. Although the lower figure of 20,000,000 was accepted by W. W. Rockhill and hence gained currency, he did not accept it without hesitation. The best proof of Rockhill's uncertainty is that, for reasons entirely unexplained, he qualified his figure of 47,700,000 (a sum total of his estimates of population losses due to various causes between 1846 and 1895) with the strange but highly significant word "adults." [47]

It is more difficult to assess the effect of other wars and rebellions in the nineteenth century. The Nien wars wrought havoc in southwestern Honan, parts of western Shantung and southern Chihli, and particularly in northern Anhwei. Compared with the Taiping upheaval, the Nien wars seem to have had a much less permanent impact on the population, although the cumulative effect of battles, scorched-earth policy, famines, and epidemics could have been severe for a number of localities.

In addition to these two wars, there was the campaign against the rebellious Moslems of Shensi and Kansu in the 1860's and 1870's. The famous statesman and general Tso Tsung-t'ang, who eventually accomplished the incredible feat of reconquering Chinese Turkestan, was very much like Tseng Kuo-fan in his belief in exterminating the rebels. On the eve of the great disturbance in the northwest the Moslem population in Shensi was estimated as between 700,000 and 800,000. After more than a decade's fighting only between 20,000 and 30,000 Moslems were reported to have survived and between 50,000 and 60,000 fled to Kansu, where most of them eventually met their doom.[48] To this must be added a much larger number of Chinese who were slaughtered by the Moslems or perished otherwise. Richthofen again testified:

> The Mohammedan rebellion has reduced northern Shensi so considerably, that it would not be doing justice to that country if one were to form an opinion to it solely from its present state. Evidently, the Mohammedans have had the firm purpose to exterminate the entire pagan population and the destructible portion of their property. They made a wholesale slaughter of men, women, and children, and destroyed villages and cities. Where mountains were in the neighbourhood, the inhabitants fled to them, if they were able to do so. But the movements

of the rebels, who were on horseback, were so rapid and unexpected, that the proportion of those who were able to take refuge was small. The destruction was greatest in the central portion of the Wei Basin, on account of great distance of the hills. Solid city-walls proved an efficient barrier, because the rebels had no artillery. Si-ngan-fu [Hsi-an], Tung-chau-fu [T'ung-chou], and most of the *fu* cities, together with some *hien* [hsien], were saved. But a great many are destroyed. On the road from Tung-kwan [T'ung-kuan] to Si-ngan-fu, every city has had that fate. In the villages not a house was left standing, those of Christians excepted. The villages in the Wei Basin were large and numerous; not one of them has escaped destruction. The temples in particular were most savagely dealt with. Even the caves in the loess were not spared, and their brick fronts torn down. The destruction of life counts by millions.[49]

Here again Richthofen was a keen observer. Shensi was substantially repopulated, after the Moslem uprising and the great drought famine of 1877–78, largely by immigrants from Szechwan and Hupei, and also from various other provinces.[50]

Kansu, particularly its historically famous and strategically vital corridor, was a much worse theater of war and slaughter than Shensi. Nearly fifty of the province's seventy counties suffered heavy losses during thirteen years of turmoil.[51] Although the relative scarcity of modern Kansu local histories makes it difficult to assess the approximate loss of population, a modern biography of Tso Tsung-t'ang, based on exhaustive reading of documents and memorials, puts the death toll in Kansu at several million. Tso made every effort to attract new settlers, including the doubling of conventional rates for converting cultivated land into tax-bearing land.[52] Extensive immigration has since filled many local vacuums, but the 1953 population of Kansu, which actually included that of Ninghsia, was only 12,928,102 as compared with 15,437,000 in 1850.

Wars of the twentieth century. With the Manchu abdication and the death of Yüan Shih-k'ai in 1916 China was ushered in an era of full-fledged warlordism. For a whole decade, from 1917 to the establishment of the Nationalist government in Nanking in 1927, there were incessant civil wars in various parts of the country. Even after 1927 civil wars were still going on in certain provinces. In Szechwan, where the land tax was collected in some counties decades in advance, over 400 large and small civil wars were fought after the founding of the Republic in 1911. Between 1932 and 1934 the population of fifteen northern Szechwan counties was estimated to have been reduced by 1,100,000. In Nan-chiang county there were only five survivors in the walled city after one of these wars.[53] From 1928 on the Nationalists were repeatedly at war with the Communists, who had established a rural base in the Kiangsi–Hunan-Fukien border area. These wars were brought to an end in December 1936; then war with Japan broke out in July 1937. After the end of World War II in August 1945 civil war between the Nationalists and the Communists was soon renewed. For over a generation, therefore, China was seldom at peace.

It is impossible to reckon the aggregate number of people affected by the civil and international wars of the twentieth century. But there is a limited amount of material that would indicate roughly the extent of population loss directly attributable to the Nationalist-Communist civil wars and the Sino-Japanese war.

Concerning the civil war between the Nationalists and the Communists in the 1930's, the 1937 *Kiangsi Yearbook* gives the following figures on losses caused by the Communists up to July 1934: 279,798 houses destroyed, damages to property of $581,908,860, and 567,869 people killed.[54] The number of adult males, and sometimes females and children, who were recruited into the Red Army and were subsequently killed in battles or died of related causes must also have been considerable.[55] The ratio of the population drafted into the Red Army may be indicated by the case of Hsingkuo county, which, according to reports based on interviews with local people in 1953, contributed in 1928–1934 three divisions comprising over 60,000 men, or about 20 per cent of its total population.[56] Various journalistic reports of the 1930's were vague about numbers but insistent upon the severe loss of adult male population in Kiangsi.[57] The total of Communist military casualties between 1928 and 1934 is impossible to calculate, but it was estimated by the Nationalists that out of a total of some 300,000 men (the peak strength of the Red Army in 1934) less than 30,000 survived the Long March and eventually reached Yenan in 1935.[58]

Both Nationalist and Communist sources testify abundantly to the application of the scorched-earth policy, whose devastating effect on farmland, on food production, and hence indirectly on the loss of human life is difficult to estimate. After the Communists were uprooted from Kiangsi in 1934, the Nationalist forces carried out systematic reprisals by destroying property and by liquidating Communist adherents among the peasantry. According to some journalists who visited or revisited the old Red base in Kiangsi, which has become a national shrine since the Communist victory in 1949, the Nationalists exterminated 8,334 households in Ning-tu and killed more than 12,000 men in Jui-chin, the old Red capital, after the Communist withdrawal in 1934. Nationalist reprisals were also severe in other old Red bases. Huang-an county in Hupei, with a pre-1934 population of some 160,000, had but 66,000 in 1953. For six years after the Communist retreat the eighty-li stretch between Hsin-hsien and Lo-t'ien in Hupei remained entirely leveled by the Nationalists and turned into a no-man's-land. In 1953 the Jui-chin population belonging to old Communist martyr families exceeded 70,000. All these journalists testified to the general underpopulation of the southern and central Kiangsi area.[59] As has been shown, the 1953 population of Kiangsi was given as 31.4 per cent less than that of 1850.

The Sino-Japanese war lasted eight full years. General Ho Ying-ch'in, the chief of staff, released in 1946 the following figures on Chinese Nationalist military casualties: 1,319,958 dead, 1,761,335 wounded, 130,126 missing in action, with a total of 3,211,419.[60] These figures, however, are very incomplete. The unreasonably low figures for the wounded and missing makes them highly suspect. A far more careful and strictly scholarly study was made by a member of the Academia Sinica's Institute of Social Sciences, whose estimates, as he states in the preface, are definitely too cautious and too low. His findings are summarized in Table 42. While his revised figures sound more plausible than the official figures, his estimate of the deaths caused by sickness, want of medical care, malnutrition, etc. seems unreasonably low. This can readily be seen in comparison with foreign war statistics which he has consulted. In the American Civil War, for example, deaths due to sickness were twice the number killed in battle. In the American-Spanish War the ratio was as high as 5.2 to 1. In view of the appallingly inadequate medical, sanitary, and dietary care for the Nationalist forces in wartime, the total number for deaths resulting from sickness and want of care should have been far greater than those for the killed

Table 42. Nationalist military casualties, 1937–1943.

Killed	1,500,000
Wounded	3,000,000
Missing in action	750,000
Deaths caused by sickness, etc.	1,500,000
Total	6,750,000

Source: Han Ch'i-t'ung, *Chung-kuo tui-Jih chan-shih sun-shih chih ku-chi, 1937–1943* (Institute of Social Sciences, Monograph no. 24, 1946).

and wounded. Since his statistics only cover up to the end of 1943, it would probably not be unreasonable to put the total military casualties for various categories in the neighborhood of 10,000,000.

The figures of Table 42 include only casualties in the regular forces. The number of conscripts, though overlapping in part with the above figures, should be briefly studied. According to official announcements, altogether 14,053,988 adult males were drafted into service during the eight-year period. This figure may well have been somewhat exaggerated. Officers sometimes kept the size of a division small so that they could pocket the difference in pay. The incidence of death among the draftees must have been high relative to their numbers. An ex-officer and historian of the modern Chinese army recalls:

> Draftees were chiefly from the lowest classes, and they were often treated as military coolies. Many conscripts had to walk hundreds of miles to join their units . . . and suffered from lack of food, shelter, warm clothing and medical attention. It is not surprising that thousands upon thousands were lost, through desertion or death, before they could reach their units. General Stilwell found that in 1943 only 56 per cent of all recruits reached their assigned units. The rest died or "went over the hill" on the way.[61]

Anyone who lived in the southwest during the war witnessed the inhuman treatment such hapless draftees received from the authorities. Even in the suburb of Kunming, an air base and terminus of the Burma Road where Americans congregated, one saw the conscripts in chains, trudging painfully like moving human skeletons. It would be surprising if the total number of deaths among the recruits was lower than the official figures for killed and wounded combined.

The loss in civilian life up to the end of 1943, as estimated by the statistician Han Ch'i-t'ung, is shown in Table 43. He computes the ratios of civilian casualties as follows: 40 per 1,000 households in areas of repeated major battles; 25 in areas of one major battle or several small battles; 19 in areas of one secondary battle; and 3 in areas of light skirmishes or areas that fell to the Japanese without resistance. Statistics of 43 localities indicate that the average casualty rate for males was three times as high as for females. There is one group of figures that was unavailable to Mr. Han. This relates to

Table 43. Loss in civilian life, 1937–1943.

Killed	1,073,496
Wounded	237,319
Captured by Japanese	71,050
Killed in air raids	335,934
Wounded in air raids	426,249
Total	2,144,048

Source: Han Ch'i-t'ung, *Chung-kuo tui-Jih chan-shih sun-shih chih ku-chi, 1937–1943* (Institute of Social Sciences, Monograph no. 24, 1946).

the casualties of the wartime puppet forces and of the Communist guerrillas. The figures were released by the Communists shortly after the end of World War II. During the eight-year period between 1937 and 1945 altogether 960,000 puppet and 446,736 Communist troops were reported killed and wounded.[62] Even with these necessarily incomplete statistics it may be said that the Sino-Japanese war was different in nature from China's domestic wars of the nineteenth century in two aspects. First, the loss in civilian life was proportionately very small. Secondly, barbarous and cruel as it was, the Sino-Japanese war was by and large fought under a far more "civilized" war code. Even so, a conservative estimate would put the total human casualties in China directly caused by the war of 1937–1945 at between 15,000,000 and 20,000,000.

Regarding the renewed civil war between the Nationalists and the Communists from 1946 to 1950, statistics are meager. So far only military casualties have been released by the new government. The fact that these Communist figures have been tacitly accepted by the Nationalist government on Formosa would indicate that they are somewhat superior to any statistics the latter may have and that they are probably not too far from the truth. The official Communist figures show that the number of Nationalist troops killed and wounded between July 1, 1946 and June 30, 1950 was 1,711,110, with an additional 85 officers killed in action. During the same period 263,800 Communist soldiers were killed, 1,048,900 wounded, and 196,000 missing in action.[63] With only fragmentary newspaper reports on civilian sufferings accompanying a few major battles, the total effect of these four years of intense civil war on the civilian population cannot even be surmised.

IV

The efforts of modern scholars to inquire into the cumulative effects of China's natural calamities and human catastrophes upon her population have been hampered by the sketchy Chinese historical and twentieth-century statistics. A further handicap is the lack of reliable data on the birth and death rates which fluctuate greatly according to the amount of disorder and malnutrition. The only possible course is to summarize in chronological order the natural calamities and human disasters of a few sample areas, so that their cumulative effect on the populations of these areas can be more clearly visualized.

The Ching-chou area in Hupei, where the Yangtze has reached the lowlands from the gorges between eastern Szechwan and western Hupei, has been selected. Here floods which struck only a limited number of counties actually wrought far greater havoc in the area than the statistics presented in this chapter show. The results of three years from the long annals of natural calamities in this area are shown.

> 1830. Great flood struck Chih-chiang county, with untold numbers drowned. Considerable area of rich rice paddies was buried in sand.
>
> 1831. The dike in Shih-shou county collapsed. Half of its population starved to death.
>
> 1832. Chiang-ling, the prefectural city, suffered serious famine. There was a serious outbreak of epidemic, probably cholera, in Kung-an county during spring and summer. Its autumn flood further brought about countless deaths. Consequently one *tou* [one-tenth of a shih, a Chinese bushel] of rice cost more than

500 cash and cannibalism occurred. Shih-shou county, where the dike broke, had further epidemics in summer which caused innumerable deaths. Chien-li and Sung-tze counties likewise suffered from epidemics and serious famine. I-tu county was visited by epidemics in spring.[64]

Although the death toll brought about by floods themselves may not in most cases have been very high, this account of calamities crowded into a span of three years suggests that famines and epidemics following in the wake of floods accounted for more deaths than the floods. All these disasters occurred in peacetime, when ordinary relief could be mobilized.

Feng-yang prefecture in northern Anhwei in the Huai River valley lists the following long series of disasters [65] during the second half of the nineteenth century:

1852. The Nien and other bandits roamed at large in Lu-chou, Feng-yang, and Ying-chou. Flood hit Shou-chou, Su-chou, and Ling-pi.

1853. Serious famine occurred in Ling-pi. Battles with both Taipings and Niens in the prefectural area.

1854. Continued wars with Taipings and Niens.

1855. Continued wars with Niens. Shou-chou suffered first from unusually heavy rainfall and then from severe drought and locusts, with almost total loss of crops.

1856. Wars on smaller scale went on as usual with Niens. In the fourth month Feng-t'ai and Ling-pi suffered from drought and locusts.

1857. Taipings, after besieging Shou-chou for nine days, withdrew. Ling-pi suffered serious famine in spring. In summer Shou-chou was visited by great flood.

1858. Feng-yang, Lin-huai, and Huai-yüan fell to the rebels for the second time. More than 700 officials and gentry died. Locusts invaded Shou-chou in autumn.

1859. Government forces made significant gains by killing hundreds of rebels, but government lost more than 1,400 men.

1860. Intense battles with Taipings because of the arrival of rebel reinforcements led by Ch'en Yü-ch'eng; dogged fights with Miao P'ei-lin, the Nien leader who was allied with Taipings.

1861. In the first month there was serious famine in Feng-yang, where one *tou* of rice cost 1,000 cash. Most of the fortresses north of the Huai River fell to Miao P'ei-lin and other Nien groups. People within these fortresses were generally exterminated.

1862–65. With the arrival of strong government reinforcements the war tide gradually changed. There were intensified battles.

1866. Shou-chou was visited by great flood in the sixth month, with many houses destroyed and large area of farmland submerged.

1867. Continued flood in Shou-chou, with countless men and domestic animals drowned. Ling-pi suffered serious famine.

1869. Shou-chou was hit by typhoon and hailstorm. Numerous houses were uprooted and loss of crops was heavy.

1877–79. Shou-chou was hit by great flood and Ling-pi was invaded by locusts, which brought about serious famine.

1883. Great flood in Ling-pi.

1888. Great flood in Shou-chou.

1889. Continued flood in Shou-chou.

1897. Extensive flood and famine in the whole prefectural area.

1898. Continued famine in the prefectural area. 30,000 people rioted, demanding food.

1899. Hundreds of thousands of taels earmarked for famine relief in the prefectural area north of the Huai River.

The crowded events of the 1850's and 1860's must be read against the background of the scorched-earth policy adopted by both government and Nien forces. Recovery must have been extremely slow, and there must have been real substance in the famous folk song of the Feng-yang area that there were nine years of famine out of every ten, during which large households became bankrupt and small households could only flee to other areas or sell their children.

Shensi province during the early Republican period offers[66] another example:

1912. There was fierce civil war between Shensi and Kansu which lasted eight months.

1913. Civil war affected more than ten counties.

1914. Civil war was enlarged and affected one-half of the province.

1915. Widespread drought famine.

1916. A series of minor civil wars.

1917–20. Continued civil wars which affected dozens of counties. During this period of enhanced economic difficulties Shensi became notorious for its bandit-kidnappers.

1921. Civil wars were temporarily halted by the arrival of Feng Yü-hsiang, but the province was soon involved in the war between him and the Chihli clique. This was also a year of severe drought famine.

1926. During the ten months' siege of Sian between 70,000 and 80,000 people within the walled city starved to death. Many other counties also suffered severely from civil wars.

1928. The well-known great drought famine.

When the cumulative effect of natural calamities, wars, rebellions, and epidemics is visualized, the Malthusian checks seem to hold true, particularly with regard to China. It may further be said that the emphasis of William Godwin (1756–1836) on the injustice and irrationality of social institutions as a deterrent to population growth, against which Malthus wrote his famous *Essay* in 1798, is also relevant to China. While natural calamities have wrought havoc from time to time, other drastic potent checks to the growth of Chinese population have been wars and rebellions, caused not by nature but by defects in social, economic, and political institutions. It is for this reason that periodization is so important to understand the history of Chinese population. It is important not only because each period yielded different types of data but also because the general economic and political conditions varied from one period to another. A famine, which is basically a natural phenomenon, in the era of K'ang-hsi and Yung-cheng may have had a much less crippling effect on the population than a famine of similar magnitude in an age of wars and civil disorder.

CHAPTER XI

Conclusion

An attempt will be made here to suggest certain ways of reconstructing China's historical population data — a most difficult undertaking. For one thing, none of the long series of Ming, Ch'ing, and modern population figures was based on censuses in the technical sense of the term. Data for the years 1776–1850 seem better than the data of any other periods, but some annual totals contain regional omissions which are not clearly explained. Secondly, unlike European historians who can approach eighteenth-century population problems in the light of detailed modern censuses, students of Chinese history at most can have only the summaries of the imperfect 1953 census for reference. A third difficulty lies in the long intervals between the three periods that yield the comparatively more reliable population figures — the Ming T'ai-tsu era, the mid-Ch'ing period between 1776 and 1850, and the census year of 1953–54. Furthermore, traditional China failed to produce political arithmeticians of the caliber of Gregory King, let alone founders of economic and demographic science like Adam Smith and T. R. Malthus. With regard to figures and economic matters the traditional Chinese scholar-official is often vague, and his testimony impressionistic rather than quantitative.

For these reasons it is impossible for any modern student to suggest definite numbers in his attempt at historical reconstruction. The best that he can do is to suggest ranges and limits within which, to the best of his knowledge, China's population is likely to have grown or fluctuated throughout the last five and three-quarter centuries. Such ranges should be ascertained by correlating all major economic and institutional factors which had important bearing on population movements and which varied from one period to another. Broad demographic theories must therefore be resisted until all available facts and factors peculiar to each period have been examined.

I

The official Ming population data, though indicating relatively little change in the national totals for households and mouths throughout the two and three-quarter centuries of the dynasty, nonetheless show important changes in the geographic distribution of population which are summarized in Table 44. These data show

Table 44. Changes in the geographic distribution of population.

Province	1393	1542	Increase or decrease (number)	Increase or decrease (percent)
Nan-Chihli	10,755,938	10,402,198	—353,740	—3.3
Pei-Chihli	1,926,595	4,568,259	2,641,664	137.1
Chekiang	10,487,567	5,108,855	—5,378,712	—51.3
Kangsi	8,982,481	6,098,931	—2,883,550	—32.1
Hu-kuang	4,702,660	4,436,255	—266,405	—5.7
Shantung	5,255,876	7,718,202	2,462,326	46.8
Honan	1,912,542	5,278,275	3,365,733	176.0
Shansi	4,072,127	5,069,515	997,388	24.5
Shensi	2,316,569	4,086,558	1,769,989	76.4
Fukien	3,916,806	2,111,027	—1,805,779	—46.1
Kwangtung	3,007,932	2,052,343	—955,589	—31.8
Kwangsi	1,482,671	1,093,770	—388,901	—26.2
Szechwan	1,466,778	2,809,170	1,342,392	91.5
Yunnan	259,270	1,431,017	1,171,747	452.0
Kweichow	....	266,920	....	....
Total	60,545,812	62,531,295	1,985,483	3.3

Source: *Hou-hu chih.*

that the five northern provinces registered an increase of 11,230,000 mouths, or 73 per cent, in the course of 150 years, while the population of the southern provinces, excepting Szechwan, Yunnan, and Kweichow, declined by 12,000,000. When the increase of the southwestern provinces is included, the total population of China increased by a mere 3.3 per cent.

These changes as shown in the official data, however, are probably more apparent than real. The gains in population in the northern provinces cannot be explained simply by the moving of the national capital from Nanking to Peking in 1421, nor can the extraordinarily large percentage gains registered by Yunnan and Szechwan be accounted for by immigration alone.[1] It seems reasonable to assume that these regional increases reflected a general growth of population throughout China. But, because some of the northern provinces, particularly those of the low plain area, had suffered most from the devastation of the wars that attended the downfall of the Mongols, they were at first assigned much smaller landtax and labor-service quotas, and at lower rates than those of the southeastern provinces. The abundance of land, together with domestic peace and government retrenchment in the early fifteenth century, presumably stimulated the growth of population. The increasing population of the north was required to share a greater portion of taxes and labor services, since the demands for the lightening of the fiscal burden by the people and officials of the southeast had always been highly vocal.[2]

In addition, the soil of north China, being less fruitful than that of the rice area, could not be expected to bear much more than its own share of the combined fiscal burden; whereas in the rice area there had been a gradual, and indeed inevitable, shift of the incidence of the labor services from the people to the land. In north China services for the most part had to be assessed directly on the people; therefore the original system of population registration was maintained longer than in the south. For these reasons the natural growth of the population in the northern provinces was substantially reflected in the official population returns. Immigration and the choice of Peking as the national capital were other, but probably comparatively minor, factors contributing to the growth of the population in north China.

The reasons for the high percentage-increase of recorded population in Yunnan and Szechwan may have been political and cultural. That there had been some natural growth of population in these two provinces cannot be doubted, but the unusually high percentage gains probably can be attributed to unusual under-enumerations in early Ming times. It was not until 1420 that Yunnan and Kweichow were made into provinces. It was not until Chinese cultural influence gradually spread to the rich basins and valleys of Yunnan that the ordinary civil administration was set up and population was registered in these districts.[3] Despite the abnormally high percentage-increases, the 1542 returns for these southwestern provinces were still far too low, for some districts in these provinces remained outside the pale of Chinese civilization down to the eighteenth century. Because of very large aboriginal populations all pre-1953 population figures for the southwestern provinces of Yunnan, Kweichow, and Kwangsi were too low.

While the northern and southwestern provinces registered population gains, the official population data show a general decline of population in the southern provinces, particularly Chekiang and Fukien. When the prefectural breakdowns are studied, it is found that, although the recorded population of Nan-Chihli declined but slightly, the populations of Su-chou and Sung-chiang, two prefectures that bore the heaviest fiscal burden in the country, substantially declined.[4] Northern Chekiang and parts of Fukien were also among the heavily taxed areas. Since a decline of population meant a reduction of fiscal burden, the decline of registered population in these heavily taxed areas was often connived at, and at times even endorsed by, provincial and local authorities. The memorials urging tax reduction for the southeast are voluminous. The compiler of a famous sixteenth-century encyclopedia commented: "South of the Yangtze the household often does not have actual ting, for the household is assessed to its [landed] property. The laws punishing the evasion of household and ting registration cannot be strictly enforced." [5] A scholar and fiscal expert of the late sixteenth century also said: "When the labor services and taxes are heavy, there are bound to be evasions." [6] K'uai-chi county, a part of modern Shao-hsing in Chekiang, testified that among the local population only those liable to labor services were entered into the official registers and that those who were outside the registers were probably three times the official population figure.[7]

A Fukien local history is most revealing:

From 1391 to 1483, in a period of 92 years, the number of households in our locality had declined by 8,890 and that of mouths by 48,250. This means a loss of between 60 and 70 per cent. Why is it that after a long period of peace and recuperation there should be such a decline? It was probably due to the fear [on the part of local officials] that an increase would mean a further burden to the people and they therefore connived at the evasions and omissions without bothering to check.[8]

It is clear that in the heavily taxed areas a decline of registered population was not infrequently regarded as a convenient means of redressing fiscal inequities.

While most counties in the southeastern provinces registered a declining population, some recorded a slightly increasing, and others a stationary population. Whatever the type of return, few if any can have reflected the actual changes in population. Under-registration was probably common even in those prefectures and counties that registered an increasing population. For instance, the population of Ch'ang-chou prefecture in southern Kiangsu nearly doubled between 1377 and 1602, yet a well-informed native scholar and fiscal expert remarked that the prefectural population had been in fact considerably under-registered because of the common practice by the well-to-do of avoiding the division of clan property, in order to lighten or evade the labor-service assessment.[9] The editor of the 1629 edition of the history of Fukien went so far as to say that "neither the numbers of households nor those of mouths are authentic." [10]

The 1613 edition of the history of Fu-chou prefecture not only explains the reasons behind under-registration but throws light on the general question of population growth, at least in the southeastern provinces:

> I have gathered from my studies that usually towns and villages were partially deserted at the beginning of a dynasty, because the people had just gone through a period of turmoil and suffering. During a long period of peace, population could not but grow and multiply. Our old prefectural history recorded that in the Cheng-te period (1506–1521) the population had increased but two-tenths or three-tenths since the beginning of the reigning dynasty. Our present population remains more or less the same as that of the Cheng-te period. The empire has enjoyed, for some two hundred years, an unbroken peace which is unparalleled in history. During this period of recuperation and economic development the population should have multiplied several times since the beginning of the dynasty. It is impossible that the population should have remained stationary. This may have been due to the intrigues of government underlings and the failure of the authorities to get at the truth, but more probably to the old convention that at the time of decennial assessment little should be done beyond the fulfilment of the old quotas.[11]

The evidence from southeastern local histories accords well with the impressions of contemporary observers. Unfortunately, although a fair amount of qualitative evidence can be gathered to bear witness to a continual growth of population, none yields a reliable quantitative statement. To mention an extreme statement, an official said in a memorial of 1614, "the present population is probably five times that at the beginning of the dynasty." [12] Exaggerated as this and other vague guesses are, they seem to substantiate the observations and theory of population put forth by Hsieh Chao-che, who served as the governor of Kwangsi in the early seventeenth century and was one of the most widely traveled and best informed persons in the country. In his famous *Wu-tsa-tsu*, a miscellany on China in five parts which was widely read in Tokugawa Japan for its valuable information on Ming China, he advanced the theory that population growth is slow and difficult except under very favorable human and material conditions. The slow increments during periods of peace were often more than offset by the effects of wars and epidemics that attended the end of certain dynasties. He was therefore of the opinion that population movements had always been cyclical from ancient times to the end of the Yüan dynasty, a generalization with which most traditional commentators on China's population concurred. But, he was convinced through his travels and observations that, since the founding of the Ming dynasty in 1368, the population had had a continual, and perhaps linear, growth. He said of the population of the Ming period: "During a period of 240 years when peace and plenty in general have reigned, people have no longer known what war is like. Population has grown so much that it is entirely without parallel in history." If a leader's merit was to be judged by the conditions which he had created for the multiplication of the human race, then in Hsieh's opinion the founder of the Ming dynasty deserved equal ranking with P'an-ku, the Creator in ancient Chinese mythology; and if the greatness of an age was to be measured by the extent to which mankind was allowed to rest, nourish and reproduce, then he was certain that the Ming period should outclass all the previous glorious dynasties — Shang, Chou, Han, and T'ang.[13]

Although such statements argue for a continual and more or less linear and uninterrupted growth of population during Ming times, we have yet to find a relatively safe clue whereby we can reconstruct the population of Ming China. The only clue we have so far is that the aggregate recorded population of the five northern provinces increased from roughly 15,500,000, recorded in 1393, to about 26,700,000 in 1542 — an apparent gain of 73 per cent in almost 150 years, at an average annual rate of growth of 0.34 per cent. But, since the growth of population in these northern provinces was not entirely reflected in the official returns, this is not a satisfactory statistical guide. In all likelihood the under-registration of northern population in the early Ming was proportionately less serious than that in the middle and late Ming periods. The actual growth of the population of north China, therefore, must have been somewhat greater than is revealed in the official data. Even estimated at the above rate, perhaps the population of north China should have at least doubled itself by 1600.

North China, of course, is not representative of the whole country. For one thing, an unusually large extent of land was available for agricultural settlement in the low plain of north China. It was so large that the founder of the Ming dynasty granted permanent tax exemption to settlers of certain areas in Pei-Chihli, Shantung and Honan, which was revoked in the fifteenth century. For another, in north China there seemed to be fewer institutional impediments to population growth. Generally speaking, north China was a land of small landowning peasants who bore a lighter fiscal burden than their southern counterpart. The effects of landlordism, if not entirely absent in the north, were certainly far weaker than those in the lower Yangtze area, northern Chekiang, and greater parts of Fukien and Kwangtung.[14] These are the main arguments for the occurrence of a more rapid growth of population in north China.

On the other hand, various factors seem to favor the opposite view. In the first place, if the south suffered more from a heavier fiscal burden and from the evils of landlordism, it benefited greatly by an expanding and more varied economy. The effects of landlordism on population growth, particularly during a period when the total population of the country was small as compared with the modern population, were probably far less important than some have judged. Concentration of landed property in an age of increasingly labor-intensive agriculture had little adverse bearing on employment. The southeast had been allotted the heaviest fiscal burden precisely because of its ability to bear it. Hsieh Chao-che was right in saying that it was the gainful opportunities created by a many-sided economy that made the lower-Yangtze area prosperous, despite a fiscal load so heavy that it would have crushed any other region.[15] Agriculturally, the inland-Yangtze provinces remained to be fully developed and plenty of good land was reported as available all along the Huai River and the lowlands of Hupei and Hunan. Kwangtung, Yunnan, and central and southern Hunan, with their natural abundance and small populations, made up what Hsieh called a "land of paradise." [16] Even the common Ming impression of congestion in the southeast must be read in its context and measured by the standard then prevailing.

When all factors are weighed, it seems that the population of south China was increasing somewhat more rapidly than the population of north China. It was by no means coincidental that the writers who testified to a continual and fairly rapid growth of population were in general southerners, while northern local histories were often full of complaints about the heavy labor-service burden which, though lighter than that of the lower Yangtze, was a cause of rural depopulation in some northern localities.[17] On these assumptions one may guess that China's population had increased from some 65,000,000 in the late fourteenth century to the neighborhood of 150,000,000 by 1600. Even assuming that the southern population had been increasing at the same moderate rate revealed in official figures for the northern population, it may be hazarded that China's population had exceeded 130,000,000 by the turn of the sixteenth century.

There are strong reasons for favoring the higher estimate. In the first place, the extent of officially registered land reached 176,000,000 acres in 1602, which is 86 per cent of the lowest, or 75.8 per cent of the highest, estimate made by Professor J. L. Buck for the entire cultivated area in China Proper in the 1930's. Secondly, the continual dissemination of early-ripening rice throughout the Ming must have contributed heavily to the growth of population, particularly the population of the rice belt. It was during the Ming that the lowlands of Hupei and Hunan, along with the lower-Yangtze delta, ranked as the nation's main rice baskets. In fact, as early as 1102, within one century of the introduction of the Champa rice into China, the Northern Sung government registered over 20,000,000 households. Since evasion of population registration was rampant in the Sung period, 20,000,000 households seems to indicate a national population on the order of 100,000,000 at a time when the clan and compound family system had just been strengthened. On this basis, had it not been for the subsequent political division of China, the serious agricultural retrogression suffered by the Huai River region and Hupei (which became a much contested theater of war between the Chinese and the Jurchen), and the unusually oppressive government and vested interests of the Mongol period (1260–1368), the population of the country would probably have reached our assumed height of 150,000,000 much earlier.

By about 1600, however, the effects of Ming misgovernment apparently outweighed the benefits of commercial and agricultural expansion. Although the fiscal burden of the nation had been steadily increasing since the middle of the sixteenth century, the increases had been moderate compared with those of the early seventeenth century. There is no way of knowing how the nation adjusted itself, during the first quarter of the seventeenth century, to the apparently worsening economic and political conditions. In all likelihood the adjustment was slow and painful and the population was still increasing. But the second quarter of the seventeenth century witnessed the outbreak of great peasant rebellions. It was reported that the population of the densely settled Red Basin of Szechwan was nearly exterminated and that relatively few counties in Szechwan escaped from the scourge of Chang Hsien-chung and the conquering Manchus. Chang and other bandit leaders also exacted a heavy toll in human lives in Shansi, Shensi, Honan, Hupei, northern Hunan, the Huai River valley, and parts of Shantung and Hopei. It is impossible to estimate even aproximately the number of people who perished directly in the two decades of peasant wars and indirectly from famine, pestilence, and economic dislocation. But one thing is clear: in terms of life destruction these wars must be compared with the Thirty Years' War in Europe and the Taiping wars in nineteenth-century China. Overpopulation, a factor believed by some Chinese demographers to have caused the social upheavals toward the end of every dynastic cycle,[18] is conspicuously absent in the recorded discussions and data on late Ming China.

II

Despite the abolition of all late Ming surtaxes after the enthronement of the first Manchu emperor in 1644, the benefit of a reduced fiscal burden was not immediately evident. Wars against the various groups of Ming loyalists were still going on in central and south China and occasionally the conquering Manchus vented their wrath by widespread slaughter of loyalist Ming forces and civilian populations in a number of localities. It was not until the Prince of Kuei, the last Ming pretender, was driven to Burma in 1659 that the conquest of the Chinese mainland was completed. Formosa remained in the hands of Ming loyalists who from time to time harassed the southeast coast. Consequently, there was considerable economic dislocation along the coast; some of the coastal population was forced to move to inland districts and was thus deprived of its normal agricultural pursuit and its lucrative trade with foreigners. In 1673 the new Ch'ing empire was shaken by the rebellion of the powerful southern feudatories, which were pacified only after seven years of bitter fighting and heavy taxation. Not until the conquest of Formosa in 1683 did China see the beginning of an era of real peace, government retrenchment, and prosperity. It is true that the military campaigns, with the exception of the war against the feudatories, were of limited scale; possibly the population of the areas unaffected by wars began to increase at moderate rates shortly after the change of the dynasty, for some contemporaries were of the impression that by the 1690's the population of the provinces, except Szechwan, had exceeded that of the middle seventeenth century.[19] Yet the second half of the seventeenth century as a whole must be regarded as a period of slow and gradual recovery from previous heavy losses. One cannot be certain whether by 1700 the population of China was as large as that of 1600, for the return of favorable economic and political conditions may have been too late for the seventeenth century to register any net gain in population.

We may further hypothesize as follows: combined economic and political conditions became most favorable after 1683. The last years of the seventeenth century were a prelude to the unique chapter of population growth that did not end until the outbreak of the

Taiping Rebellion in 1851. The tone of early eighteenth-century observers on population differs drastically from that of late seventeenth-century writers. In fact, one of the main reasons for the K'ang-hsi Emperor's freezing of the national ting payment in 1711 was his idea of a rapidly increasing population, an idea that was the outcome of his personal tours of the empire and interviews with the common people.[20] The plains, valleys, and easily worked agricultural land of the greater parts of the country must have been densely settled by the Yung-cheng period (1723–1735), when a series of edicts exhorted the nation to practice more intensive farming, to improve the crops, and to discourage the cultivation of non-cereal commercial plants like tobacco.[21] Li Fu, while governor of Kwangsi in 1724–25, was of the impression that the population of the whole country had nearly doubled during the sixty years of the reign of K'ang-hsi.[22] Although his statement cannot be taken as an accurate estimate, there can be little doubt as to the unusual rate of population growth during the last years of the seventeenth century and the first quarter of the eighteenth century.

One phenomenon of this period was the steady rise in the price of rice throughout the first half of the eighteenth century. By 1743 the young Ch'ien-lung Emperor was so puzzled by this phenomenon in years of bumper harvests that he ordered the provincial authorities to discover the reason for the high prices. Neither the exemption of taxes on rice nor the reduction of public purchase of rice could halt this upward swing in agricultural prices.[23] Finally in 1748 Yang Hsi-fu, governor of rice-rich Hunan and a native of Kiangsi, who had a firsthand knowledge of rice production and marketing, offered perhaps the most satisfactory explanation. He attributed the rising rice prices to the higher standard of living for the nation as a whole, the increasing extravagance of the people, rice hoarding and price manipulation by the rich and cunning, large-scale public purchase of rice, and the rapid multiplication of the population. The last factor he considered the most basic. He said in a much-cited memorial:

I was born and brought up in the countryside and my family had been for generations engaged in farming. I can recall from personal memory that prices of one *shih* of rice ranged between two-tenths and three-tenths of a tael during the K'ang-hsi period. By the Yung-cheng period such low prices were no longer possible and they rose to four-tenths or five-tenths of a tael. Nowadays the prices can never be lower than five-tenths or six-tenths of a tael. This is because a large population consumes large quantities of rice. Despite the considerable cultivated area that has been developed during the past few decades, in many regions there is no more room for agricultural expansion. It is inevitable that a rapidly increasing population should have caused a steady rise in the price of rice.[24]

Although Yang referred probably only to the changing price structure in Kiangsi and Hunan, his generalization is supported by scattered official reports on rice prices for various provinces. By about 1750 the surplus rice of lowland Hupei and those districts of Hunan and Kiangsi easily reached by rivers was no longer sufficient to meet the demand of the southeast coast. At times even Hupei became partially dependent on Szechwan for the supply of rice, although in the remoter districts of Hupei, Hunan, and Kiangsi, inaccessible to cheap water transportation, there was still surplus rice.[25] It is helpful in estimating the eighteenth-century population to recall that when the population was presumably in the neighborhood of 150,000,000 around the turn of the sixteenth century, Hupei, Hunan, and Kiangsi, the traditional rice-bowl of China, were far from fully developed agriculturally. The filling-up of the important rice-producing lowlands and low hills is one of the indications that China's population may have considerably exceeded 150,000,000 in the early decades of the eighteenth century.

To mitigate the increasing regional pressure of population, the central and provincial officials repeatedly exhorted the nation to experiment with various new crops, particularly the sweet potato. A series of laws issued in the late 1730's granted permanent tax exemption to small plots of cultivable land, at varying sizes for various provinces, to prospective cultivators. It was during the first half of the eighteenth century that extensive interregional migrations took place and dry hills and mountains began to be systematically developed into maize and sweet-potato fields. Szechwan, the inland Yangtze highlands, the Han River drainage, and even southwestern aboriginal districts were invaded by millions of migrant farmers and the forested mountains were opened up on a truly national scale. By the end of the eighteenth century the hilly and mountainous areas had been so intensively exploited that soil erosion and diminishing agricultural returns were reported to have become a serious problem.

The population growth throughout the eighteenth century assumed above was presumably connected with a standard of living. The benefit of the freezing of the national ting payment and of the merger of the ting payment with the land tax seems to have been passed on to the common people. The famous fiction writer P'u Sung-ling (1640–1715) testified that the average peasant household in his native Shantung had ample storage of grains, fresh and salted meats, vegetables and dried fruits, chickens and eggs, and home-brewed wine for the lunar new year.[26] This description need not have been an exaggeration or a portrayal of the standard of living for a comparatively well-to-do family. Although he himself was from a small landowning family, he received only twenty mou of land at the time of division of the family property,[27] while many tenants in his novel worked on at least forty mou. Some of the peasants in Shantung, by no means one of the rich provinces, were so well off in the late K'ang-hsi period that they spent relatively freely and wasted a great deal of grain in good years.[28]

The rising standard of living and general increase in wealth are attested to by various eighteenth-century local histories. A rapidly increasing population and a prolonged economic prosperity were said to be responsible for a manifold increase in land values in northeastern Hunan.[29] The hilly and backward southeastern corner of Hunan bordering Kwangtung had never been reported as richer than during the second half of the eighteenth century, when the area could boast of a number of fortunes ranging from scores of thousands to over a million taels.[30] Whereas Ying-ch'eng county in Hupei had been poor and sparsely populated in the 1660's, the county was not only densely populated but had several hundreds of inhabitants with a fortune of 10,000 taels by the turn of the eighteenth century.[31] Of even a relatively poor district in Shensi it was testified in 1762:

> Our old local history [1557 edition] said that the people were so frugal that sometimes they failed to comply with the customary standards of propriety. Previously even at a banquet there were but few courses, now people vie with each other in offering more. Wine, silk and meat have become common articles of local consumption . . . .

The prices of cotton and silk fabrics, vegetables and fruits, meats and fuel have consequently increased several fold.[32]

The lower Yangtze area probably established a new standard of extravagance during the eighteenth century. Such cities as Nanking, Yang-chou, Su-chou, and Hang-chou became great centers of consumption and luxurious living. Many contemporaries deplored this trend toward more wasteful and extravagant living.[33] There must have been a substantial gap between the newly acquired standards of living in the late seventeenth and early eighteenth century and the subsistence living for a large part of the population during the first half of the nineteenth century.[34]

A continuous rapid growth of population was reflected in the official figures for the years 1779–1794, the first years of improved pao-chia population registration without noticeable regional omissions. The average rate of increase during these fifteen years is 0.87 per cent, as compared with 0.63 per cent for the entire period 1779–1850 and 0.51 per cent for the years 1822–1850. This may be comparable with the average rate of 0.771 per cent for the population of preindustrial Eastern Europe in the period 1800–1850.[35] The population of China, which was presumably in the neighborhood of 150,000,000 around 1700 or shortly after, probably increased to 275,000,000 in 1779 and 313,000,000 in 1794. If so, it had more than doubled itself within a century that offered uniquely favorable conditions for growth.

III

Although it is often difficult to determine exactly when and where population pressure increases, there is reason to believe that in the case of Ch'ing China, the optimum condition (the point at which "a population produces maximum economic welfare"[36]) at the technological level of the time, was reached between 1750 and 1775. Up to the third quarter of the eighteenth century contemporaries had viewed the continual rapid increase of population as an almost unqualified blessing; the generation of Chinese who reached maturity during the last quarter of the century, however, began to be alarmed by the noticeable lowering of the standard of living that had become "customary" since the earlier decades of the eighteenth century. If the transition from high prosperity to growing economic strain was relatively abrupt, it may be accounted for by the fact that the optimum population, probably in the neighborhood of 250,000,000 in the middle of the third quarter of the eighteenth century, was already large and any further proportionate growth brought about a formidable increase in the total numbers. Granted that the deterioration of economic conditions occurred in various places at different times,[37] the generalizations and the analysis of the economic situation made by Hung Liang-chi (1746–1809), "the Chinese Malthus," toward the end of the eighteenth century seem applicable to large areas of the nation. In his two famous essays, "Reign of Peace" and "Livelihood,"[38] written in 1793 while serving as educational commissioner of Kweichow, five years before the appearance of the first edition of Malthus' *Essay on Population*, he expounded many ideas similar to those of Malthus. Hung's theory of population is summarized and paraphrased by a modern Chinese scholar:

1. The increase in the means of subsistence and the increase of population are not in direct proportion. The population within a hundred years or so can increase from five-fold to twenty-fold, while the means of subsistence, due to the limitation of the land-area, can increase only from three to five times.
2. Natural checks like flood, famine and epidemic, cannot diminish the surplus population.
3. There are more people depending on others for their living than are engaged in productive occupations.
4. The larger the population, the smaller will be the income *pro rata*; but expenditure and the power of consumption will be greater. This is because there will be more people than goods.
5. The larger the population, the cheaper labor will be, but the higher will be the prices of goods. This is because of the over-supply of labor and over-demand for goods.
6. The larger the population, the harder it will be for the people to secure a livelihood. As expenditure and power of consumption become greater than the total wealth of the community, the number of unemployed will be increased.
7. There is unequal distribution of wealth among the people.
8. Those who are without wealth and employment will be the first to suffer and die from hunger and cold, and from natural calamities like famine, flood and epidemics.[39]

As to the possible remedies for overpopulation and their effectiveness, Hung says:

> Some may ask: "Do Heaven and Earth have remedies?" The answer is that their remedies are in the form of flood, drought, sicknesses and epidemics. But those unfortunate people who die from natural calamities do not amount to more than 10 or 20 per cent of the population. Some may ask: Does the government have remedies? The answer is that its methods are to exhort the people to develop new land, to practice more intensive farming, to transfer people from congested areas to virgin soils, to reduce the fiscal burden, to prohibit extravagant living and the consumption of luxuries, to check the growth of landlordism, and to open all public granaries for relief when natural calamities strike . . . . In short, during a long reign of peace Heaven and Earth could not but propagate the human race, yet their resources that can be used to the support of mankind are limited. During a long reign of peace the government could not prevent the people from multiplying themselves, yet its remedies are few.

Like Malthus, Hung could not foresee the effect of technological inventions and scientific discoveries on agricultural and industrial production. Unlike Malthus, who, while expounding his abstract theory, analyzed practically all available material concerning the population problem, Hung misread, or at least did not bother with, the exact meaning of official population data. Whereas Malthus in the revised edition of his *Essay* and especially in his later economic writings succeeded in formulating a system, Hung's ideas are fragmentary and his quantitative statements irresponsible. But by far the most serious drawback in Hung's theory of population is his failure to understand the law of diminishing returns, which Malthus belatedly understood and which saves his arithmetical ratio of food increase from being a pure fallacy. To do him justice, however, Hung's ideas are original; among his bookish contemporaries he had the keenest grasp of the changing economic condition.

Although the tempo of population growth may have slackened because of a series of regional disturbances after 1796, the nation continued to register large net gains in population after peace was restored. In 1820 Kung Tzu-chen, a gifted scholar profoundly disappointed by the traditional cultural and institutional system, viewed the situation as a lull before the storm:

Since the late Ch'ien-lung period the officials and commoners have been so distressed and slipping fast. Those who are neither scholars and farmers nor artisans and traders constitute nearly one-half of the population . . . . In general the rich households have become poor and the poor hungry. The educated rush here and there but are of no avail, for all are impoverished. The provinces are at the threshold of a convulsion which is not a matter of years but a matter of days and months.[40]

Kung's alarmist view seems largely justified in the light of evidence from nineteenth-century local histories. Many local histories of Hunan and Kiangsi, the rice bowl of China, give the impression that by the first half of the nineteenth century the urgent economic problem was no longer how to maintain the customary living standards but how to maintain bare subsistence. Some traditional rice-exporting areas had only a small surplus in years of bumper harvests and became partially dependent on other regions for their food supply in bad years.[41] Although an area of intensive farming, southern Kiangsi by the 1840's could not cope with the problem of rural unemployment, which had become more and more serious.[42] With Szechwan almost entirely filled up, there was no more room for agricultural expansion in China Proper. Even the comparatively newly developed Yangtze highlands and the Han River drainage had begun to suffer from diminishing returns and soil erosion. Because of the silting-up of the Yangtze, its tributaries, and the lakes, which acted as its reservoirs, there was also an increasing incidence of flood. Yet the growing population of the Hupei and Hunan lowlands defied the law and the public interest and reclaimed and expanded the newly formed sandbanks and islets. A Hupei local history stated that the locality had become so congested by about 1850 that even after the successive Taiping and Nien wars, which somewhat relieved the local pressure of population, "still there can only be idle adults in the households but no untilled land in the fields." [43] There is reason to believe that in the middle of the nineteenth century the Yangtze region was even more crowded than it is today.[44]

"It was precisely in this situation," says a modern economic historian, "when a growing population was straining at resources and a large part of it living near the margin of subsistence, that a society was particularly vulnerable." [45] So vulnerable was the

Chinese society that in 1851 the world's greatest civil war, the Taiping Rebellion, broke out. It lasted fourteen years and affected nearly all provinces of China Proper, but most of all the densely populated central and lower Yangtze regions. Although the factors contributing to the rebellion were many, there can be little doubt that the pressure of population was one of the most basic.

The most frighteningly realistic discussion of the population question was made by Wang Shih-to (1802–1889), whose *I-ping jih-chi*, a diary written in 1855–56 while he was a Taiping captive, is an important source for the study of the background and early events of the rebellion. He was one of the millions who were innocent victims of the process of progressive impoverishment brought about by continual increase of population. Although his great-grandfather had been a rich merchant, by his lifetime the family had become so poor that he was sent out as a boy to apprentice in stores. The necessity for putting aside a little in order to marry off his daughters made it impossible for him to provide sufficient medical care for his consumptive first wife, which caused him deep sorrow for the rest of his life. Among his five daughters and three sons only two girls reached adulthood, and both of these met with tragic deaths in the 1850's.[46] His bitter experience made him a misanthropist. He said of overpopulation in general:

> The harm of over-population is that people are forced to plant cereals on mountain tops and to reclaim sandbanks and islets. All the ancient forestry of Szechwan has been cut down and the virgin timberland of the aboriginal regions turned into farmland. Yet there is still not enough for everybody. This proves that the resources of Heaven and Earth are exhausted.[47]

As far as he could diagnose, the ills of nineteenth-century China were due neither to misgovernment nor to the nation's lack of ingenuity and diligence. They were primarily accounted for by an increasing disproportion between population and economic resources. The remedy that he suggested was to relax the prohibition against female infanticide, or rather to encourage such a practice en masse, to establish more nunneries and to forbid the remarriage of widows, to propagate the use of drugs that would sterilize adult women, to postpone the age of marriage for both sexes, to impose heavy taxation on families having more than one or two children, and to drown all surplus infants of both sexes except the physically fittest.[48] The problem of overpopulation and mass poverty which has plagued modern China had come into full existence by 1850.

In retrospect, it seems a great irony that the benevolent despotism of K'ang-hsi and Yung-cheng, one of the stimulants to population growth, sowed the seeds of the decline and fall of the Manchu empire and to a large extent accounted indirectly for the economic difficulties of modern China.

IV

Although it is difficult to suggest any definite figure for the net loss in population during the fourteen years of Taiping wars, our detailed local evidence testifies very clearly that a figure of twenty or thirty million, estimated by contemporary Western residents of the treaty ports, was too low. The Nien wars, which ravaged the Huai River area of northern Anhwei and parts of the north China plain, did not come to an end until the early 1870's. The Moslem rebellion, which greatly reduced the population of Shensi and Kansu, lasted until the late 1879's. It seems as though a readjustment between population and land resources could not be effectively brought about by nature's routine revenge alone but had to be supplemented by unprecedented man-made disasters.

The readjustment between population and land resources may best be evidenced by post-Taiping interregional migrations. Ironically it was the historically most densely settled lower Yangtze area that played the role of the pre-1850 Szechwan as the leading recipient of immigrants for about a generation. The sudden and unexpected lessening of population pressure could not entirely fail to have a certain beneficent, if transient, effect on the living standard of the people in the lower Yangtze region. The abundance and cheapness of good agricultural land, together with the government's eagerness to attract immigrants, helped many tenants to become small landowners. Even without the opportunity of climbing up the agricultural ladder, tenant farmers benefited for a while from the lenient tenurial terms that the acute shortage of labor had forced owners to offer.

The unusually observant Hu Ch'uan, father of Dr. Hu Shih, testified:

I personally witnessed in 1866–67 that within the several-hundred-*li* stretch between Hui-chou and Ning-kuo and T'ai-p'ing [in southern Anhwei] the survivors had surplus grain in their storerooms, meat in their kitchens, and wine in their pitchers. They were well fed, occasionally got drunk, and enjoyed to the full what a restored peace could offer. On the roads practically no one would care to pick up something dropped by others and at night one did not have to shut his gate.[49]

But the civil wars of the third quarter of the nineteenth century at best conferred upon the nation a brief breathing spell and failed to redress the old population-land balance. While the growth of the lower Yangtze population was definitely halted by the Taiping wars, the population of the low plain provinces of north China seems to have been increasing even faster than it had been before 1850. Emigration must have had comparatively little effect on provinces like Honan, whose population, as Richthofen observed, was apparently increasing rapidly due chiefly to the custom of early marriage. We may surmise that in all likelihood China's total population surpassed the 1850 peak sometime in the last quarter of the nineteenth century. The opening-up of Manchuria and overseas emigration, though having regional alleviating effects, failed to bring about a more favorable population-land ratio for the nation at large.

The absence of a major technological revolution in modern times has made it impossible for China to broaden the scope of her land economy to any appreciable extent. It is true that, after the opening of China in the 1840's, the moderately expanding international commerce, the beginnings in modern money and banking, the coming of steamers and railways, the establishment by both Chinese and foreigners of a number of light and extractive industries have made the Chinese economy somewhat more variegated, but these new influences have been so far confined to the eastern seaboard and a few inland treaty ports. There has not been significant change in the basic character of the national economy. In fact, for a century after the Opium War the influence of the West on the Chinese economy was as disruptive as it was constructive.

While the Chinese economy failed to make a break-through, the ideal of benevolent despotism was dead once and forever. The reduction of the land tax in the lower Yangtze region was only temporary, and very soon the financial straits of the central and provincial governments made large increases in surtaxes and perquisites inevitable. The people's fiscal burden was made even heavier by nationwide official peculation. With the lower Yangtze region partially recovered from war wounds, land tenure once more became a serious factor aggravating the peasants' economic difficulties. But the worst days did not come until China was ushered into an era of warlordism, at a time when the nation had so little economic reserve that natural calamities exacted disproportionately heavy tolls of human lives. Even under the twenty-two years of Nationalist rule the nation hardly enjoyed a year without war. All in all, therefore, the combined economic and political condition in post-1850 China was such that the nation seems to have barely managed to feed more mouths at the expense of further deterioration of its living standard.

Small wonder, then, that China's population has been computed as having increased only 35.5 per cent between 1850 and 1953, giving an average annual rate of growth of 0.3 per cent. If we take into account the probable net loss in population during the greater part of the third quarter of the nineteenth century and take 1865 instead of 1850 as the datum, then the average assumed rate of growth was between 0.4 and 0.5 per cent, which is substantially lower than the average of 0.63 per cent assumed for the period 1776–1850. It is even slightly lower than the average rate of 0.51 per cent assumed for the period 1822–1850, when the basic population-land pattern had fully assumed its modern characteristics. Whatever the margin of error in the pre-1850 population figures, the evidently persistent trend of a declining rate of growth tallies well with our knowledge of the changing economic and political conditions.

In conclusion, it may be guessed that China's population, which was probably at least 65,000,000 around 1400, slightly more than doubled by 1600, when it was probably about 150,000,000. During the second quarter of the seventeenth century the nation suffered severe losses in population, the exact extent of which cannot be determined. It would appear that the second half of the seventeenth century was a period of slow recovery, although the tempo of population growth was increased between 1683 and 1700. The seventeenth century as a whole probably failed to register any net gain in population. Owing to the combination of favorable economic conditions and kindly government, China's population increased from about 150,000,000 around 1700 to perhaps 313,000,000 in 1794, more than doubling in one century. Because of later growth and the lack of further economic opportunities, the population reached about 430,000,000 in 1850 and the nation became increasingly impoverished. The great social upheavals of the third quarter of the nineteenth century gave China a breathing spell to make some regional economic readjustments, but the basic population-land relation in the country as a whole remained little changed. Owing to the enormous size of the nineteenth-century Chinese population, even a much lowered average rate of growth has brought it to its reported 583,000,000 by 1953.

Today the population of China is again increasing rapidly, even more rapidly than during the eighteenth century. The return of peace and order, the removal of some institutional barriers, the beginnings of large-scale industrialization, and especially the nationwide health campaign cannot fail to stimulate population growth. Historically the Chinese population has been responsive to favorable economic and political conditions and has had a tendency toward prolonged growth even at the expense of progressive deterioration in the national standard of living. Whether history will repeat itself or whether the new China can achieve a rate of economic growth greater than her current rate of population growth remains to be seen. But the existence of a population of 600,000,000 — which is both China's strength and weakness — has already compelled the pragmatic Communist state to adopt a policy of limiting future population growth.[50]

APPENDIXES

APPENDIX I

OFFICIAL POPULATION DATA, 1741–1850.

YEAR	POPULATION	YEAR	POPULATION	YEAR	POPULATION
1741	143,411,559	1762	200,472,461	1783	284,733,785
1742	159,801,551	1763	204,209,828	1784	286,331,307
1743	164,454,616	1764	205,591,017	1785	288,863,974
1744	166,808,604	1765	206,693,224[a]	1786	291,102,486
1745	169,922,127	1766	208,095,796	1787	292,429,018
1746	171,896,773	1767	209,839,546	1788	294,852,189
1747	171,896,773[a]	1768	210,837,502[a]	1789	297,717,496
1748	177,495,039[a]	1769	212,023,042	1790	301,487,114
1749	177,495,039	1770	213,613,163	1791	304,354,160
1750	179,538,540	1771	214,600,356	1792	307,467,279
1751	181,811,359	1772	216,467,258	1793	310,497,210
1752	182,857,277[a]	1773	218,743,315	1794	313,281,795
1753	183,678,259	1774	221,027,224	1795	296,968,968
1754	184,504,493	1775	264,561,355	1796	275,662,044
1755	185,612,881	1776	268,238,181	1797	271,333,544
1756	186,615,514	1777	270,863,760[a]	1798	290,982,980
1757	190,348,328	1778	242,695,618	1799	293,283,179
1758	191,672,808	1779	275,042,916	1800	295,273,311
1759	194,791,859	1780	277,554,413	1801	297,501,548
1760	196,837,977	1781	279,816,070	1802	299,749,770
1761	198,214,555	1782	281,822,675	1803	302,250,673

[a] Annual national total missing in Lo's table, added by Ping-ti Ho in accordance with the *Ch'ing shih-lu.*

[b] Figure revised by Lo Ehr-kang in accordance with the archives of the Board of Revenue.

YEAR	POPULATION	YEAR	POPULATION	YEAR	POPULATION
1804	304,461,284	1820	353,377,694	1836	404,901,448
1805	332,181,403	1821	355,540,258	1837	406,984,114[b]
1806	335,309,469	1822	372,457,539	1838	409,038,799
1807	338,062,439	1823	375,153,122	1839	410,850,639
1808	350,291,724	1824	374,601,132	1840	412,814,828
1809	352,900,024	1825	379,885,340	1841	413,457,311
1810	345,717,214	1826	380,287,007[a]	1842	416,118,189[b]
1811	358,610,039	1827	383,696,095	1843	417,239,097
1812	361,691,431	1828	386,531,513	1844	419,441,336
1813	336,451,672	1829	390,500,650	1845	421,342,730
1814	316,574,895	1830	394,784,681	1846	423,121,129
1815	326,574,895	1831	395,821,092	1847	425,106,201[b]
1816	328,814,957	1832	397,132,659	1848	426,928,854[b]
1817	331,330,433	1833	398,942,036	1849	428,420,667[b]
1818	348,820,037	1834	401,008,574	1850	429,931,034[b]
1819	351,260,545	1835	403,052,086[b]		

APPENDIX II

CHANGES IN THE GEOGRAPHIC DISTRIBUTION OF POPULATION, 1787–1953

Province	Population (A) (1787 in '000)	Population (B) (1850 in '000)	Population (C) (1953 in '000)	(B) as a % of (A)	(C) as a % of (B)
Kiangsu	31,427	44,155	47,457[a]	140.4	107.4
Chekiang	21,719	30,027	22,866	138.2	76.2
Anhwei	28,918	37,611	30,344	130.1	81.7
Kiangsi	19,156	24,515	16,773	128.2	68.5
Hupei	19,019	33,738	27,789	177.0	82.4
Hunan	16,165	20,614	33,227	128.0	161.2
Hopei	22,957	23,401	38,678[b]	102.2	165.2
Shantung	22,565	33,127	48,877	147.1	147.5
Honan	21,036	23,927	44,215	113.8	184.7
Shansi	13,232	15,131	14,314	122.0	94.6
Shensi	8,403	12,107	15,881	144.0	131.1
Kansu	15,162	15,437	12,928	101.8	83.8
Fukien	12,012	....	13,143	....	109.4[c]
Kwangtung	16,014	28,182	34,770	175.6	123.3
Kwangsi	6,376	7,827	19,561	122.9	249.9
Yunnan	3,461	7,376	17,473	213.1	236.9
Kweichow	5,158	5,434	15,037	105.3	276.7
Szechwan	8,567	44,164	62,304	515.8	141.7
Liao-ning	811	2,571	18,545	317.0	721.3
Kirin	150	327	11,570	218.0	3638.2
Heilungkiang	....	....	11,897	....	....

[a] Includes population of Shanghai.

[b] Includes population of Tientsin but not the population of Peking; all through the Ch'ing period the figures for Peking were never included in Chihli.

[c] The 1850 Fukien figure has been proven unauthentic. The percentage is of (C), with (A) as 100.

The changes in the geographic distribution of China's population in Ming, Ch'ing, and modern times cannot be studied in detail, since reliable provincial figures are lacking for the greater part of the period covered in this study. The problem is of sufficient importance, however, to justify a brief discussion of a necessarily tentative nature.

One of the most striking features revealed in the table is the decline of the populations of four Yangtze provinces in the past century. The reasons for the decline of the lower Yangtze provinces have already been discussed, but critics of traditional Chinese population data not unnaturally raise the question: were some of the pre-1850 figures exaggerated? This question has been anticipated by the editors of the latest and uncompleted history of Anhwei province which is unusually informative. (*Anhwei t'ung-chih-kao*, latest uncompleted edition, Book on Civil Affairs, ch. on "population.") Based on their intimate knowledge of the changing economic and political conditions in the province, they not only uphold the pre-1850 figures for Anhwei but explain convincingly how the cumulative effects of the Taiping and Nien wars, destruction of farmland, famines, epidemics, the evils of opium, and banditry have made a full recovery of the Anhwei population to the 1850 level impossible.

Kiangsi, in the light of its arbitrary post-1850 figures, appears highly suspect. The drastic changes in the pao-chia registration after 1851 have already been explained. From an exhaustive sampling of nineteenth-century Kiangsi local histories it may be generalized that the province in the first half of the past century was not as important a rice-surplus area as it is in the twentieth century. Some Kiangsi local histories, like the 1848 edition of the history of Kan-chou prefecture, impress upon modern readers the problem of serious rural unemployment, which is in sharp contrast to recent accounts of that area. Some correspondents who toured the old Red shrines of southern Kiangsi in 1953 testify that Chi-an county alone has an annual rice surplus of over a million shih, because its population is small (150,000 in 1953) and available agricultural land is plentiful. (Ch'en Mu, *Nan-fang lao-ken-chü-ti fang-wen chi* [Wuhan, 1953], esp. p. 121.) Everywhere in central and southern Kiangsi today the problem is an abundance of idle land and a shortage of farm labor.

Moreover, in 1933 the Economic Commission of the Kiangsi Provincial Government made a careful investigation of the production and distribution of rice. In the 1920's and the early 1930's, before the Nationalist-Communist civil war reached an acute stage, the province on the average could secure annually a rice surplus of 10,000,000 shih, which made Kiangsi, along with Hunan and Anhwei, a leading rice-exporting area. No such impression can be gathered from mid-Ch'ing editions of Kiangsi local histories. The Commission finally estimated that the size of the Kiangsi population at which all surplus rice would be consumed by the province would probably be in the neighborhood of 25,000,000, which slightly exceeds the 1850 figure. (*Chiang-hsi chih mi-mai wen-ti* [Nan-ch'ang, 1933], *passim.*) How could the 1850 Kiangsi figure have been falsified when it tallies so well with other facts and figures, while the 1850 Fukien figure of 19,987,000, which has been proven false, contradicts almost every known fact about modern Fukien?

Concerning Chekiang, we know that the loss of population was unusually heavy in the three most densely settled northern prefectures of Hang-chou, Chia-hsing, and Hu-chou, and fairly heavy in two central prefectures of Yen-chou and Chin-hua. It is also known that a full agricultural recovery in Chekiang was indefinitely postponed by the opposition of native landlords to further immigration after 1883, if not earlier. The bulk of the devastated farmland in northern Chekiang was resettled by immigrants from other parts of Chekiang. The Taiping wars, therefore, brought about mainly a readjustment of land resources and of the population of the province itself. Moreover, in the last century there must have been a considerable number of people from Chekiang who were in government service, or established in trade and professions outside the province.

But one basic factor, aside from wars, that may have led to a decline of population in the lower Yangtze provinces is that rice culture reached its saturation in this area much earlier than in other rice areas. Ever since the Sung period early-ripening rice had been disseminated widely and uninterruptedly in this region. The efforts of Lin Tse-hsü and Li Yen-chang in the 1830's to persuade Kiangsu peasants to grow such extremely early-ripening but low-yielding varictics as the "forty-day" and "thirty-day" certainly defied the law of diminishing returns. Since the growth of the lower Yangtze population was halted by the Taiping wars and since in the last century the choice of food and cash crops was wider than before, it is probable that much of the inferior rice land has gone to other crops. This change, though not easily documented, may have had some bearing on the size of the population of so "overdeveloped" a province as Chekiang.

In the last hundred years the population of Hupei has also reportedly declined by 17.6 per cent. This may have been brought about by the Taiping wars and by large-scale emigration. Prior to 1850 Hupei was already an important emigrating province, but it attracted many more millions than it sent out, chiefly because of its central location and the development of the western Hupei highlands. There is reason to believe that after 1865 Hupei became a net emigrating province. A number of modern Hupei local histories show that post-Taiping emigration to the lower Yangtze and occasionally to Shensi was accounted for less by dire economic necessity than by the desire of Hupei tenant farmers to improve their economic lot.

Among the central and lower Yangtze provinces only Kiangsu and

Hunan are reported as having gained in population since 1850. But for one century of commercial and industrial development and the rise of Shanghai to a metropolis of 6,000,000, Kiangsu would not have been able to make this small gain of 7.4 per cent. Hunan, however, has gained 61.2 per cent; it is the only province in the Yangtze region to show a robust increase since 1850. There are several possible explanations. First, unlike other Yangtze provinces, Hunan escaped serious war devastation, as the Taipings only passed through the province without gaining a foothold. As shown in Chapter VII, the conclusions of some modern scholars that Hunan in the Ch'ing period was a net emigrating province do not agree with detailed local evidence which indicates that immigration in modern times has been considerable. Secondly, the 1850 figure of 16,165,000 may have been somewhat low because of the exclusion of the Miao people, who were then more numerous in the western hilly districts than they are now. The population increases during the past century should therefore be somewhat smaller than 61.2 per cent. Thirdly, Hunan alone of the Yangtze provinces actually benefited both politically and economically from the Taiping wars. Owing to the vital role played by the Hunan Army, the province in the late Ch'ing produced more high-ranking officials, generals, and officers than any other province. Remittances from Hunan people who prospered elsewhere must have contributed substantially to the relative wealth and plenty that the province has enjoyed in modern times. The vitality of the people and the availability of land in the hilly peripheral districts have also favored population growth.

The second striking feature revealed in the above table is the rapid population growth in the low plain provinces of north China during the past hundred years. The relative immunity from major wars has been an important factor. But, more basically, it would appear that the rapid increase in Hopei, Shantung, and Honan may have been due to a relatively late agricultural development. Historically, northern agriculture was characterized by low yield and lack of labor-intensive methods of cultivation. As late as 1738 Yin Hui-i, governor of Hunan and himself a native of that province, said in a famous memorial (cited in Hsü Hsi-lin, *Hsi-ch'ao hsin-yü* [*Ch'ing-tai pi-chi ts'ung-k'an* ed.], 9.3a–4b): "In the south the produce per mou is measured by the shih, but in the north the produce per mou is measured by the tou (one-tenth of a shih)." To encourage more intensive farming he suggested a law limiting the size of a tenant farm to 30 mou. Detailed local evidence shows that for centuries northern peasants were handicapped by the lack of means to cultivate drylands that were unsuited to wheat and other native northern cereals, and that the dissemination of maize, sweet potatoes, and peanuts did not become truly extensive until the nineteenth century. This belated revolution in land utilization cannot have failed to have an important bearing on the growth of north China's population.

On the loess high plain, Kansu and Shansi have shown some decline and Shensi has registered some net gain; the aggregate northwestern population has remained stationary during the past century. Actually there have been sharp fluctuations in the population in this region. Kansu has yet to recover from the Moslem wars of the 1860's and 1870's and from repeated drought famines. Shansi suffered from the drought famine of 1877–78 and from subsequent less severe ones. Because of its poor soil and scanty rainfall, it could not compete with Shensi in attracting immigrants. Although Shensi has been stricken by several severe drought famines and wars, the Ch'in-ling range and particularly the fertile Wei River valley were usually refilled fairly promptly by immigrants from the congested neighboring provinces of Szechwan and Hupei.

As to the unusually high percentage-increase in the populations of Yunnan, Kweichow, and Kwangsi, the chief explanation is that their pre-1850 populations were very much under-registered, because large numbers of various non-Chinese races were not included. There can be little doubt, however, that their populations have substantially increased since 1850, because the main trend in land utilization in modern times has been a continual devlopment of mountains. The unusually large percentage-gain in Szechwan population up to 1850, in contrast to a 41.1 per cent increase after that date, confirms my conclusion that Szechwan relinquished its role as a leading recipient of immigrants by the middle of the nineteenth century. The increase since 1850 seems to have been mainly, if not exclusively, accounted for by natural growth.

Some erratic features are shown in the population figures of two southeastern coastal provinces, Kwangtung and Fukien. Without extensive overseas emigration after 1850, the population of Kwangtung would have increased more than 23.3 per cent in the last century. The most disconcerting feature in the table is the 1850 Fukien figure. Since the Fukien pao-chia personnel became reluctant to carry out their duties in the early nineteenth century, the provincial administration had to resort to an arbitrary compilation of population figures; the 1850 figure of 19,987,000 must therefore be rejected. The net increase of about a million, or 9.4 per cent, between 1787 and 1953 suggests that much of the moderate growth in Fukien population has been absorbed by emigration to Formosa and southeast Asia.

Although the pre-1850 population figures for Manchuria are undoubtedly very incomplete, it is certain that in the last hundred years Manchuria has registered big increases in population.

Of the geographic redistribution of China's population as a whole, it may be said that in modern times the population of northern provinces has been increasing rapidly and that of the Yangtze region has been declining. As the country becomes more industrialized, the shift of the demographic center of gravity from the Yangtze region, whose large population and past glory were based largely on rice and domestic commerce, to north China, which is endowed with richer natural resources, will become every decade more apparent.

APPENDIX III

CHINA'S IMPORTS OF MAJOR FOODS, 1867–1937.

Year	Rice (piculs)	Wheat (piculs)	Wheat flour[a] (piculs)
1867	713,494		
1868	349,167		
1869	346,573		
1870	141,298		
1871	248,394		
1872	658,749		
1873	1,156,052		
1874	6,293		
1875	84,612		
1876	576,279		
1877	1,050,901		
1878	297,567		
1879	248,939		
1880	30,433		
1881	197,877		
1882	233,149		
1883	253,210		
1884	151,952		
1885	316,999		
1886	518,448		
1887	1,944,251		
1888	7,132,212		
1889	4,270,879		
1890	7,574,257		
1891	4,684,675		
1892	3,948,202		

Sources: *Statistics of China's Foreign Trade during the Last Sixty-Five Years* (Institute of Social Sciences, Academia Sinica, 1931); figures from 1929 to 1937 are from *Statistics of China's Foreign Trade, 1912–1930* (Research Department, Bank of China, 1931) and *China, The Maritime Customs, Foreign Trade of China*, 1929 to 1937 inclusive.

[a] Between 1887 and 1901 figures for wheat flour are given only in value in taels and are here omitted.

[b] Although the unit from 1934 to 1937 is quintal (1.654 piculs) the figures for these years have been converted to piculs.

Year	Rice (piculs)	Wheat (piculs)	Wheat flour [a] (piculs)
1893	9,474,562		
1894	6,440,718		
1895	10,096,448		
1896	9,414,568		
1897	2,103,702		
1898	4,645,360		
1899	7,365,217		
1900	6,207,226		
1901	4,411,609		
1902	9,730,654		
1903	2,801,894		766,324
1904	3,356,830		937,946
1905	2,227,916		931,761
1906	4,686,542		1,784,681
1907	12,765,189		4,414,383
1908	6,736,616		369,445
1909	3,797,705		596,777
1910	9,409,594		740,841
1911	5,302,805	3,197	2,183,042
1912	2,700,391	2,564	3,202,501
1913	5,414,896	2,064	2,596,821
1914	6,774,266	998	2,166,318
1915	8,476,058	2,586	158,273
1916	11,284,023	59,555	233,464
1917	9,837,182	36,169	678,849
1918	6,984,025	16	4,551
1919	1,809,749	20	271,328
1920	1,151,752	5,425	511,021
1921	10,629,245	81,346	752,673
1922	19,156,182	873,142	3,600,967
1923	22,434,962	2,595,190	5,826,540
1924	13,198,054	5,145,367	6,657,162
1925	12,634,624	700,117	2,811,500
1926	18,700,797	4,156,378	4,285,124
1927	21,091,586	1,690,155	3,824,674
1928	12,656,254	903,088	5,984,903
1929	10,820,950	5,676,144	11,951,743
1930	19,921,918	2,762,324	5,150,307

Year	Rice (piculs)	Wheat (piculs)	Wheat flour[a] (piculs)
1931	9,213,643	22,835,996	4,746,912
1932	21,386,444	15,084,723	6,636,658
1933	12,128,036	10,714,634	1,957,113
1934[b]	10,728,195	7,691,919	985,367
1935	11,066,403	8,615,835	844,360
1936	3,084,003	1,932,026	512,852
1937	4,747,246	711,992	502,593

APPENDIX IV

NATURAL CALAMITIES IN HUPEI, 1644–1911
(numbers of counties affected)

Year	Famine	Drought	Flood, excessive rain, typhoon, etc.[a]	Epidemic	Locusts, pests, etc.	Total number of counties affected
1644	5(e)	1	—	1	—	7
1645	5	—	3	1	—	9
1646	—	3	—	3	1	7
1647	6	6(d)[b]	3	—	—	9
1648	—	1	—(2)	—	—	3
1649	2	—	4(1)	—	—	7
1650	—	—	1(2)	—	—	3
1651	—	1	—(7)	—	1(d)	8
1652	—	25(e)	3	—	—	28
1653	—	—	3(2)	—	—	5
1654	—	—	3(1)	—	—	4
1655	—	—	3(1)	—	—	4
1656	—	—	—(2)	—	—	2
1657	—	1	—(1)	—	—	2
1658	—	—	27(e)	—	1(d)	27

Source: *Hupei TC* (1921 CP ed.), 76, *passim.* Sometimes the source only states that certain prefectures were visited by natural calamities without mentioning the exact number of counties affected. Since a prefecture usually consisted of from five or six to over ten counties, we take the lowest number (5) for each prefecture. In such cases there is (e), which means "estimated," after the number of counties. For example, "25(e)" under drought in the year 1652 means that 5 prefectures suffered from drought. Or, as in the year 1658 "27(e)" under flood means that in addition to 5 prefectures hit by flood there were two other counties specifically mentioned.

[a] The figures in parentheses indicate the number of counties hit by excessive rainfall, typhoon, etc. and hence suffering from considerable crop loss. On rare occasions a county might have been hit by typhoon, etc. without a significant loss of crops. Such cases are omitted.

[b] Sometimes a county had more than one kind of natural calamity in the same year. For example, in the year 1658 among the estimated 27 counties suffering from flood one suffered from locusts as well. Although this occurrence of locusts is listed under locusts, pests, etc. it is followed by (d), which means "duplicate," and is not added to the total number of calamity-struck counties for that year.

Year	Famine	Drought	Flood, excessive rain, typhoon, etc.[a]	Epidemic	Locusts, pests, etc.	Total number of counties affected
1659	—	1	4(2)	—	—	7
1660	—	1	1	—	—	2
1661	—	10	1	—	—	11
1662	—	2	7	—	—	9
1663	—	1	18(1)	—	—	20
1664	—	—	7	—	—	7
1665	—	—	2(1)	—	—	3
1666	—	1	—(3)	—	—	4
1667	1	5(e)	—(1)	—	—	7
1668	1	1	6	—	—	8
1669	—	—	1(7)	—	—	8
1670	—	3	8(1)	—	—	12
1671	2	8	3	—	—	13
1672	—	2	4	—	1	7
1673	—	6(e)	—	—	—	6
1674	—	6	2	—	—	8
1675	—	1	—	—	—	1
1676	—	—	13	—	—	13
1677	2	—	3(1)	—	1	7
1678	—	1	1	—	—	2
1679	—	18	2	—	—	20
1680	3	2	4(1)	—	1	11
1681	—	1	5	—	1	7
1682	—	—	6	1	—	7
1683	—	—	3(1)	1	1	6
1684	—	—	1	—	—	1
1685	—	—	17(1)	—	—	18
1686	—	—	—	—	—	0
1687	—	—	—	—	—	0
1688	—	—	—	—	—	0
1689	1	9	—	—	—	10
1690	36	36(d)	—(1)	—	—	37
1691	1	—	—	—	—	1
1692	—	—	—	1	—	1

Year	Famine	Drought	Flood, excessive rain, typhoon, etc.[a]	Epidemic	Locusts, pests, etc.	Total number of counties affected
1693	—	—	2	2	—	4
1694	—	—	—	—	—	0
1695	—	—	1(1)	—	—	2
1696	—	—	11(1)	—	1	13
1697	7	—	1(1)	—	—	9
1698	—	1	—	1	—	2
1699	—	9	1	—	—	10
1700	—	—	1	1	—	2
1701	—	—	1(1)	—	—	2
1702	—	—	2(1)	—	—	3
1703	—	—	4(1)	—	—	5
1704	—	1	6	—	1	8
1705	—	1	13(1)	—	—	15
1706	10	—	6	2	—	18
1707	—	1	3	2	—	6
1708	1	—	3	1	1	6
1709	—	—	12	—	—	12
1710	—	1	1	1	—	3
1711	—	—	—	—	—	0
1712	—	3	1	—	—	4
1713	—	—	1	—	—	1
1714	—	3	3	—	—	6
1715	—	1	7	—	1	9
1716	—	—	15	—	—	15
1717	—	—	—(1)	—	1	2
1718	—	12	—(1)	—	—	13
1719	—	—	1	—	—	1
1720	—	—	4	—	—	4
1721	—	1	—	—	—	1
1722	—	5	—	—	—	5
1723	—	3	2	—	—	5
1724	—	—	19	—	—	19
1725	—	1	3	—	—	4

Year	Famine	Drought	Flood, excessive rain, typhoon, etc.[a]	Epidemic	Locusts, pests, etc.	Total number of counties affected
1726	—	—	18	—	—	18
1727	1	—	17(2)	2	—	22
1728	—	—	2(1)	5	—	8
1729	—	1	—	1	1	3
1730	—	—	—(1)	—	—	1
1731	—	—	1	—	—	1
1732	—	—	—(1)	—	—	1
1733	—	—	—	—	—	0
1734	—	—	1	1	—	2
1735	—	5	—(2)	—	—	7
1736	—	—	6	—	—	6
1737	—	—	5(2)	—	—	7
1738	—	6	2(1)	—	—	9
1739	—	4	1	—	—	5
1740	—	—	4	—	—	4
1741	—	—	5	—	—	5
1742	1	—	20	—	—	21
1743	—	—	—(1)	—	—	1
1744	—	—	—	—	—	0
1745	—	—	10	—	—	10
1746	—	—	8	—	—	8
1747	—	—	—(1)	—	—	1
1748	—	—	8(2)	—	—	10
1749	—	—	8	—	—	8
1750	—	—	1(2)	—	—	3
1751	—	2	—	—	—	2
1752	1	9	17	—	—	27
1753	1	—	5(1)	—	1	8
1754	1	—	—	—	—	1
1755	1	—	3(3)	—	—	7
1756	2	—	6	—	—	8
1757	—	—	1	—	—	1
1758	—	—	—	—	—	0

Year	Famine	Drought	Flood, excessive rain, typhoon, etc.[a]	Epidemic	Locusts, pests, etc.	Total number of counties affected
1759	—	—	4(1)	—	—	5
1760	—	—	—(1)	—	—	1
1761	2	—	7	—	—	9
1762	—	—	—	—	—	0
1763	—	—	—(4)	—	—	4
1764	—	—	14	—	—	14
1765	—	—	2	—	—	2
1766	—	—	1	—	—	1
1767	—	—	16	—	—	16
1768	—	10	—	1	—	11
1769	—	—	16	—	—	16
1770	—	—	1	—	—	1
1771	—	1	1	—	—	2
1772	—	—	1	—	—	1
1773	—	—	1	—	—	1
1774	—	4	—	—	—	4
1775	—	2	—	—	—	2
1776	—	—	—	—	—	0
1777	—	2	—	—	—	2
1778	8	30	—(1)	—	—	39
1779	4	—	3(1)	—	—	8
1780	1	—	6	—	—	7
1781	—	—	10	—	—	10
1782	—	—	—	—	—	0
1783	—	—	7	—	—	7
1784	1	1	—(1)	—	—	3
1785	4(d)	46	—	—	—	46
1786	—	1	—	—	1	2
1787	—	—	—	—	—	0
1788	—	—	16	—	—	16
1789	—	—	4	—	—	4
1790	—	—	—(1)	—	—	1
1791	—	—	3(1)	—	—	4
1792	1	2	—	—	—	3

Year	Famine	Drought	Flood, excessive rain, typhoon, etc.[a]	Epidemic	Locusts, pests, etc.	Total number of counties affected
1793	—	—	—	—	—	0
1794	—	—	4(1)	—	—	5
1795	—	—	2(2)	—	—	4
1796	—	—	5(e)	—	—	5
1797	—	2	—(1)	—	—	3
1798	—	—	—	—	—	0
1799	—	—	—	—	—	0
1800	—	1	—(1)	—	—	2
1801	1	—	—	—	—	1
1802	—	20(e)	—	—	—	20
1803	—	—	9	—	—	9
1804	—	—	2	—	—	2
1805	—	—	3	—	—	3
1806	1	—	—	—	—	1
1807	4	—	1(1)	—	—	6
1808	2	—	—	—	—	2
1809	—	—	1(5)	—	—	6
1810	—	1	1	—	—	2
1811	—	2	1(1)	—	—	4
1812	—	1	4	—	—	5
1813	6	3	5	—	—	14
1814	7	6	1(1)	1	—	16
1815	—	—	2	1	—	3
1816	—	—	2	—	—	2
1817	—	—	1(1)	—	—	2
1818	—	—	2	—	—	2
1819	—	1	3	1	—	5
1820	—	4	1(1)	—	—	6
1821	—	—	3	—	—	3
1822	2	—	2	—	—	4
1823	2	—	2	—	—	4
1824	—	—	—	—	—	0
1825	—	1	2(3)	—	—	6

Year	Famine	Drought	Flood, excessive rain, typhoon, etc.[a]	Epidemic	Locusts, pests, etc.	Total number of counties affected
1826	—	1	11(1)	—	—	13
1827	—	—	8	—	—	8
1828	—	—	4	1	—	5
1829	—	—	2	—	—	2
1830	1	—	5(2)	—	—	8
1831	—	—	19	—	—	19
1832	3	—	11(2)	10	2	28
1833	4	—	11(1)	—	3	19
1834	3	1	6(4)	—	1	15
1835	2	14	4	—	2	22
1836	—	—	4	—	4	8
1837	1	1	1	—	—	3
1838	—	—	3	—	1	4
1839	—	1	9	—	—	10
1840	—	—	8(1)	—	—	9
1841	—	—	8(4)	—	—	12
1842	2	—	6(1)	1	—	10
1843	—	4	(2)	1	—	7
1844	—	5	6	1	—	12
1845	—	—	2(1)	1	1	5
1846	—	—	3	—	—	3
1847	—	1	4	—	—	5
1848	—	—	13(2)	—	—	15
1849	4	—	14	1	—	19
1850	—	—	3(3)	—	—	6
1851	—	—	5(2)	—	—	7
1852	—	—	6(1)	1	—	8
1853	—	—	5	—	—	5
1854	—	—	6(5)	—	—	11
1855	—	3	7	—	—	10
1856	1	7	—	—	—	8
1857	3	10	3	—	—	16
1858	—	—	3(1)	—	4	8

Year	Famine	Drought	Flood, excessive rain, typhoon, etc.[a]	Epidemic	Locusts, pests, etc.	Total number of counties affected
1859	—	1	2(1)	—	—	4
1860	—	—	5(1)	—	—	6
1861	—	—	5(5)	1	—	11
1862	4	2	5(2)	—	—	13
1863	1	1	5(1)	—	—	8
1864	—	—	3(2)	1	—	6
1865	1	—	—(3)	—	—	4
1866	—	1	7(2)	—	—	10
1867	1	4	11(2)	—	—	18
1868	4	—	5(5)	—	—	14
1869	1	—	2(7)	1	—	11
1870	1	—	9(2)	1	—	13
1871	—	—	2(1)	—	—	3
1872	—	—	3(1)	1	—	5
1873	—	1	2	—	—	3
1874	—	3	1(2)	—	—	6
1875	1	1	3(1)	—	—	6
1876	—	2	5	—	—	7
1877	—	4	4	1	1	10
1878	1	2	3(2)	2	1	11
1879	—	—	10(2)	1	1	14
1880	—	—	—(3)	—	—	3
1881	—	—	—	—	—	0
1882	—	—	4	—	—	4
1883	—	—	8	—	—	8
1884	—	—	2	—	—	2
1885	—	—	3(1)	—	—	4
1886	—	2	3	—	—	5
1887	1	2	20(3)(e)	—	—	26
1888	1	1	1	1	—	4
1889	1	—	1(1)	—	—	3
1890	—	—	—	—	—	0
1891	—	—	—	—	—	0
1892	—	—	6(1)(e)	—	—	7

Year	Famine	Drought	Flood, excessive rain, typhoon, etc.[a]	Epidemic	Locusts, pests, etc.	Total number of counties affected
1893	—	—	—	—	—	0
1894	—	—	1	—	—	1
1895	—	—	—	—	—	0
1896	1	—	3	—	—	4
1897	—	—	8(e)	—	—	8
1898	—	—	15(e)	—	—	15
1899	—	—	—	—	1	1
1900	—	1	—	—	—	1
1901	—	—	1	—	—	1
1902	—	—	—	—	—	0
1903	—	—	—	—	—	0
1904	—	—	—	—	—	0
1905	—	—	—	—	—	0
1906	—	—	—	—	—	0
1907	—	—	20(e)	—	—	20
1908	—	—	30(1)(e)	—	—	31
1909	4	—	4(2)	—	—	10
1910	2	—	1	—	—	3
1911	—	—	—	—	—	0

NOTES

ABBREVIATIONS

CC	*chou-chih*
CCHWHTK	*Ch'ing-ch'ao hsü-wen-hsien t'ung-k'ao* (CP ed.)
CCWHTK	*Ch'ing-ch'ao wen-hsien t'ung-k'ao* (CP ed.)
CP	Commercial Press
FC	*fu-chih*
HC	*hsien-chih*
HCCSWP	*Huang-ch'ao ching-shih wen-pien* (Edited by Ho Ch'ang-ling, 1886 ed.)
HJAS	*Harvard Journal of Asiatic Studies*
HWHTK	*Hsü-wen-hsien t'ung-k'ao* (CP ed.)
Liang, "Single-Whip System,"	Liang Fang-chung, "I-t'iao-pien fa," *Chung-kuo she-hui ching-chi-shih chi-k'an*, Vol. IV, no. 1.
Liang, "Ming Yellow Registers,"	Liang Fang-chung, "Ming-tai huang-ts'e k'ao," *Ling-nan hsüeh-pao*, Vol. X, no. 1.
Liang, "Chronological Chart,"	Liang Fang-chung, "Ming-tai i-t'iao-pien fa nien-piao," *Ling-nan hsüeh-pao*, Vol. XII, no. 1.
SPPY	Ssu-pu pei-yao
SPTK	Ssu-pu ts'ung-k'an
TC	*t'ung-chih*
THCKLPS	Ku Yen-wu, *T'ien-hsia chün-kuo li-ping shu* (SPTK ed.)
TSCC	Ts'ung-shu chi-ch'eng
YCHP	*Yen-ching hsüeh-pao*

CHAPTER I. The Nature of Ming Population Data.

1. *Ming-shih* (SPPY ed), 77.2b.

2. Wang Ch'ung-wu, "Ming-tai hu-k'ou ti hsiao-chang," *YCHP*, December 1936, pp. 366–367.

3. The proclamation of 1370 and the original form of the household certificate are found in two rare local histories, *Hsing-hua-ts'un chih* (1685 ed.), 11.1a-2b, and *P'u-chen chi-wen* (1787 ed., Ch'ing manuscript). The texts of the proclamation in these two works differ but slightly. Dr. Arthur W. Hummel's translation, which has been used here in part, is in *Annual Report of the Library of Congress, 1940*, pp. 158–159. It is cited in full, together with a facsimile reproduction of the Chinese text, in George Sarton, *Introduction to the History of Science*, vol. III, part II (Baltimore, 1948), pp. 1268–1270. Sarton's comment on the trustworthiness of the early Ming census system is particularly interesting.

4. The Chung-shu-sheng, which was the real Cabinet of the Yüan period and of the first twelve years of the Ming dynasty, was abolished in 1380 in consequence of the alleged treason of the Prime Minister, Hu Wei-yung. See *Ming-shih*, 72.1a and 308.1a–3b. For a brief discussion of the evolution of this office, see Wu Han, *Chu Yüan-chang chuan* (1949), pp. 166–167.

5. *Ta-Ming hui-tien* (1502 ed.), 20.14a; also the 1587 edition and various editions of *Ta-ch'ing hui-tien*.

6. *Hui-chou FC* (1502 ed.), 2.34b ff.

7. *Su-chou FC* (1379 ed.), 10, *passim*.

8. *HWHTK*, 13.2891.

9. Although some later local histories have preserved the totals of households and mouths for the early Ming period, the lack of relevant breakdowns prevents us from making a more systematic study of the late fourteenth century population data.

10. *Shao-hsing FC* (1586 ed.), 14.3a–7a.

11. *Fu-chou FC* (1613 ed.), 26.12–1b.

12. Wang Ch'ung-wu, "Ming-tai hu-k'o ti hsiao-chang," pp. 336–337.

13. *HWHTK*, 13.2892.

14. Chao Kuan, *Hou-hu chih*, the Board of Revenue manual on population first compiled in 1513. The copy used is a microfilm reproduction of the 1621 revised edition originally in the possession of the National Peiping Library. Chao's comment is in chapter 2. See also Liang Fang-chung, "Ming Yellow Registers," p. 166.

15. The statutes are given in *Hou-hu chih*, 4.1a–3b; also Liang "Ming Yellow Registers," pp. 159–160.

16. For an excellent summary of these illegal practices, see Liang, "The Single-Whip System," Introduction.

17. Liang, "Chronological Chart"—a masterly discussion of the general social and economic conditions during the late Ming period.

18. *Hou-hu chih*, 4.2a–2b.

19. *Shang-hai HC* (1814 ed.), 4.1a–1b, gives the number of ting, or male adults liable to labor services, as 81,961 for the period 1621–1644.

20. *Shao-hsing FC* (1586 ed.), 14.3a–7a.

21. *Mien-yang CC* (1531 ed., 1926 reprint), 9.1b–5b.
22. *Hua-chou chih* (1572 ed., 1882 reprint), 8.2a.
23. *Ch'in-an HC* (1535 ed.), 7.1b.
24. *Shun-te HC* (1585 ed.), 3.4b.
25. Chao I, *Nien-erh shih tsa-chi* (1947, Shih-chieh-shu-chu ed.), ch. 34, pp. 495–496.
26. The problem of landownership will be further discussed in Chapter IX.
27. Chou Chen's famous letter to the Board of Revenue, cited in Wang Ch'ung-wu, "Ming-tai hu-k'o ti hsiao-chang," pp. 372–373; also cited in *Ming-shih*, 77, and *HWHTK*, 2.
28. Chao I, *Nien-erh-shih tsa-chi*, ch. 32, pp. 473–474.
29. For an appraisal of the administration of the Ming T'ai-tsu period, see Wu Han, *Chu Yüan-chang chuan* (1949).
30. *Liu-yang HC* (1551 ed.), B.8a–9a.
31. Liang, "Chronological Chart," discussion on pp. 46–47.
32. Ho Ch'iao-yüan, *Min-shu* (1629 ed.), 39.2a.
33. *Shun-te HC* (1585 ed.), 3.1; also Ku Yen-wu, *THCKLPS*, 40.52a–53a, cites the 1633 edition of *Chao-ch'ing FC*.
34. *Ta-Ming hui-tien* (1502 ed.), 20.15a.
35. Hu Shih-ning, *Hu-tuan-min tsou-i* (Chekiang shu-chü ed.), 3.16a–16b; also *Shensi TC* (1542 ed.) 33.29b: "For every *chu-hu* (literally 'host household') in Szechwan there are several tens of *k'o-hu* ('guest households'); so that when one man answers the labor-service call, he is actually assisted by several tens of people. This is why in spite of a large number of services people do not feel burdened."
36. See Table 3.
37. *HWHTK*, 13.2893.
38. *Ibid.*; *Ta-Ming hui-tien* (1502 ed.), 20.15a–15b; *Hou-hu chih*, 4.9a–10b.
39. A sixteenth-century scholar-official's essay on *chün-chi* people, cited in *Chiang-ning HC* (1598 ed.), 3.28b–29b.
40. Wang Ch'ung-wu, "Ming-tai hu-k'o ti hsiao-chang," p. 363.
41. This was again Chou Chen's impression, cited in *Ming-shih*, 77.3a, and *HWHTK*, 13.2896.
42. Ku Yen-wu, *THCKLPS*, 9.64a and 33.123b–124a; Liang, "Chronological Chart," pp. 44–45.
43. *Ch'ang-shan HC* (1585 ed., printed in 1660), 8.62–6b.
44. *I-hsing HC* (1590 ed.), 4.3a–3b.
45. *Shao-hsing FC* (1586 ed.), 14.3a–7a.
46. *Hai-ning HC* (1557 ed., 1898 reprint), 2.1b–2b.
47. The ting figure is given in *HWHTK*, 13.2895. *Shun-t'ien FC* (1593 ed.), 3, *passim.*, only gives the number of households and mouths.
48. Both are originally the property of the National Peiping Library. They are the only available works of their kind outside China and Japan.
49. *Kiangsi fu-i ch'üan-shu* (1611 ed.), book on Jui-chou; also *Jui-chou FC* (1628 ed.), 10.2a–4a.
50. *Ch'ien-t'ang HC* (1609 ed., 1893 reprint), 1.22a.
51. Otto B. van der Sprenkel, "Population Statistics of Ming China," *Bulletin of the School of Oriental and African Studies, University of London*, vol. XV, part 2, 1953. The author is most impressed by the fact that the

average numbers of households per li for some provinces, prefectures, and counties agree with one another within a narrow range and that when the averages are multiplied by the actual numbers of li the totals agree within a narrow range with the totals of households of such provinces, prefectures, and counties. This has proved the consistency of the data themselves, but is far from proving their "remarkably high level of accuracy" as sources of demographic information. In a sense, his reasoning is caught in a self-drawn circle. For how could the provincial totals for households and li disagree with the sum totals of households and li of those subdivisions within their jurisdiction, when the former are computed from the latter? His study therefore is valuable only in providing raw figures, and falls far short of the goal of "interpreting in their fullest administrative context" the Ming population data. The weakness of his study lies in his failure to consult modern Chinese writings in the field of Ming economic and institutional history, particularly those of the leading authority, Liang Fang-chung.

52. Liang, "Ming Yellow Registers," p. 168.

CHAPTER II. The Nature of Ting.

1. *Nei-ko-ta-k'u hsien-ts'un Ch'ing-tai huang-ts'e mu-lu* (Peiping Palace Museum, 1936). In addition to the 6602 volumes preserved by the Palace Museum, the National University of Peking has more than 7,300 volumes and the Institute of History and Philology of the Academia Sinica has more than 1,900. See also *Ch'ing-nei-ko chiu-ts'ang han-wen huang-ts'e lien-ho mu-lu* (Peiping Palace Museum, 1947). It must be pointed out that, although during the period 1651–1740 the only population data available to modern students are the annual ting returns, different kinds of population figures were no doubt reported by various localities in *Ch'ao-chin hsü-chih ts'e*, a handbook that provided information about the locality for the reference of the court. It is not known whether this handbook was regularly compiled or compiled only on special occasions, but its compilation seems to have been an old practice. Some Ming local histories, for instance *Fu-chou FC* (1613 ed.), 30.1b–2a, record that the expense of compiling this handbook was appended to the local labor-service payment. In Ch'ing times Tung-yang county in central Chekiang, for example, though carrying a quota of 22,577 ting, which was first fixed in 1581, listed 26,571 households and 80,085 mouths in its 1671 *Ch'ao-chin hsü-chih ts'e*. The editor thoughtfully commented: "This is the sum total of all the local population, including the military rank and file, artisans, females and children and is therefore similar to what used to be called population in ancient times. It is entirely different from the ordinary fiscal population based on ting enumeration." (*Tung-yang HC* [1832 ed.], 2.6b.) Unfortunately, such handbooks do not seem to have been preserved in any significant number.

2. The writings on Ch'ing population are legion; we need only mention a few which have been influential: W. W. Rockhill, "An Inquiry into the Population of China," *Annual Report of the Smithsonian Institution*, 1905 (the above quotation is from p. 665); E. H. Parker, "Notes on Some Statistics Regarding China," *Journal of the Royal Statistical Society*, 1899; W. F. Wilcox, "A Westerner's Effort to Estimate the Population of China

and its Increases since 1650," *Journal, American Statistical Association*, 1930; also in the book edited by him, *International Migrations* (New York, 1931), vol. II, part 1; Ta Chen, *Population in Modern China*, published as a special monograph in *American Journal of Sociology*, part 2 (July 1946); and Otake Fumio, *Kinsei Shina keizaishi kenkyu* (Tokyo, 1942), pp. 271–282. Otake, like Rockhill and Wilcox, believes that the eighteenth-century population returns of China were largely guesswork and were magnified to please the Ch'ien-lung Emperor. The best, though not entirely satisfactory, study to date on Ch'ing population is Lo Erh-kang, "The Population Pressure in the Pre-Taiping Rebellion Years," *Chung-kuo she-hui-ching-chi-shih chi-k'an* (January 1949) (in Chinese).

3. Ta Chen, *Population.*

4. That the ting tax is not a poll tax becomes very obvious when we compare it with the poll taxes of the Former Han period. In Han times there were the *k'ou-fu* (poll tax on children under 14) and *suan-fu* (poll tax on adults between 15 and 56), in addition to the land tax, and compulsory labor and military services. See L. S. Yang, "Notes on Dr. Swann's *Food and Money in Ancient China*," *HJAS* (December 1950).

5. Li Fu, *Mu-t'ang ch'u-kao* (1831 reprint), 31.6a.

6. Our discussion of the Ming tax system is essentially based on the authoritative studies of Liang Fang-chung, particularly on his most useful article, "The Single-Whip System."

7. Hu Shih-ning, *Hu-tuan-min tsou-i*, 3.16b–17a.

8. *Wu-chin HC* (1605 ed.), 3.62b–63a.

9. Except in areas where the illegal practice of merging households for tax and labor-service evasion was common, as in Szechwan, Hupei, Hunan, and possibly some northern provinces (see Chapter I), the average household was likely to have only one adult male who tilled the land and bore the burden of labor services. It is interesting that from a few pages of the 1565 Yellow Register of a southern Fukien village, in the possession of the Library of Kyoto University, we find that out of 36 households, 35 had only one male adult tilling the land. These few pages are photographically reproduced as an appendix to Liang, "Ming Yellow Registers." What was true in southern Fukien may have held for other parts of the densely populated southeast. In their chapters on social customs, *Su-chou FC* (1379 ed.), *Wu-hsien chih* (1642 ed.), *Shao-hsing FC* (1586 ed.), and *Hsin-ch'ang HC* (1477 ed., printed in 1521) all testify to the old custom whereby grown brothers as a rule divided up the common property and lived separately even when their father was alive. All this bears out the fact that the clan was but a superstructure, while the most important factor determining the family unit was human labor.

10. *Shun-teh HC* (1585 ed.), 3.4a–4b.

11. *Yang-chou FC* (1685 ed.), 10.48a–48b.

12. *K'uai-chi HC* (1575 ed.), 6.3a–4b.

13. *Wu-chin HC* (1605 ed.), 3.7a and 3.31b–32a.

14. *Ch'ang-shan HC* (1584 ed., printed in 1660), 8, *passim.*

15. A late-Ming edition of the history of Hai-yen county in Chekiang, cited in *THCKLPS*, 32.6a–6b.

16. Such as Yang-chou in Kiangsu and a few counties in Kwangtung and Hunan. See Liang, "The Single-Whip System," and "Chronological Chart."

17. *THCKLPS*, 17.109a–112a; *Wu-chin HC* (1605 ed.), 3.83b, discussion of the initial difficulties encountered by the single-whip system in Shantung, by its well-informed editor, T'ang Ho-cheng. Also Liang, "Chronological Chart," pp. 37–38.

18. Cited in *THCKLPS*, 11.24b.

19. *Ibid.*, 17.109a–112a; Liang, "Chronological Chart," pp. 37–38.

20. For instance, in Honan approximately every 100 mou of land bore only four-tenths of a tael of ting tax, while an upper-grade household was assessed 1.2 taels. In some Shantung counties the land bore only between 10 and 20 per cent of the ting tax at the initial stage. See *THCKLPS*, 17.111a–111b; 20.161a–161b.

21. Such as Ts'ao-hsien of western Shantung, where approximately 40 mou of land bore the tax of one ting. *Ibid.*, 20.177a–184b.

22. *Ibid.*, 17.109a–112a.

23. *Ibid.*, 20.177a–184b.

24. Chang Yü-shu, *Chang-wen-chen-kung chi* (1772 ed.), 7.23b.

25. This was the general impression of high officials in the central government in 1741, when the ting assessment had been discontinued. *Ch'ing-kao-tsung shih-lu*, 133.5b–6a.

26. *CCWHTK*, 19.5027; *Ta-Ch'ing hui-tien* (1764 ed.), 9.2b.

27. *CCWHTK*, 19.5023; 21.5044.

28. Liang, "The Single-Whip System," pp. 43–44.

29. Ho Ch'ang-ling, ed., *HCCSWP*, 30.11a.

30. *Ma-ch'eng HC* (1877 ed.), 10.1a–1b.

31. *Tang-yang HC* (1866 ed.), 4.1a–1b.

32. *Lo-ch'uan HC* (1806 ed.), 9.1a–2b.

33. *Hunan TC* (CP ed.), 48.1289.

34. *Shang-hai HC* (1813 ed.), 4.1a. Both the late-Ming and the "original" Ch'ing quotas were at 81,961 ting.

35. Lu Lung-ch'i, *San-yü-t'ang wen-chi* (1868 reprint), 3.13a–15a.

36. Li Fu, *Mu-t'ang ch'u-kao*, 31.6a.

37. *Chekiang TC* (1735 ed., CP reprint), 71–74, *passim.*

38. *Lo-t'ien HC* (1876 ed.), 4.5a–6a.

39. *Ta-Ch'ing hui-tien* (1732 ed.), 30.1a. In the rest of the country the merger was completed in the early nineteenth century.

40. Given in E. H. Parker, "Notes." The *Ch'ing-shih-lu* and *Tung-hua-lu* are of course the sources. In view of the nature of the ting it is little use to collect the annual ting totals.

41. *Pao-ning FC* (Szechwan, 1826 ed.), 22.

42. *Ch'iung-chou chih* (Szechwan, 1818 ed.), 17.

43. *Chao-ch'ing FC* (Kwangtung, 1833 ed.), 3.16b–17a.

CHAPTER III. Population Data, 1741–1775.

1. The important exception is Ta Chen, who believes, entirely without foundation, that the population data before and after 1711 differs qualitatively as well as quantitatively.

2. Rockhill, "An Inquiry," p. 665; for his discussion on the rates of population increase during the eighteenth century, see p. 673.

3. Li Fu, *Mu-t'ang chu-kao*, 31.6a–7a; also *Yung-cheng chu-p'i yü-chih*, 8th *ts'e*, pp. 78a–78b. For the true nature of the pao-chia system, see K. C. Hsiao, "Rural Control in Nineteenth-Century China," *Far Eastern Quarterly*, February 1953; also his forthcoming book.

4. *Ch'ing-kao-tsung shih-lu*, 130.1a–3a.

5. *Ibid.*, 131.4b–5a.

6. *Ibid.*, 133.5b–6a.

7. *CCWHTK*, 19.5028–5029.

8. Liu Jui-t'u, *Tsung-chih che-min wen-hsi* (1672 ed.), 2.75a–77a testified that the pao-chia system was set up only in the mountainous border districts between Chekiang and Fukien, where disturbances were likely to occur. Chu Ch'ang tso (see his *Fu-che hsi-ts'ao* [1664 ed.], 1.75a–76b) acquired the impression while serving as governor of Chekiang in 1662–1664 that the pao-chia system was a mere formality in the province and never concerned itself with population canvassing. It seems that up to 1740 the pao-chia had functioned efficiently as a local policing force only under the close surveillance of some unusually energetic provincial officials. See Wang Shih-chün, *Li-chih hsüeh-ku pien* (preface dated 1723–1724, printed later by his son), A.14b–15a; and especially Yüan Mei, *Hsiao-ts'ang shan-fang wen-chi* (SPPY ed.), 27.12b–14b, which says that the pao-chia system was so successful under T'ien Wen-ching, governor of Honan in 1724–1728 and governor general of Honan and Shantung in 1728–1732, and Li Wei, governor and then governor general of Chekiang in 1725–1732, that wayfarers could travel at night without fear of highwaymen.

9. *Ta-Ch'ing hui-tien tse-li* (1764 ed.), 33.17a–17b.

10. Yang Hsi-fu, *Ssu-chih-t'ang wen-chi* (1806 ed.), 4.5a.

11. *Hu-nan wen-cheng* (1873 ed.), Ch'ing works, 2.18a–21b.

12. *Ch'ing-kao-tsung shih-lu*, 560.14a–15b.

13. Wang Hui-tsu, *Hsüeh-chih i-shuo* (1823 ed.), B.42a.

14. *Huang-Ch'ing tsou-i* (1805 ed., 1936 reprint), 51.1a–3b.

15. Yüan Mei, *Hsiao ts'ang shan-fang wen-ehi*, 15.3a–4b, mentions Ch'en's orders that local officials in Kiangsu should prepare the *hsün-huan-ts'e*, or the rotating registers, for double-checking the pao-chia population returns. The fact that Ch'en took such an energetic step seems to suggest that he may have been reprimanded for his suggestion that women and children be omitted from the pao-chia registration.

16. *Ibid.*

17. *Yung-p'ing FC* (1880 ed.), 45.

18. For example, *Ch'i-hsien chih* (Honan, 1788 ed.), 7.13a–15b, and *Heng-chou FC* (Hunan, 1774 ed., 1875 reprint), 13.

19. *Lin-shui HC* (1835 ed.), 2.3a–3b.

20. *Ch'i-shui HC* (1880 ed.), 14.2b–3a.

21. *P'ing-ting CC* (1790 ed.), ch. on "Food and Money."

22. *Ta-Ch'ing i-t'ung-chih* (1812 ed., CP reprint), Hsiang-yang prefecture.

23. Chang Yü-shu, president of the Board of Revenue in 1691–1711, stated that while the abstract of the Yellow Registers should give only the total of the registered ting in a locality, the Registers themselves should contain the figures for all the male and female adults and children of the taxpaying households. See *Chang-wen-chen-kung chi*, 7.23a–23b. However, there is

evidence that the methods used in compiling the Yellow Registers during the early Ch'ing period were by no means uniform. For example, *Hun-yüan CC* (Shansi, 1661 ed.), A.40b–41a, states that, in compiling the Yellow Registers during the early years of the Ch'ing, the female population was entirely excluded. It seems possible that the mid-eighteenth-century Lo-ch'uan Yellow Registers included not only the adults and children of the ting-taxpaying households but those of the land-tax payers as well, for the ting could not substantially exceed the quota of 1711, which was only 1,642.

24. *CCWHTK*, 19.5033.

25. *Ibid.*, 19.5032–5033

26. Wei Yuan, *Ku-wei-t'ang wai-chi* (1878 ed.), 6.5a.

27. *Ch'ing-nei-ko chiu-ts'ang han-wen huang-ts'e lien-ho mu-lu.*

28. *Ta-Ch'ing hui-tien shih-li* (1818 ed.), 233.8b–9a, gives a statute of 1810 in which the local officials were reminded not to exaggerate the number of people in need of famine relief, which sometimes appeared to exceed the total population in certain localities.

29. This is invariably the case in the initial years of each of the three subperiods in Ch'ing population data. In 1651 the national total for ting was a little over 10,000,000. It had jumped to over 19,000,000 by 1660. The increase in the registered population between 1741 and 1745 was 26,500,000. Likewise, the increase between 1776 and 1779 was also as large as 10,000,000. It is surprising that Rockhill and some other writers on Ch'ing population should have chosen the year 1651 as the datum.

CHAPTER IV. Population Data, 1776–1850.

1. *Ch'ing-kao-tsung shih-lu*, 993.17a–8a, 20b–22a; 995.15a–16b.

2. *Hu-pu tse-li* (1776 ed.), 3.1a–1b.

3. See Chapter III, note 8.

4. Chu Yün-chin, *Yü-ch'eng shih-hsiao lu* (1826 preface, 1873 ed.), 1.17a–21b; Wang Ch'ing-yün, *Hsi-ch'ao chi-cheng* (1898 ed.), 3.14a–16a.

5. M. Huc, *A Journey through the Chinese Empire* (New York, 1855), vol. II, p. 98.

6. The text of Yeh's original proclamation is given in Yang Ching-jen, *Ch'ou-chi pien* (preface 1824, 1879 ed.), 27.6a–15a, which also contains a brief account of the adoption of Yeh's method by various provinces; also in Hsü Tung, ed., *Pao-chia shu* (preface 1837, 1848 ed.), 2.1a–8a.

7. For the various forms of the pao-chia door placard, see Wen Chün-t'ien, *Chung-kuo pao-chia chih-tu* (CP, 1939), pp. 235–248.

8. The best source of information is Wang Feng-sheng, *Yüeh-chou ts'ung-cheng lu* (1823 ed.), *passim.* For a brief self-appraisal of Wang's success in pao-chia work in P'ing-hu county in Chekiang, see his *Sung-chou ts'ung-cheng lu* (preface 1825, 1826 ed.), p. 1a.

9. *Ta-Ch'ing hui-tien shih-li* (1818 ed.), 233.4a.

10. S. Wells Williams, *The Middle Kingdom* (New York, 1907), vol. II, p. 281. This book was first published in 1848.

11. *Fu-shun HC* (1777 ed., 1882 reprint), 48.4a–4b.

12. Tsung Chi-ch'en, *Kung-ch'ih-chai wen-ch'ao* (1851 ed.), 4.1a–6a.
13. Preface to Li's "New Regulations Concerning the *Pao chia* Organization in En-ssu Prefecture," cited in Hsü Tung, *Pao-chia shu*, 2B.20a–40a.
14. *Ta-yao HC* (1845 ed.), 3, *passim.*, gives detailed information for every subunit within the walled city and suburb and for every peripheral village, including the number of occupants of the odds-and-ends households. The returns for these subunits run to 48 double pages.
15. Sheng K'ang, ed., *Huang-ch'ao ching-shih-wen hsü-pien* (1897 ed.), 80.19a–24a.
16. Theodore Shabad, *China's Changing Map, A Political and Economic Geography of the Chinese People's Republic* (New York, 1956), p. 42; *Chung-kuo fen-sheng ti-t'u* (Ta-chung Book Company, 1956).
17. Shabad, *China's Changing Map*, pp. 39–48.
18. For a detailed account of the clash between the encroaching Chinese and the Miao in western Hunan and the successive slaughters of the Miao by Chinese government forces in the late eighteenth and early nineteenth centuries, see Tan Hsiang-liang, *Hu-nan Miao-fang t'un-cheng k'ao* (1882 ed.), *passim.*
19. For details see Chapter X.
20. Cited in Hsü Tung, *Pao-chia shu*, 2B.41a–50b.
21. *Kuang-chou FC* (1879 ed.), 70.
22. *Tung-kuan HC* (1911 ed.), 22.
23. Ho Ch'ang-ling, ed., *Huang-ch'ao ching-shih-wen pien* (1886 ed.), 82, *passim.*
24. *Ibid.*
25. Li Chao-lo, *Yang-i-chai wen-chi* (1878 ed.), 2; and *Feng-t'ai HC* (1814 ed.), 2.6a.
26. *Yin-hsien t'ung-chih* (compiled since the late 1930's, still uncompleted), "Geography," "Population," *passim.*
27. *Kuang-tung ching-chi nien-chien* (1941), ch. 3.
28. *Ch'ing-hsüan-tsung shih-lu, passim.* National population totals and brief explanations of regional omissions can be found at the end of each year.
29. Provincial breakdowns for the period 1786–1898, with certain gaps, based on the archives of the Board of Revenue, are given in Yen Chung-p'ing, *et al.*, eds., *Chung-kuo chin-tai ching-chi-shih t'ung-chi tzu-liao hsüan-chi* (Peking, 1955); Appendix prepared by Lo Erh-kang.
30. For a comparison of the pre-1850 and post-Taiping populations of these four provinces and detailed evidence on the loss of population in the lower Yangtze region, see Chapter X.
31. *CCHWHTK*, 25.7759–7760.
32. Wang Ch'ing-yün, *Hsi-ch'ao chi-cheng*, 3.16a.
33. Huc, *A Journey*, vol. II, p. 99. It must be pointed out, however, that the latest population data known to him were those given in the 1818 edition of the *Ta-Ch'ing hui-tien.*
34. Ta Chen, "Population in Modern China," in *American Journal of Sociology*, part ii, July 1946.
35. See Chapter I, notes 38, 39.
36. For a detailed account of immigration to Szechwan see Chapter VII.
37. *Shih-ch'üan HC* (Shensi, 1849 ed.), 2.
38. Hupei had the largest average household in 1812, but we have no

means of checking the youth percentages of the populations in Hupei counties, whose local histories give no relevant breakdowns.

39. *P'u-ch'i HC* (1864 ed.), 1.1b.

40. *An-lu HC* (1872 ed.), 2.51b.

41. *Ching-an HC* (1870 ed.), 13.35a.

42. Cited in Sheng K'ang, ed., *Huang-ch'ao ching-shih-wen hsü-pien*, 74.46a–47b.

43. *Ch'eng-pu HC* (1866 ed.), 4.49a–49b.

44. Huc, *A Journey*, vol. II, pp. 327ff., 338–339.

45. *Hsia-men chih* (1839 ed.), 15.13a–13b; *Chien-yang HC* (1874 ed.), 56.21a–23a.

46. Wang Shih-to, *I-ping jih-chi* (Peiping, 1935), pp. 13b, 25b, 28b, 29a–31a.

47. *Hsiao-kan HC* (1883 ed.), 5.5a.

48. *Chiu-chiang FC* (1873 ed.), 13.36a–36b.

49. Cited in Sheng K'ang, ed., *Huang-ch'ao ching-shih-wen hsü-pien*, 74.44a–45a.

50. *T'ung-an HC* (1929 ed.), 23.2b.

51. Hu Shih, *The Chinese Renaissance* (Chicago, 1935), ch. 5.

52. See Chapter X.

53. Surveys of five sample localities in Hopei conclude that the percentage of married males to males of marriageable age ranges from 66.7 to 71.9; that of married women to women of marriageable age ranges from 88 to 92.3. *Chung-kuo jen-k'o wen-t'i chih t'ung-chi fen-hsi* (Directorate of Statistics, 1946), pp. 37–38. The disproportion between male and female populations may in a sense be mitigated by the fact that women have a much higher marriage rate than men, the poorest of whom are likely to be doomed to remain single for life.

54. *Shih-liao hsün-k'an*, no. 21.

55. Ta Chen, "Population," p. 21. He regards the population of the K'unming Lake area, whose youth percentage is 33.7, as a "truly stationary population."

56. *CCHWHTX*, 25.7760.

57. Yen Chung-p'ing *et al.*, *Chung-kuo*, appendix prepared by Lo.

58. For a good brief account of Wang Lun's revolt in western Shantung in 1774, see Teng Chih-ch'eng, *Ku-tung so-chi ch'üan-pien* (Peking, 1955), pp. 97–98.

59. *CCHWHTK*, 25.7756.

CHAPTER V. Population Data, 1851–1953.

1. Mei Tseng-liang, *Po-hsien-shan-fang wen-chi* (1856 ed.), 2.1a.

2. Cited in Shen Nai-cheng, "On the Powers of the Viceroys and Governors of the Provinces in the Last Years of the Ch'ing Dynasty," *She-hui k'e-hsüeh* (Social Sciences, Tsing Hua University), vol. II, no. 2 (January 1937).

3. Lo Erh-kang, *Hsiang-chün hsin-chih* (CP, 1939), contains a very scholarly discussion of the vital institutional changes in the post-Taiping period. For post-Taiping finance one of the most detailed and useful studies

is Lo Yü-tung, *Chung-kuo li-chin shih* (CP, 1936), a two-volume monograph on the history of likin.

4. The most authoritative studies of the changes in the power of military command are Lo Erh-kang's *Hsiang-chün hsin-chih*, and his article, "Ch'ing-chi ping-wei-chiang-yu ti ch'i-yüan," *Chung-kuo she-hui-ching-chi-shih chi-k'an*, vol. V, no. 2 (June 1937).

5. Wen Chün-t'ien, *Chung-kuo pao-chia chih-tu* (CP, 1935), p. 262.

6. Ke Shih-chün, ed., *Huang-ch'ao ching-shih-wen hsü-pien* (1888 ed.), 68.9a–10a.

7. *Ibid.*, 68, *passim.* Also *Huang-ch'ao Tao Hsien T'ung Kuang tsou-i* (1902 ed.), 56, *passim.*

8. Ke Shih-chün, *Huang-ch'ao ching-shih weu hsü-pien*, 68.16b–18a.

9. *Sung-chiang FC* (1882 ed.), 14.22a–23b.

10. *Chü-yung HC* (1904 ed.), 4.4b–8a, gives detailed pao-chia regulations as of 1897.

11. *Hupei TC* (1921 CP ed.), 43.1227.

12. The post-1850 provincial totals in the Board of Revenue archives are given in Yen Chung-p'ing *et al.*, eds., *Chung-kuo chin-tai ching-chi shih-t'ung-chi tzu-liao hsüan-chi* (Peking, 1955), appendix compiled by Lo Erh-kang.

13. *Hunan TC* (1885 ed.), 49.37a.

14. On the severe population loss suffered by central and lower Yangtze provinces during the Taiping period, see Chapter X.

15. *Shensi t'ung-chih-kao* (1934 ed.), 31, introduction.

16. *Shansi TC* (1885 ed.), 65.22a, editor's comments.

17. Cited in *T'ung-ch'uan FC* (1897 ed.), 18.7b–9a.

18. Hsü Shih-ch'ang, ed., *Tung-san-sheng cheng-lüeh* (1911 ed.), 6.45a–45b.

19. *Huang-ch'ao ching-shih-wen hsü-pien*, 68, *passim.*

20. *Huang-ch'ao Tao Hsien T'ung Kuang tsou-i*, 56.6a.

21. Ibid., 9.25a–36a.

22. *Ch'ing-teh-tsung shih-lu*, 505.12a.

23. For a good brief account of the early history of the police system, see *Nei-cheng nien-chien* (1935), "Police Administration," ch. 6.

24. For a factual summary of the reforms of the late Ch'ing period, see M. E. Cameron, *The Reform Movement in China, 1898–1912* (Stanford University Press, 1931).

25. Wang Shih-ta, "The Min-cheng-pu Census of 1909–1911: A New Study Based on Recently Discovered Documents," *She-hui k'o-hsüeh tsa-chih* (Quarterly Review of Social Sciences), vol. III, no. 3 (September 1932) and vol. IV, no. 1 (March 1933).

26. *Ta-Ch'ing fa-kuei ta-ch'üan, 1901–1909*, "Book on Constitutional Affairs," ch. 6.

27. *Anhwei t'ung-chih-kao* (latest uncompleted ed.), "Book on Civil Affairs," chs. on "Police," *passim.*

28. *Nei-cheng nien-chien* (1935), part C, p. 418.

29. *Tung-san-sheng cheng-lüeh*, 6.45a–45b.

30. *Yin-hsien t'ung-chih*, compiled during the late 1930's and early 1940's, still uncompleted, "Geography," "Population," pp. 293a–294b.

31. *Hua-yang HC* (1934 ed.), 4.1a–7b.

32. Lou Yün-lin, *Szechwan* (Shanghai, 1941), pp. 48–61.
33. *Hsia-p'u HC* (1929 ed.), 9.1a, editor's comments.
34. *Fukien TC* (1942 ed.), 14, *passim.*
35. The 1911 Shanghai figure is from Wang Shih-ta, "Min-cheng-pu Census"; the 1932–33 figure is from *Nei-cheng nien-chien* (1935), C, 403–404.
36. *Nei-wu fa-ling chi-lan* (Peking, 1918), book 11, pp. 147–148.
37. *Ibid.*, book 11, p. 112.
38. Wang Shih-ta, "Min-cheng-pu Census."
39. *Nei-cheng nien-chien* (1935), "Population Administration," p. 404.
40. *Ti-fang tzu-chih ch'üan-shu* (Shanghai, 1929), a collection of statutes and laws concerning local self-government proclaimed by the Nationalist government, vol. II, p. 63.
41. *Nei-cheng nien-chien* (1935), vol. II, ch. 4; also Ch'eng Mao-hsing, *Hsien-hsing pao-an chih-tu* (Shanghai, 1936), chs. 3 and 4, *passim.*
42. *Nei-cheng nien-chien* (1935), vol. II, ch. 1, *passim.*
43. *Ibid.*, ch. 2.
44. *Ibid.*, ch. 1, p. 50.
45. *Ti-fang tzu-chih ch'üan-shu*, preface; also vol. II, p. 63.
46. Complete text given in Wen Chün-t'ien, *Chung-kuo pao-chia chih-tu*, appendix, pp. 547–549. Italics mine.
47. *Kiangsi nien-chien* (1937), part IV, pp. 97–99.
48. Ch'eng Mao-hsing, *Hsien-hsing pao-an chih-tu*, ch. 4; also Lang Ching-hsiao, *Pao-chia yün-tung chih li-lun yü shih-chi* (Shanghai, 1930), pp. 63–64.
49. Yang's report is cited in Ch'eng Mao-hsing, *Hsien-hsing pao-an chih-tu*, pp. 83–95.
50. *Nei-cheng nien-chien* (1935), "Population Administration," ch. 4.
51. *Ibid.*, ch. 5.
52. *Hu-pei-sheng nien-chien* (1937), "Population," pp. 106–109; Chang Hsiao-mei, *Kuei-chou ching-chi* (Shanghai, 1939), part D, p. 5, and *Kuei-chou-sheng t'ung-chi tzu-liao hui-pien* (Kweichow Provincial Statistical Bureau, 1942), p. 17.
53. For observations and comments on the actual working of the pao-chia system in various provinces up to 1935, see Wen Chün-t'ien, *op. cit.*, *Chung-kuo pao-chia chih-tu*, chs. 23–25, *passim.*
54. S. Krotevich, "Vsekitayskaya perepis' naseleniya 1953 g." (The All-China Population Census of 1953), *Vestnik Statistiki* (Journal of Statistics), no. 5, p. 39 (September-October 1955).
55. These are the only precensus official figures cited in the atlas, *Chung-hua jen-min kung-ho-kuo fen-sheng ti-t'u* (Ti-t'u ch'u-pan-she, 1953).
56. Y. L. Wu, *An Economic Survey of Communist China* (New York, 1956), p. 14.
57. Krotevich, "Vsckitayskaya," pp. 36–37.
58. Wu, *Economic Survey*, p. 14; *Chung-hua jen-min kung-ho-kuo fen-sheng ti-t'u*. It should be noted that the preface to the atlas is dated April 1953.
59. Krotevich, "Vsekitayskaya," pp. 36–37.
60. *Ibid.*, p. 40. This is so far the most systematic account of the census procedures. For an excellent condensation, see Theodore Shabad, "Counting 600 Million Chinese," *Far Eastern Survey*, April 1956.

61. Irene B. Taeuber and Leo A. Orleans, "A Note on the Population Statistics of Communist China," *Population Index*, October 1956, p. 274.

62. *Ibid*., p. 276.

63. Theodore Shabad, *China's Changing Map* (New York, 1956), p. 37.

64. *Jen-min jih-pao* (People's Daily), Peking, November 1, 1954, cited in Wu, *Economic Survey*, pp. 15–19.

65. The new regulations and comments by Lo Jui-ch'ing, Minister of Public Safety, are given in *Hsin-hua pan-yüeh-k'an*, 1958, no. 3, pp. 46–49.

66. See Chapter X.

67. Wu, *Economic Survey*, pp. 19–20.

CHAPTER VI. Population-Land Relation: Ming, Ch'ing, and Modern Land Data.

1. *Sung-shih* (SPPY ed.), 17.2a–2b; Ma Tuan-lin, *Wen-hsien t'ung-k'ao* (CP ed.), p. 61.

2. *Ch'ang-chao ho-chih* (1783–93 ed.), 3.6a.

3. T'ang Shun-chih, *Ching-ch'uan hsien-sheng wen-chi* (SPTK ed.), 9.29a–29b. The *Chou-li* principle, which was incorrectly paraphrased by T'ang, is: "Each household [should be alloted] 100 mou of nonfallow land, or 200 mou of once-fallow land, or 300 mou of twice-fallow land." During the feudal Chou period agricultural technique was relatively primitive and in the three-field system land, excepting the most fertile, had to be fallowed once or twice in a three-year cycle. *Chou-li* (*Shih-san-ching chu-shu* ed.), 10.7a–7b, text and commentary.

4. *Ming-shih* (SPPY ed.), 77.3b–4a; Ku Yen-wu, *Jih-chih lu* (SPPY ed.), 10.2a–3b.

5. *Yü-chou chih* (1546 ed.), 3.la–3b.

6. *Li-ch'eng HC* (1771 ed.), 4.6a. The "uniform large taxpaying land" was common in many parts of Shantung. See *Yen-chou FC* (1768 ed.), 13; *Lai-chou FC* (1740 ed.), 3; *I-chou FC* (1760 ed.), 8–9; *Tung-ch'ang FC* (1777 ed.), 8; and *Chi-nan FC* (1839 ed.), 14.

7. *Yung-ch'eng HC* (1840 ed.), 3.5a–5b.

8. Wang Ying-chiao, *Wang-ch'ing-chien-kung tsou-shu* (late Ming ed.), 8.15a.

9. *Wen-shang HC* (1608 ed.), 4, *passim*.; *Ssu-shui HC* (1661 ed.), 3.4b–5b.

10. *Chi-yang HC* (1763 ed.), 3.4a–7a.

11. *Sung-hsien chih* (1767 ed.), 19.1b–2a.

12. *Fen-chou FC* (1609 ed.), 5, *passim*.

13. The unusually well-documented *Lo-ch'uan HC* (1944 ed.), 14.1a–2b, has much information on the entire north central Shensi area.

14. *Ch'ing-kao-tsung shih-lu*, 8.20a–20b.

15. *Kansu TC* (1909 ed.), 16, *passim*.

16. Huang Tso, *Nan-yung chih* (1544 ed.), 1.36a–36b. Huang's exact date for the submission of the *Yü-lin t'u-ts'e* to the emperor is confirmed by *Ming-tai-tsu shih-lu*, 184.4a–4b.

17. The *Ming-shih* and *HWHTK* both err in saying that the survey team was dispatched in 1398, the very year in which the *Yü-lin t'u-ts'e* were

completed. It is a credit to Huang Tso's careful scholarship that no date for the dispatch of the survey team is given. Obviously it took considerable time for the team to prepare the survey maps and registers carefully.

18. *Chia-shan HC* (1894 ed.), 10.6a–7a; *Hsiu-shui HC* (1684 ed.), 3, appendix.

19. Feng Kuei-fen, *Hsien-chih-t'ang kao* (1876 ed.), 9, *passim.*; Ch'en Han-sheng, *Mu-ti ch'a-i* (CP, 1929), *passim.*

20. Cited as commentary on Ku Yen-wu, *Jih-chih lu chi-shih* (SPPY ed.), 10.3b. Yen's famous commentary is paraphrased from an important memorial by Liu Kuo-fu, cited in n. 40 below.

21. *Huai-an FC* (1748 ed., 1852 reprint), 12.26b–27a.

22. *Hai-chou chih* (1811 ed.), 15.10b–11a.

23. For a good brief criticism of modern Anhwei statistics on land and cultivated area, see Wu Ch'uan-chün, *Chung-kuo liang-shih ti-li* (CP, 1948), ch. 3.

24. *Nan-ch'eng HC* (1873 ed.), 3B.1a–1b.

25. *P'ing-hsiang HC* (1872 ed.), 3.16b–19a.

26. *CCHWHTK*, p. 7550.

27. *Chia-hsiang HC* (1869 ed.), 3.5b.

28. For Ning-hua county, see Liang, "Single-Whip System"; *P'u-t'ien HC* (1758 ed., 1879 reprint), 5.9b–10b; *Hsien-yu HC* (1770 ed., 1873 reprint), 4.4a–4b.

29. *Ku-t'ien HC* (1606 ed.), 4.21a–32b.

30. *Hui-chou FC* (1881 ed.), 14.11a–11b.

31. *Kwangtung TC* (1840 ed.), 162–164, *passim.* For example, Kuang-chou prefecture returned: "The total of public and private irrigated and non-irrigated land, hilly land, and ponds, etc. is *shui* (tax) 10,623,170.437 mou."

32. For example, *Ch'ing-kao-tsung shih-lu*, 194.3b.

33. *Ibid.*, 45.18b–19a.

34. Cited in *Anhwei TC* (1877 ed.), 69.13b–14a, and *Pao-ying HC* (1840 ed.), 8.4a–5a.

35. *Hsiu-shui HC* (1684 ed.), 3.24a–86b; *Chia-shan HC* (1800 ed.), 8, appendix.

36. *Li-ling HC* (1871 ed.), 1.24b.

37. *I-yang HC* (1874 ed.), 2.6b.

38. For Kwangsi, see *Kuang-hsi-sheng nung-ts'un tiao-ch'a* (Executive Yüan, CP, 1935), chs. 1–2; for Szechwan, see *Chung-kuo ching-chi nien-chien* (1935), F, p. 261.

39. *Ming-shih*, 77.3b–42; *HWHTK*, p. 2786.

40. *Ch'ing-kao-tsung shih-lu*, 134–170, *passim.* A summary of regulations on tax-exempt small plots for various provinces is given in *CCWHTK*, pp. 4884–4885.

41. *Ming-shih*, 77.42; *HWHTK*, p. 2792.

42. *Yung-cheng chu-p'i yü-chih*, 45.12b–21a.

43. *CCHWHTK*, p. 7512.

44. Cited in Ch'en K'uei-lung, ed., *Ch'ien-shih chi-lüeh hou-pien* (1911 ed.), 13.4b.

45. J. L. Buck, *Land Utilization in China* (Shanghai, 1937), vol. III, *Statistics*, table 7.

46. For a discussion of this fiscal principle, see L. S. Yang, "Notes on Dr. Swann's Food and Money in Ancient China," *Harvard Journal of Asiatic Studies*, December 1950, esp. pp. 527–528.

47. *Kuang-p'ing FC* (1894 ed.), 25; also *Kuang-p'ing HC* (1608 ed.), 1.11a.

48. Liang, "Chronological Table," introduction.

49. *Wen-shang HC* (1608 ed.), 4, *passim*. Its 1764 quota of taxed land is from *Yen-chou FC* (1768 ed.), 13.77b–78a.

50. Feng Kuei-fen, *Hsien-chih-t'ang kao*, 9.

51. Ch'en Han-sheng, *Mu-ti ch'a-i*, *passim*.

52. *Jen-min shou-ts'e* (Ta-kung-pao, 1952), p. 341.

53. *Ch'ang-shan HC* (1585 ed.), 8.6b–7a.

54. *Ti-cheng yüeh-k'an*, vol. IV, nos. 3–4, pp. 133–134.

55. Cited in Liang, "Single-Whip System."

56. *Ning-p'o FC* (1560 ed.), 13, *passim*.; and *Shao-hsing FC* (1586 ed.), 14, *passim*.

57. *K'uei-chi HC* (1575 ed.), 6.3a–4b.

58. *Ti-cheng yüeh-k'an*, vol. IV, nos. 2–3, p. 134.

59. For reminiscences on the land self-reporting movement in Kiangsu, see Ch'en Kuo-fu, *Su-cheng hui-i* (Taipei, 1951), pp. 37–38.

60. For a detailed explanation of the method used in land self-reporting and summarizing findings, see *An-hui-sheng Tang-t'u-hsien t'u-ti ch'en-pao kai-lüeh*, *Chiang-su-sheng Chiang-tu-hsien t'u-ti ch'en-pao kai-lüeh*, and *Chiang-su-sheng Hsiao-hsien t'u-ti ch'en-pao kai-lüeh*, all published as reports of the Local Taxation Reform Committee of the Ministry of Finance in 1935–36.

61. Fukien self-reported figures are given in *Chung-kuo t'u-ti wen-t'i chih t'ung-chi fen-hsi*; Szechwan figures are given in *Ssu-ch'uan-sheng t'u-ti hsing-cheng kai-k'uang* (Szechwan Bureau of Land Administration, 1940).

62. Wu Ting-ch'ang, *Ch'ien-cheng wu-nien* (Kweichow Provincial Government, 1943), pp. 29, 59.

63. *Kansu TC* (1909 ed.), 16, *passim*.

64. *The Manchoukuo Yearbook, 1942*, p. 419.

65. Han Ch'i-t'ung, *Chung-kuo tui-jih chan-shih sun-shih chih ku-chi* (Institute of Social Sciences, Academia Sinica, 1946).

66. Hsiao Ch'ien, *How the Tillers Win Back Their Land* (Peking, 1951), pp. 112–113.

67. Yen Chung-p'ing *et al.*, eds., *Chung-kuo chin-tai ching-chi-shih t'ung-chi tzu-liao hsüan-chi* (Peking, 1955).

68. See Chapter VIII, Section III.

CHAPTER VII. Population-Land Relation: Interregional Migrations.

1. *Shang-yüan HC* (1594 ed.), cited in *THCKLPS*, *ts'e* 10, p. 54a. This account of the forced migration of wealthy households to Nanking is more detailed and specific than that given in *Ming-shih*.

2. All the above facts about government-supervised migrations are summarized from *Ming-shih*, 77, *passim*; and *HWHTK*, 13, *passim*.

3. Fujii Hiroshi, "A Study of the Hsin-an (Hui-chou) Merchants, I," *Toyo Gakuho*, June 1953.

4. *Ming-shih*, 77.
5. Hsieh Chao-che, *Wu-tsa-tsu* (1795 Japanese ed.), 4.25a–25b.
6. T'an Ch'i-hsiang, "Chung-kuo nei-ti i-min-shih — Hu-nan-p'ien" (History of Internal Migrations in China, chapter on Hunan), *Shih-hsüeh nien-pao* (Historical Annual), vol. I, no. 4, 1932.
7. *Tzu-chung-hsien hsü-hsiu Tzu-chou chih* (1928 ed.), 8.6b.
8. See Chapter VI, section II.
9. Huang Hsün, ed., *Huang-Ming ming-ch'en ching-chi lu* (1549 ed.), 8.7b–8a, a famous collection of Ming discussions on various economic and administrative problems.
10. *Wu-tsa-tsu*, 4, *passim.*
11. Li Kuang-t'ao, "Chang Hsien-chung shih-shih" (Certain Facts about Chang Hsien-chung), *Bulletin of the Institute of History and Philology, Academia Sinica*, vol. 25, pp. 21–30 (June 1954).
12. *CCWHTK*, 1.4858.
13. *Ibid.*, 2.4865.
14. *Ibid.*, 2.4868.
15. *Hsin-fan HC* (1907 ed.), 5.2b.
16. *Chien-wei HC* (1937 ed.), "Ethnic Table," pp. 6a–6b, 51a–52a.
17. *An-hsien chih* (1932 ed.), 56.1a–2b.
18. *Tzu-chung-hsien hsü-hsiu Tzu-chou chih* (1928 ed.), 8, *passim.*, "Dialects."
19. Lou Yün-lin, *Szechwan* (Shanghai, 1941), ch. 2.
20. For example, *P'u-ch'i HC* (1868 ed.), 1.1b–2a, has preserved two poems depicting the dependence of a substantial portion of the local population upon earning their living in Szechwan and the human sorrows attending long absence from home. They were written by a *chü-jen* of 1747 and a *kung-sheng* of 1815. *Wu-ch'ang HC* (1885 ed.), Appendix, 2.19a, has preserved a similar song written at the beginning of the nineteenth century.
21. *CCWHTK*, 1.4860.
22. *Ibid.*, 3.4873.
23. For land surveys in Szechwan in the 1730's, see *ibid.*, 3.4776-4777; and *Yung-cheng chu-p'i yü-chih* (undated late Ch'ing ed.), *ts'e* 34.
24. Yen Chung-p'ing *et al.*, eds., *Chin-tai chung-kuo ching-chi-shih t'ung-chi tzu-liao hsüan-chi* (Peking, 1955), Appendix.
25. *San-t'ai HC* (1814 ed.), 8.1a.
26. *Baron Richthofen's Letters, 1870–1872* (2nd ed., Shanghai, 1903), pp. 177–178.
27. Wei Yüan, *Ku-wei-t'ang wai-chi* (1878 ed.), 6.5a–6b.
28. T'an, "Internal Migrations in China."
29. For example, Shao-yang, capital city of Pao-ch'ing prefecture, a well-known emigrating area in Ch'ing times, had 8 prominent Ch'ing immigrant clans, and Ju-ch'eng, at the southeastern corner of the province, had no less than 43 Ch'ing immigrant clans listed. See *Shao-yang-hsien hsiang-t'u-chih* (1907 ed.), 2, *passim*; and *Ju-ch'eng HC* (1932 ed.), 29, *passim.*
30. *Ling-hsien chih* (1873 ed.), 7.2a–3a.
31. *Wu-ling HC* (1863 ed.), 16.45a; and 1867 edition, 29, *passim.*
32. *Yu-hsien chih* (1871 ed.), 7.6a–6b.
33. *Ju-ch'eng HC* (1932 ed.), 18.4b–5a.
34. *Tao-chou chih* (1877 ed.), 3.5a

35. Fu Chiao-chin, *Hu-nan ti-li chih* (Ch'ang-sha, 1933), pp. 207–212, 586–587, 342–351, 641.

36. Cited in *Yüan-chou FC* (1874 ed.), ch. 9, part 7, p. 73b.

37. For further detail on maize and sweet potatoes, see Ping-ti Ho, "The Introduction of American Food Plants into China," *American Anthropologist*, April 1955; and "American Food Plants in China," *Plant Science Bulletin*, January 1956.

38. *CCWHTK*, 19.5027.

39. Cited in *P'ing-hsiang HC* (1784 ed.), 2.66b.

40. *Yü-shan HC* (1873 ed.), 1B.24b.

41. *Wu-ning HC* (1870 ed.), 8.3a–4b.

42. Tan Hsiang-liang, *Hu-nan Miao-fang t'un-cheng k'ao* (1882 ed.), chs. 3–4, *passim*.

43. For more systematic accounts of the development of mountains in Hunan, see *Hunan TC* (1757 ed.), 50.1b; *Ch'ang-sha FC* (1747 ed.), 36.1b–2b; *Yüeh-chou FC* (1763 ed.), 10.8b; *Ch'en-chou FC* (1765 ed.), 15.5a–5b; *Feng-huang-t'ing chih* (1758 ed.), 12.1b; *Lung-shan HC* (1818 ed.), 8.5b–6a; *Yung-chou FC* (1825 ed., 1866 reprint), 5A.12b–13a; *Pao-ch'ing FC* (1849 ed.), appendix, 2.4a; *Ching-chou chih* (1837 ed.), 11.28a–30a; *Yüan-chou* FC (1790 ed.), 20.2b (Yüan-chou of Hunan, not Kiangsi).

44. *An-chi HC* (1750 ed.), *Wu-k'ang HC* (1748 ed. and 1829 ed.), cited in *Hu-chou FC* (1874 ed.), 32.3b, 10a, 14b.

45. *Hui-chou FC* (1827 ed.), 4B.43a.

46. *Wei-yüan-t'ing chih* (1839 ed.), 3.50a–51b, cites the imperial decree of 1826, which ordered mountain farmers to be brought under the pao-chia.

47. Lin Tse-hsü, *Lin-wen-chung-kung cheng-shu* (1879 ed.), 3rd series, 18a–22a.

48. *Ch'ing-yüan FC* (1828 ed.), 8.14a.

49. Cited in *Tsun-i FC* (1841 ed.), 16; and *Li-p'ing FC* (1892 ed.), 30.

50. *Wu-ning HC* (1870 ed.), 8.3a–3b, 27.17a, 22.20a.

51. On soil erosion, see *Hui-chou FC* (1827 ed.), 4B.42a–43a; *Ch'ang-hua HC* (1822 ed.), 3.6a–6b. On the silting of rivers and lakes and more frequent inundation, see Wang Feng-sheng, *Che-hsi shui-li pei-k'ao* (1823 ed., 1878 reprint), 1.11b–12b, 1.28b–29b, 4.4a.

52. T'ao Chu, *T'ao-wen-i-kung ch'üan-chi* (late Ch'ing ed.), 26.1a–3b, 26.7a–8a. The aboriginal and Chinese mountain farmers of Kweichow had wisely planted maize in between young *Cunninghamias* in order to prevent soil erosion. It seems that southwestern mountain farmers had had more experience in cultivating mountains than eastern mountain tenants. For a description of an ingenious method of growing maize in Kweichow, see *Li-p'ing FC* (1845 ed.), 12.3a.

53. See Chapter X, section III.

54. *I-ch'ang FC* (1864 ed.), 11.24b.

55. *Fang-hsien chih* (1866 ed.), 11.9a–9b.

56. *Ch'ang-lo HC* (1870 ed.), 16.14b.

57. *Shensi TC* (1735 ed.), 42.5b, cites the 1694 edition of *Shang-yang HC*

58. *Fang-hsien chih* (1866 ed.), 11.11a.

59. *Tzu-yang HC* (1843 ed.), 3.12b–13b.

60. *I'ch'ang FC* (1864 ed.), 11.6b, and *Ho-feng CC* (1822 ed.) 6.1b.

61. Lin Tse-hsü, *Lin-wen-chung-kung cheng-shu*, 2nd series, 2. Local histories of this area containing accounts of soil erosion and frequent floods are too numerous to be listed. This phenomenon was found in southeastern Szechwan as well. See *Shih-chu-ting chih* (1843 ed.), 6.4b.

62. *Tang-yang HC* (1867 ed.), 4.

63. *Kuang-hua HC* (1882 ed.), 3.

64. *Chu-hsi HC* (1869 ed.), 6.2a, cites the 1829 edition and adds new remarks.

65. *Ch'ang-lo HC* (1870 ed.), 16.14b. The majority of immigrants were not from other provinces but from lowland districts of Hupei.

66. *Shih-ch'üan HC* (1849 ed.), 2.25a–25b.

67. *Hsün-yang HC* (1870 ed.), 11.13b–14a.

68. *Hsiao-i-t'ing chih* (1883 ed.), 3.2a.

69. See Chapter X, section III.

70. Cited in *Huang-ch'ao ching-shih-wen t'ung-pien* (Shanghai, 1901), 10.11a–12a.

71. *An-lu HC* (1872 ed.), B.64a.

72. Baron von Richthofen, *Letter on the Provinces of Chekiang and Nganhwei* (Shanghai, 1872), p. 12.

73. *Yü-hang hsien-chih-kao* (1906 ed.), ch. on land tax and population, p. 3b.

74. *Chiang-su chien-fu ch'üan-an* (1865 ed.), *passim*; *CCHWHTK*, 3, *passim*.

75. *Chü-yung HC* (1904 ed.), 4.8a–8b.

76. *Huang-ch'ao Tao Hsien T'ung Kuang tsou-i* (1902 ed.), 29.20a–20b.

77. *Huai-an FC* (1884 ed.), 2.4b.

78. Baron von Richthofen, *Reports of Hunan, Hupeh, Honan and Shensi* (Shanghai, 1870), p. 23.

79. *Kuang-shan HC* (1936 ed.), 1.1a–1b.

80. *Shang-chiang liang-hsien chih* (1874 ed.), 7.7b.

81. *Chü-yung HC* (1904 ed.), 6B, "Social Customs." The novel and intelligent method or rice culture introduced by Honan immigrants is described in *Liu-ho-hsien hsü-chih-kao* (1920 ed.), 14.2a–3a.

82. Information supplied the author by Dr. Hu Shih.

83. *Kuang-teh CC* (1881 ed.), 24.15b.

84. *Ch'u-chou chih* (1897 ed.), 2A.1a.

85. *Meng-ch'eng HC* (1915 ed.), 12.4b; and *Su-chou chih* (1889 ed.), 35.23a.

86. *Yü-i HC* (1903 ed.), 16.68b; and *Ch'üan-chiao HC* (1920 ed.), 4.1b.

87. *Hang-chou FC* (compiled between 1879 and 1919, 1923 ed.), 57, *passim*; and *Chia-hsing HC* (1906 ed.), 11.48a.

88. For details, see Chapter X, section III.

89. The permanent decline of Su-chou, which in the Ming and Ch'ing produced more men of letters and advanced degree-holders than any other prefecture, is the subject of an excellent modern study by P'an Kuang-tan, "Chin-tai Su-chou ti jen-ts'ai" (Su-chou as an Aristogenic Center), *She-hui k'o-hsüeh* (Social Sciences, Tsinghua University), vol. I, no. I, 1935.

90. *Hsin-teng HC* (1922 ed.), 10.1a.

91. Hsieh Hsing-yao, *Ch'ing-ch'u liu-jen k'ai-fa tung-pei shih* (Shanghai, 1948), appendix.

92. One of the best historical studies of the early colonization of Manchuria is Liu Hsüan-min, "Ch'ing-tai tung-san-sheng ti i-min yü k'ai-k'en" (The Migrations to and Colonization of Manchuria during the Ch'ing Period), *Shih-hsüeh nien-pao*, vol. II, no. 5, 1938. My brief historical account of Manchurian colonization is based substantially on Liu's well-documented article.

93. Hsü Shih-ch'ang, ed., *Tung-san-sheng cheng-lüeh* (1911 ed.), 6, *passim.*

94. Sir Alexander Hosie, *Manchuria, Its People, Resources and Recent History* (New York, 1904), p. 122.

95. For a brief discussion of natural conditions in Manchuria, see E. E. Ahnert, "Manchuria as a Region of Pioneer Settlement," in *Pioneer Settlement Cooperative Studies* (New York, 1932). On crops, see *Fengtien TC* (1934 ed.), *Kirin TC* (1891 ed.), *Heilungkiang tung-chih-kao* (1932 ed.), chs. on crops and products.

96. *The Soya Beans of Manchuria* (Chinese Maritime Customs, Special Series, no. 31, 1911).

97. T. H. Shen, *Agricultural Resources of China* (Ithaca, 1951), ch. 27 and appendix, table 2.

98. On emigrating communities of Shantung, see Walter Young, "Chinese Immigration and Colonization in Manchuria," *Pioneer Settlement Cooperative Studies.*

99. A good account of this changing attitude among emigrants is given in *Ch'ing-ho HC* (1934 ed.), 3.

100. *Ch'ang-ke HC* (1931 ed.), preface.

101. F. C. Jones, *Manchuria since 1931* (London, 1949), pp. 206–207.

102. A. J. Grajdanzev, "Manchuria as a Region of Colonization," *Pacific Affairs*, March 1946.

103. Lan Ting-yüan, *Lu-chou ch'u-chi* (1732 ed.), 11.33b–34a.

104. *Lu-chou tsou-shu* (1732 ed.), 1.5b.

105. Ting Yüeh-chien, *Chih-T'ai pi-kao lu* (1867 ed.), 2.73a, 78b.

106. *Fukien TC* (1829 ed., 1871 reprint), 48.

107. Chou Yin-t'ang, *T'ai-wan chün-hsien chien-chih shih* (Chungking, 1943).

108. Lien Heng, *T'ai-wan t'ung-shih* (Shanghai, 1947), p. 109.

109. James W. Davidson, *The Island of Formosa, Past and Present* (New York, 1903), p. 398.

110. *Ibid.*, p. 591.

111. This section on the Hakkas is based exclusively on Lo Hsiang-lin, *K'e-chia yen-chiu tao-lun* (Hsing-ning, Kwangtung, 1933).

112. On the great drought-famine of 1877–78, see Chapter X, section II.

113. *Shensi t'ung-chih-kao* (1934 ed.), 31, *passim.*

114. See Chapter X, section III.

115. Ta Chen, *Chinese Migrations, with Special Reference to Labor Conditions* (Washington, D.C., 1923); H. F. MacNair, *The Chinese Abroad* (New York, 1933); Victor Purcell, *The Chinese in Southeast Asia* (London, 1951).

116. *Ssu-hui HC* (1896 ed.), I, "Social Customs"; *Chia-ying CC* (1898 ed.), 8.54a–55b; *K'ai-p'ing HC* (1933 ed.), 1.20a–21b; *En-p'ing HC* (1934 ed.), 4.4b–5a.

117. C. F. Remer, *Foreign Investments in China* (New York, 1933), ch. x.

118. Yao Tseng-yin, *Kuang-tung-sheng ti hua-ch'iao hui-k'uan* (Institute of Social Sciences, Chungking, 1943).

119. Wu Ch'eng-hsi, "A Survey of Financial Conditions in Amoy with Special Reference to Overseas Chinese Remittances," *She-hui k'e-hsüeh tsa-chih* (Quarterly Journal of Social Sciences), vol. 8, no. 2, June 1937.

CHAPTER VIII. Land Utilization and Food Production.

1. Ta Chen, *Population in Modern China*, a special monograph in *American Journal of Sociology*, part 2, July 1946.

2. For a brief discussion of the agricultural improvements during the Ming period, see Liang, "Chronological Chart," p. 41.

3. *Ch'i-min yao-shu* (SPPY ed.), 2.9b.

4. For full documentation and detailed treatment, see Ping-ti Ho, "Early-Ripening Rice in Chinese History," *Economic History Review*, December 1956. Most of the statements in this section will not be documented.

5. There are a number of good studies on Sung population by Japanese scholars. The more balanced views are presented in Kato Shigeshi's three articles, all conveniently reprinted in his posthumous *Studies in Chinese Economic History* (Toyo Bunko, 1953), vol. ii, pp. 317–403.

6. Matteo Ricci, *China in the Sixteenth Century: Journals of Matteo Ricci, 1583–1610* (New York, 1953), p. 12.

7. Lin's preface to Li Yen-chang, *Chiang-nan ts'ui-keng k'o-tao pien* (preface dated 1834; 1888 ed.).

8. Michael Hagerty, "Comments on Writings concerning Chinese Sorghums," *Harvard Journal of Asiatic Studies*, 1940.

9. Only millet was indigenous to China. See N. I. Vavilov, *Selected Writings of N. I. Vavilov*, translated by K. S. Chester, *Chronica Botanica*, xiii, nos. 1–6, 1949–50.

10. Major regional food crops are mentioned only very briefly in *Yü-kung*, one of the earliest geographic treatises, probably compiled during the fourth or third century B.C.; *Chou-li*, the famous Confucian classic on rites and institutions, probably compiled in early Han times; the well-known chapter 129 on ancient capitalists and industrialists in *Shih-chi*; and the chapter on geography in *Han-shu*. More specific although also incomplete information is available in two early regional geographic works: *Yüeh chüeh-shu* (SPPY ed.), 4.3b, one of the earliest of extant treatises on Chekiang, probably compiled by Yüan K'ang during the early first century A.D., has preserved some valuable material of the fifth century B.C., in which wheat, barley, and millet were said to have been grown in the modern Shao-hsing area. Hua Chü, *Hua-yang kuo-chih* (SPPY ed.), 3.9b, 3.11a, 4.7a, a comprehensive geographical treatise on Szechwan compiled during the late third century A.D., mentions that all five major cereal crops were grown in Szechwan.

11. During the Sung, when double-cropping was not common in most parts of the rice area, Szechwan was said to grow three or four crops a year. *Sung-shih* (SPPY ed.), 89.12b.

12. Translation in L. S. Yang, "Notes on the Economic History of the Chin Dynasty," *Harvard Journal of Asiatic Studies*, June 1946, pp. 163–164.

13. For example, Kuo Wen, a native of Lo-yang, settled in a hilly district in northwestern Chekiang as a hermit, practicing wheat and legume farming and helping needy neighbors with his surplus produce. See Lü Ssu-mien, *Liang-Chin nan-pei-ch'ao shih* (Shanghai, 1948), p. 1076.

14. For example, *Wu-hsing chih* (1201 ed., 1914 reprint), 20.3b: "During the Liang period (502–556) a prefect of Wu-hsing, Chou Min, exhorted the people to grow wheat."

15. Sung-shih, 173.3a, mentions only that the edict was issued during the period 973–983 without giving the exact year.

16. *Ibid.*, 173.12a–12b. The decrees are dated 1171, 1179, and 1215.

17. Ch'iu Chün, *Ta-hsüeh yen-i pu* (1931 reprint), 24.9b.

18. Various local histories of southern Kiangsu and Chekiang; also *Chung-kuo ching-chi nien-chien* (CP, 1935), part VII, "Land Tenure."

19. Li Yen-chang, *Chiang-nan ts'ui-keng k'o-tao pien*, Lin's preface.

20. *Sung-shih*, 173.12b.

21. *Yung-cheng chu-p'i yü-chih*, *ts'e* 54, p. 75b.

22. *Ibid.*

23. *Pa-min t'ung-chih* (1490 ed.), 25, *passim*.

24. Liu Hsün, *Ling-piao lu-i* (TSCC ed.), B.9.

25. Hsü Kuang-ch'i, *Nung-cheng chüan-shu* (1843 ed.), 25.15b.

26. Hsü Kuang-ch'i, *Hsü-wen-ting-kung chi* (Shanghai, 1933), pp. 3–5.

27. *Yung-cheng chu-p'i yü-chih*, *ts'e* 54, p. 75b.

28. In the 1930's Hupei ranked after Shantung, Honan, Kiangsu, and Hopei. *Chung-kuo ching-chi nien-chien* (1935), part V, pp. 53–54.

29. Fang Kuan-ch'eng, *Fang-ch'üeh-min-kung tso-i* (1851 ed.), 1.18a–19a; 5.42b–44a.

30. Ch'en Hung-mou, *P'ei-yüan-t'ang ou-ts'un kao* (late eighteenth century ed.), 40.1a–2a, 30a–31b; 41.3a–4a; 46.24a–25b.

31. *Yung-cheng chu-p'i yü-chih*, 25.64a.

32. *Ch'en-chou FC* (1746 ed.), 11.5a.

33. *Wu-yang HC* (1834 ed.), 6.8a.

34. *Chen-chiang FC* (1685 ed.), 42.2a; *Tan-tu HC* (1879 ed.), 17.2a.

35. The introduction of millet into lakeside Hunan was first recorded by a Sung scholar, Chang Shun-min, *Hua-man lu* (*Pai-hai* ed.), 1.28a. It has been made known more widely in modern times by Wu Ch'i-chün, *Chih-wu ming-shih t'u-k'ao* (CP reprint), pp. 16–17.

36. *Hsiang-t'an HC* (1553 ed.), A.14b; *Ch'ang-teh FC* (1813 ed.), 18.3b.

37. *Kuang-hsin FC* (1783 ed.), 2.67a; *Hsing-tzu HC* (1871 ed.), 1.15a; *Po-yang HC* (1749 ed.), 7.20b; *Yü-kan HC* (1823 ed.), 6.7b.

38. Hagerty, "Chinese Sorghum."

39. *Ch'i-min yao-shu*, last chapter.

40. Vavilov, *selected writings*.

41. P. T. Ho, "The Introduction of American Food Plants into China," *American Anthropologist*, April 1955.

42. The importance of the India-Burma-Yunnan route in world cultural history has yet to be appreciated by scholars. For an interesting discussion, see Professor Carrington Goodrich's chapter, "Trades Routes to China,"

in J. Labatut and W. J. Lane, eds., *Highways in Our National Life* (Princeton University Press, 1950).

43. Lo Yüan, *Hsin-an chih* (1175 ed., 1888 reprint), ch. 2.

44. See P. T. Ho, "Introduction of American Food Plants," and "American Food Plants in China," *Plant Science Bulletin* (Botanical Society of America), January 1956.

45. While local histories are too numerous to be listed, an especially good account of the peanut in Kwangtung and Yunnan during the late eighteenth century is T'an Ts'ui, *Tien-hai yü-heng chih* (preface dated 1804, TSCC ed.).

46. The earliest reference to the peanut in northern local histories is found in a Shensi county, *Hua-chou chih* (1684 ed.), 2.44b. The other early eighteenth-century references are found in *Yung-p'ing FC* (1711 ed.), cited in *Chi-fu TC* (1732 ed., CP reprint), 73.3085, and in *Lin-ch'ing CC* (1749 ed.), 2.32a.

47. Cited in Li Chia-jui, *Pei-p'ing feng-su lei-cheng* (CP, 1937), vol. i, p. 210.

48. There is a particularly good account of the late dissemination of peanuts in *T'ai-an HC* (1929 ed.), 4.30b.

49. *Min-shu* (1629 ed.), 130.4b–6b; *Nung-cheng ch'üan-shu*, 27.29b–30a.

50. *Min-hsien hsiang-t'u chih* (date not clearly indicated, circa. 1903), *ts'e* 4, p. 323a.

51. *T'ai-ts'ang CC* (1629 ed.), 5.33a–35b.

52. Hsü Yu-ch'ü, *Chung-shu p'u* (1834 Korean ed.), *passim.* Although he states that the sweet potato was first introduced into southern Korea in 1825, the fact remains that it was not widely disseminated in Korea until the 1830's. His statement needs further checking because it seems unlikely that the plant would have taken so long to be introduced into this hermit kingdom. His best authorities on the methods of growing sweet potatoes are Ho Ch'iao-yüan and Hsü Kuang-ch'i.

53. It is regrettable that many early Ch'ing Szechwan local histories, following the example of Szechwan provincial histories, overlooked grain crops completely, while giving profuse descriptions of special products.

54. Wang Shih-to, *I-ping jih-chi* (Yenching University Press, 1935), 3.26b–28a.

55. Maize is not recorded in *Ch'üan-liao chih* (1566 ed.), 4, the first history of Manchuria, but is recorded in *Sheng-ching TC* (1684 ed., printed in 1711), 21.2a.

56. In addition to *Kung-hsien chih* (1555 ed.), other early references to maize in northern local histories are found in: *Yüan-wu HC* (1595 ed.), A.24b (Honan); the 1684 edition of *Yen-ling HC*, cited in *Yen-ling wen-hsien chih* (1862 ed.), 10.1a–1b (Honan); *Li-ch'eng HC* (1640 ed.), 5.28b (Shantung), and *Chao-yüan HC* (1660 ed.), 5.1b (Shantung). While mention of maize has not yet been found in seventeenth-century Hopei local histories, the 1732 edition of *Chi-fu TC* mentions that maize was grown in Jehol. It seems likely that a number of early Ch'ing northern local histories may have failed duly to record maize, but the fact that maize is referred to in so few seventeenth-century northern local histories indicates that maize was not extensively grown in the north until comparatively late.

57. *Shansi TC* (1892 ed.), 100.18b–19a; *Kansu TC* (1909 ed.), 12.2a; *Chahar-sheng TC* (1935 ed.), 8.29a; *Feng-t'ien TC* (1934 ed.), 109.25b.

58. J. L. Buck, *Land Utilization in China*, vol. ii. p. 271. His estimates and field data are adopted by Yen Chung-p'ing *et al.*, *Chung-kuo chin-tai ching-chi-shih tzu-liao hsüan-chi* (Peking, 1955), p. 359.

59. Chou Shun-hsin, "Financing the Economic Development of Manchuria, 1900–1945," MS. paper read at the Conference on the Modern Chinese Economy sponsored by Harvard University in September 1956.

60. *Kuang-hsi-sheng san-shih-i-nien-tu liang-shih tseng-ch'an shih-shih chi-hua kang-yao* (Kwangsi Provincial Government, 1942); also Jao Yung-ch'un, *Liang-shih tseng-ch'an wen-t'i* (Chungking, 1942).

61. *Hsin-hua pan-yüeh-k'an*, 1956, no. 2, pp. 74–75. On recent government efforts to increase the acreage under maize and sweet potatoes in the country in general, see Hua Shu, *Wo-kuo ti-i-ke wu-nien-chi-hua chung ti nung-yeh tseng-ch'an wen-ti* (Shanghai, 1956), and Ho Wei, *Wo-kuo tang-ch'ien ti liang-shih cheng-ts'e* (Peking, 1955).

62. Sung Ying-hsing, *T'ien-kung k'ai-wu* (1637 ed., 1919 reprint), 1.1b, 1.6a.

63. The six northern provinces are Hopei, Shantung, Honan, Shansi, Shensi, and Kansu. The six southern provinces are Kiangsu, Chekiang, Anhwei, Kiangsi, Hupei, and Hunan.

64. Liang Fang-chung, "International Trade and the Silver Movements in the Ming Dynasty," *Chung-kuo she-hui-ching-chi-shih chi-k'an*, vol. 6, no. 2, December 1939.

65. All the figures for cereal outputs during 1931–1937 are from T. H. Shen, *Agricultural Resources of China* (Cornell University Press, 1951), Appendix, table 2.

66. *China's Foreign Trade, 1932* (Chinese Maritime Customs), pp. 48–49. Note that the figures on rice imports for 1932, 1931, and 1930 given there slightly exceed those given in the above table.

67. Wu Ch'uan-chün, *Chung-kuo liang-shih ti-li* (Shanghai, 1948).

68. For a critical study of Communist Chinese agricultural statistics, see Alexander Eckstein and Y. C. Yin, "Mainland China's Agricultural Product in 1952," manuscript for private circulation, Russian Research Center, Harvard University.

69. *Chung-hua jen-min kung-ho-kuo fa-chan ti-i-ko wu-nien-chi-hua ti ming-tz'u chien-shih* (Peking, 1955).

70. *Hsin-hua pan-yüeh-k'an*, 1956, no. 15, pp. 47–50.

71. *Ibid.*, 1956, no. 20, pp. 11–12. For a more systematic and popular account for Western readers, see Sun Ching-chih, "Food Resources and Population Growth," *People's China*, 1956, no. 10, pp. 4–10.

CHAPTER IX. Other Economic and Institutional Factors.

1. On the rise of money economy in China, see Ch'üan Han-sheng, "Chung-ku tzu-jan ching-chi" (Natural Economy in Medieval China), *Bulletin of the Institute of History and Philology, Academia Sinica*, vol. X, and his "T'ang-Sung cheng-fu sui-ju yü huo-pi ching-chi ti kuan-hsi" (The

Relationship between Public Revenue and Money Economy during the T'ang and Sung Periods), *ibid.*, vol. XX.

2. During the period 1573–1644 the total Mexican dollars imported into China through legal trade alone is estimated by a cautious modern economic historian to have definitely exceeded 100,000,000. Japan is estimated to have exported more than 70,000,000 taels of silver to China during the period 1601–1647. See Liang Fang-chung, "Ming-tai kuo-chi mao-i yü yin ti shu-ch'u-ju" (International Trade and Silver Movements during the Ming period), *Chung-kuo she-hui-ching-chi-shih chi-k'an* (Chinese Social and Economic History Review), vol. VI, no. 2.

3. Hsieh Chao-che, *Wu-tsa-tsu* (1795 Japanese ed.), 4.25b.

4. Fu I-ling, *Ming-Ch'ing shih-tai shang-jen chi shang-yeh tzu-pen* (Peking, 1956), chs. 3, 5.

5. *Ibid.*, ch. 4.

6. Shigeshi Kato, *Shina keizaishi kenkyū* (Studies in Chinese Economic History) (Tōyō Bunko, 1953), vol. II, pp. 557–584.

7. Ping-ti Ho, "The Salt Merchants of Yang-chou: A Study of Commercial Capitalism in Eighteenth-Century China," *Harvard Journal of Asiatic Studies*, vol. XVII, nos. 1, 2, June 1954.

8. W. C. Hunter, *The "Fan Kwae" at Canton before Treaty Days, 1825–1844* (Shanghai, 1911), p. 48.

9. *Hui-chou FC* (1827 ed.), 9–12, *passim.*

10. *She-hsien hui-kuan lu* (1834 second revised ed.), list of donators to the public cemetery for the poor people of She-hsien who died in Peking, pp. 12a–15a.

11. *Tien-yeh hsü-chih lu*, undated manuscript, a manual for pawnshop apprentices, written by a native of Hui-chou, preface.

12. Various nineteenth-century western Hupei local histories.

13. P. J. B. Du Halde, *A Description of the Empire of China and Chinese Tartary* (London, 1738), vol. I, pp. 333–334.

14. Wang Shih-mao, *Min-pu-shu* (TSCC ed.), p. 12.

15. Du Halde, *A Description*, I, p. 85.

16. Ch'en Mao-jen, *Ch'üan-nan tsa-chih* (TSCC ed.), A.7.

17. Ch'u Yüan, *Mu-mien p'u* (*Shang-hai chang-ku ts'ung-shu* ed.), p. 11b.

18. Du Halde, *A Description*, I, pp. 47–53.

19. Fujii Hiroshi, "A Study of the Hsin-an (Hui-chou) Merchants, I," *Toyo Gakuho*, June 1953. This first installment is a study of the interregional trade in general and is by far the most useful work of its kind.

20. Hsieh Chao-che, *Tien-lüeh* (early seventeenth-century ed.), 4.15b.

21. Fujii, "Hui-chou Merchants," pp. 23–24.

22. G. M. Trevelyan, *English Social History* (New York, 1946), p. 216.

23. Du Halde, *A Description*, I, pp. 80–81.

24. *Ibid.*, I, p. 73.

25. Yen Chung-p'ing, *Chung-kuo mien-fang-chih shih-kao* (Peking, 1955), ch. 2.

26. Yeh Meng-chu, *Yüeh-shih pien* (*Shang-hai chang-ku ts'ung-shu* ed.), 7.5a–6a.

27. H. B. Morse, *The Trade and Administration of China* (New York, 1920), pp. 309–310.

28. Hsü Kuang-ch'i, *Nung-cheng ch'üan-shu* (1843 ed.), 35.13a–13b.

29. Fujii, "Hui-chou Merchants," p. 15.

30. This generalization is based on Nishijima Sadao, "The Extension of Cotton Cultivation during the Ming Period," *Shigaku Zasshi*, vol. 57, nos. 4–5 (1948) and the present author's sampling of Ch'ing local histories.

31. *Kiangsi TC* (1880 ed.), 49; *T'ung-ch'üan FC* (1786 ed.), 3.7b; *Kao-chou* (1827 ed.), 3.63a; *Lien-chou FC* (1756 ed.), 7.1b; *Lei-chou FC* (1811 ed.), 2.49b; T'an Ts'ui, *Tien-hai yü-heng chih* (preface dated 1804, TSCC ed.), 10.73–74, 11.85; *Po-po HC* (1832 ed.), 12.21b; *Kuei-hsien chih* (1893 ed.), 1.2a; *Mi-lo CC* (1738 ed.), 23.50a.

32. Fujii, "Hui-chou Merchants," *passim*. The memorial is cited on p. 44.

33. For a short study of the major tobacco-producing areas in the Ch'ing, see Wang Hsin, *Ch'ing-yen lu* (1805 ed.),8, *passim.*

34. Shu Wei, *P'ing-shui-chai shih-chi* (TSCC ed.), 6.63.

35. Liang Chang-chü, *T'ui-an sui-pi* (*Ch'ing-tai pi-chi ts'ung-k'an* ed.), 8.8b–9a.

36. Pao Shih-ch'en, *An-wu ssu-chung*, 6.

37. *Ch'eng-tu HC* (1815 ed.), 6.38a; *Chin-t'ang HC* (1811 ed.), 3; Huang pen-chi, *Hu-nan fang-wu chih* (1845 ed.), 3.4b–5a; *HCCSWP*, 36.22a–22b.

38. *Jui-chin HC* (1603 ed.) lacks reference to these commercial crops and also lacks the confident tone as to the state of the local economy. The change in the economy of the locality brought about by commercial crops is described in the 1872 edition, 2.37b–45b.

39. M. Huc, *A Journey through the Chinese Empire* (New York, 1855), vol. II, p. 129.

40. For a more systematic explanation of traditional China's failure to develop a capitalistic system and her difficulties in capital accumulation, see Ho, "Salt Merchants of Yang-chou," last section.

41. Ho, "Salt Merchants," and Fu I-ling, *Ming-Ch'ing shang-jen*, pp. 171–173.

42. Yen Chung-p'ing, pp. 152–155.

43. C. F. Remer, *Foreign Investments in China* (New York, 1933), part II.

44. Yen Chung-p'ing, *Chung-kuo mien-fang-chih shih-kao*, chs. 3–7.

45. Cheng, Y. K., *Foreign Trade and Industrial Development of China* (Washington, D.C., 1956), p. 211.

46. Liang Fang-chung, "Ming-tai shih-tuan-chin fa," *Chinese Social and Economic History Review*, vol. VII, no. 1, pp. 132–133 (June 1944).

47. *HWHTK*, 2.2795.

48. Ku Ying-t'ai, *Ming-shih chi-shih pen-mo*, ch. 65, *passim.*

49. *Ch'ing-sheng-tsu shih-lu*, 233.4b.

50. *Ibid.*, 245.20a.

51. P'u Sung-ling, *Hsing-shih yin-yüan*, ch. 24. I have freely rendered into prose the last half of one of his many poems depicting the general contentedness of the time.

52. Kung Tzu-chen, *Ting-an wen-chi* (SPPY ed.), second series, 2.22b.

53. *Shih-shou HC* (1866 ed.), 3.1a–2b.

54. *Ibid.*

55. The *huo-hao* was first collected to cover the loss on government-operated gold mines during the Yüan period. See Ch'ien Ta-hsin, *Shih-chia-chai yang-hsin lu* (SPPY ed.), 19.7b.

56. Cited in *HCCSWP*, 27.6a–9b.
57. Lü Ssu-mien, *Ch'in-Han shih* (Shanghai, 1947), p. 663.
58. *Min-shu* (1629 ed.), 39.1a–1b.
59. Yang Hsi-fu, *Ssu-chih-t'ang wen-chi*, 10.1a–9a.
60. *Yüeh-chou FC* (1748 ed.), 12.5a–5b.
61. J. H. Clapham, *The Economic Development of France and Germany, 1814–1914* (Cambridge, 1946), p. 41.
62. Herbert Norman, *Japan's Emergence as a Modern State* (New York, 1940), p. 21.
63. *Ho-nan FC* (1779 ed.), 24.1a.
64. Gibbon's *Autobiography*, cited in T. S. Ashton, *The Industrial Revolution, 1760–1830* (Oxford, 1948), p. 5.
65. K'ung Shang-jen, *Jen-jui lu* (*Chao-tai ts'ung-shu* ed.), second series, 13.
66. Singde, *Lu-shui-t'ing tsa-chih* (*Ch'ing-tai pi-chi ts'ung-k'an* ed.), 1.3a–3b.
67. Yü Cheng-hsieh, *Kuei-ssu lei-kao* (*An-hui ts'ung-shu* ed.), 12.22b.
68. *Kuei-yang CC* (1868 ed.), 18, *passim.*
69. It seems that the prevalence of old people in the early Ch'ing might have had something to do with the dissemination of medical knowledge. The famous book on materia medica, Li Shih-chen's *Pen-ts'ao kang-mu*, which systematized the existing pharmaceutical knowledge of China, was completed in 1578 after thirty years of intensive research and study. It has since gone through numerous editions and is still regarded as the most authoritative work of its kind. In the late Ming and Ch'ing an ever-increasing number of shorter works were based on it. Moreover, the lists of works on pediatrics, obstetrics, gynecology, smallpox, and typhoid fever given in various provincial and local histories are very impressive. Although the effect of these works on medical practice cannot be appraised by the layman, especially when the majority of them were only of regional and local significance and are not readily available now, their mere quantity would indicate a remarkable dissemination of medical knowledge in late Ming and early Ch'ing China. As to epidemics, they still wrought havoc from time to time but were seldom on an interregional scale in the late seventeenth and eighteenth centuries. One of the most destructive of the epidemics, the true Asiatic cholera, was unknown to China until 1821. This was clearly attested to by a contemporary imperial prince; see Chao Lien (Prince Li), *Hsiao-t'ing hsü-lu* (1880 ed.), 2.41b. For a modern discussion, see K. C. Wong and L. T. Wu, *A History of Chinese Medicine* (Tientsin, 1932), pp. 106–108. Local histories often contain brief mention of outbreaks of terrible epidemics in the first half of the nineteenth century — mostly cholera.
70. Hsü Hsi-lin, *Hsi-ch'ao hsin-yü* (*Ch'ing-tai pi-chi ts'ung-k'an* ed.), 12.5b–6a.
71. This generalization is based on the early years of *Ch'ing-kao-tsung shih-lu* and other documents and biographical material.
72. Article on Ho-shen, in A. W. Hummel, *Eminent Chinese of the Ch'ing Period.*
73. Cheng Ho-sheng, *Chung-kuo chin-shih shih* (Chungking, 1945), pp. 655–658.

74. Lo Yü-tung, *Chung-kuo li-chin shih* (CP, 1935), vol. I, p. 5.
75. Wang Yü-ch'üan, "The Rise of Land Tax and the Fall of Dynasties in Chinese History," *Pacific Affairs*, 1938, pp. 201–220.
76. On forced migrations of rich men in the early Ming, see Chapter VII, section I.
77. *Ku-su chih* (1506 ed.), 13.6b; *Wu-chiang HC* (1558 ed.), 13.6b–11a; *Wu-men pu-sheng* (1829 ed.), 1.1a–4a.
78. Yeh Meng-chu, *Yüeh-shih pien*, 1.18a–19a.
79. *HCCSWP*, 36.26b.
80. *Wu-tsa-tsu*, 4.36b–37a; also the excellent monograph based on old agricultural contracts and mortgages by Fu I-ling, *Fu-chien tien-nung ching-chi-shih ts'ung-k'ao* (Fukien Christian University, 1944).
81. *Shun-teh HC* (1585 ed.), 3.1a.
82. *HCCSWP*, 75.41a–42b.
83. *Ibid*., 34.31a–32a; 33a–34b
84. *Ming-shih*, 77.4a. The evasion of land tax was made easier by the absence of the fish-scale maps and books in many Kwangtung counties. See *Yung-cheng chu-p'i yü-chih*, 19.20b–21a.
85. *Pa-ling HC* (1872 ed.), 4.3a.
86. *Chiang-chin HC* (1875 ed.), 6.1b.
87. *Shan-hua HC* (1877 ed.), 16.10b, cites the 1740 edition.
88. See especially *Ch'eng-pu HC* (1863 ed.), 10.35a–44a, and *Yüan-ling HC* (1871 ed.), 10.10b.
89. Yüan Mei, *Hsiao-ts'ang shan-fang shih-wen-chi*, prose works, second series, 33.1a–2b.
90. *Ch'ing-kao-tsung shih-lu*, 201.2b–3b.
91. Yang Hsi-fu, *Ssu-chih-t'ang wen-chi*, 10.1a–9a.
92. Hung Liang-chi, *Chüan-shih-ko wen-chi* (SPTK ed.), series A, pp. 8a–9b.
93. Lo Erh-kang, *T'ai-p'ing t'ien-kuo shih-kang* (CP, 1937), ch. 1; and Wang Ying, "T'ai-p'ing t'ien-kuo ke-ming ch'ien-hsi ti t'u-ti wen-t'i" (The Land Question on the Eve of the Taiping Rebellion), *Chung-shan wen-hua chiao-yü-kuan chi-k'an*, vol. III, no. 1.
94. Wang Ying, "Tai-p'ing"; Chao Lien, *Hsiao-t'ing tsa-lu*, "Rich People of the Reigning Dynasty."
95. See Chapter X, section III.
96. Lo Erh-kang, *T'ai-p'ing t'ien-kuo shih-shih-k'ao* (Peking, 1955), "On the Question whether the Taiping Land Policy was Carried Out."
97. R. H. Tawney, *Land and Labour in China* (London, 1937), ch. 2.
98. A. K. Ch'iu, "Agriculture," in H. F. MacNair, ed., *China* (Berkeley, Calif., 1946), pp. 473–474.
99. Tawney, *Land and Labour in China*, p. 74.
100. Ch'en Kuo-fu, *Su-cheng hui-i* (Taipeh, 1951), p. 39.
101. *The Chinese Yearbook* (1937), pp. 776–778.
102. Summary results have been cited here and there by recent mainland writers; see, for example, Meng Hsien-chang, *Chung-kuo chin-tai ching-chi-shih chiao-ch'eng* (Shanghai, 1951), ch. 24, *passim*.
103. *China Handbook* (1950), pp. 581–583.
104. For a brief historical discussion, see L. S. Yang, *Money and Credit in China, A Short History* (Cambridge, Mass., 1952), ch. 1.

105. *Wu-hsien chih* (1642 ed.), 10.1b–2b.
106. *Pa-ling HC* (1872 ed.), 11.9b.
107. *HCCSWP*, 26.17a–25a; 36.14a–14b; 36.26b.
108. *Shang-yüan HC* (1593 ed.), 12.97a–98b.

CHAPTER X. Catastrophic Deterrents.

1. Walter H. Mallory, *China: Land of Famine* (New York, 1928). The well-known statistical studies on natural calamities in Chinese history are: Alexander Hosie, "Droughts in China, A.D. 620 to 1643," *Journal of the North China Branch of the Royal Asiatic Society*, 1878, pp. 51–89; Co-ching Chu, "Climatic Pulsations during Historic Time in China,' *Geographical Review*, 1926, pp. 274–282; and the more detailed two-volume work, Ch'en Kao-yung, ed., *Chung-kuo li-tai t'ien-tsai jen-huo piao* (National Chi-nan University, 1939).
2. *Chen-tsai hui-k'an* (Honan-Shensi-Kansu Famine Relief Commission, 1928).
3. The figures for droughts and floods for the Hsien-feng period (1851–1861) are missing in Chu, "Climatic Pulsations," but are duly amended according to the *Tung-hua lu* for our comparison.
4. Lin Tse-hsü's preface to Li Yen-chang, *Chiang-nan ts'ui-keng k'o-tao pien* (preface dated 1834, 1888 ed.), and his *Lin-wen-chung-kung cheng-shu* (1879 ed.), B2; Wei Yüan, *Ku-wei-t'ang wai-chi* (1878 ed.), 6.5a–6b; Wang Shih-to, *I-ping jih-chi* (Peiping, 1935), 3.26b–28a.
5. Cited in *Shensi t'ung-chih-kao* (1934 ed.), 14.
6. *Ibid.*, 31.1a.
7. *Li-ch'üan HC* (1935 ed.), 14.3a–5a.
8. Cited in Mallory, *China*, pp. 29–30.
9. On general depopulation in Shansi after the famine of 1877–78 and the provincial government's efforts to attract new settlers, see Governor Chang Chih-tung's memorial of 1882, cited in *Huang-ch'ao ching-shih-wen t'ung-pien* (Shanghai, 1901), "Geography," 10.13b–14a.
10. For details on immigration in Shensi, see Chapter VII.
11. W. W. Rockhill, "An Inquiry into the Population of China," *Annual Report of the Smithsonian Institution*, 1904, p. 673.
12. Mallory, *China*, introduction and p. 30.
13. *Shih-shih yüeh-pao*, July 1930, p. 31.
14. The total number of 20,000,000 in need of relief was estimated by the Honan-Shensi-Kansu Famine Relief Commission in 1928. See *Chen-tsai hui-k'an*. The China International Famine Relief Commission estimated in 1929 that a total population of 57,350,000 was affected, and at least 12,000,000 seriously affected by the famine. See *China International Famine Relief Commission, Annual Report, 1929*, p. 3.
15. Chung Hsin, *Yang-tzu-chiang shui-li k'ao* (CP, 1936), pp. 39–42.
16. Li Kuang-t'ao, "Chang Hsien-chung shih-shih," *Bulletin of the Institute of History and Philology, Academia Sinica*, xxv, 1954, pp. 21–30.
17. For a detailed interpretation of land statistics, see Chapter VI.
18. See Chapter VII.

19. Lo Erh-kang, ed., *Chung-wang Li Hsiu-ch'eng tzu-chuan yüan-kao chien-cheng* (Peking, 1951), p. 65. It is interesting that the recent revisionist view among historians in Mainland China is that Tseng was an executioner and a Manchu henchman. See Fan Wen-lan, *Han-chien k'uei-tzu-shou Tseng Kuo-fan ti i-sheng* (Hsin-hua Company, 1944).

20. So important was the economic phase of the war that Chang Teh-chien, *Tse-ch'ing hui-ts'uan* (original preface dated 1855), one of the most comprehensive contemporary studies on the rebels from the government standpoint, devoted a special chapter to Taiping food supplies. This work is reprinted in *T'ai-p'ing t'ien-kuo* (Peking, 1952), a collection of sources on the Taiping movement by Mainland Chinese historians.

21. Li Hsiu-ch'eng's own testimony, Lo, ed., *Chung-wang*, pp. 134, 147–148.

22. Chiang Siang-tseh, *The Nien Rebellion* (Seattle, 1954), pp. 68–69.

23. *Kuang-teh CC* (1880 ed.), 60.25a–25b.

24. *Ch'ao-p'ing yüeh-fei fang-lüeh* (1872 ed.), 401.1a–2a.

25. *She-hsien chih* (1937 ed.), 3.3b–4a.

26. Hu Ch'uan, "Tun-fu nien-p'u," manuscript loaned to the author by Dr. Hu Shih.

27. Cited in *Nan-ling HC* (1924 ed.), 41.94b.

28. *Su-chou FC* (1883 ed.), 13, *passim.*, *Wu-hsi Chin-k'uei HC* (1883 ed.), 12.4a.

29. *Kao-ch'un HC* (1881 ed.), 7.2a–2b.

30. *Chin-t'an HC* (1885 ed.), 2.3a and 3.84b.

31. *Ch'ao-p'ing yüeh-fei fang-lüeh*, years 1864–65; *Huang-ch'ao Tao Hsien T'ung Kuang tsou-i* (1902 ed.), 27, *passim.*

32. *An-lu hsien-chih pu-sheng* (1872 ed.), B.64b.

33. Baron von Richthofen, *Letter on the Provinces of Chekiang and Nganhwei* (Shanghai, 1871), pp. 12–14, written especially for the *North China Daily News.*

34. This is strongly borne out by a number of southern Kiangsu local histories, notably, *Shang-chiang liang-hsien chih* (1874 ed.), 6.12a–12b; *Chin-t'an HC* (1885 ed.), 3.84b; *Chiang-p'u p'i-sheng* (1891 ed.), 7.7b; *Chü-yung HC* (1904 ed.) 4, *passim.*

35. The acute shortage of farm and construction labor was a serious problem not only throughout southern Kiangsu but as far as central Chekiang. See Tso Tsung-t'ang, *Tso-wen-hsiang-kung tsou-kao* (1890 ed.), 11, memorial dated November 5, 1864. Also *Chien-teh HC* (1919 ed.), 4.7b–8a, 6.18b.

36. *Huang-ch'ao Tao Hsien T'ung Kuang tsou-i*, 29.20a–23b. A particularly good account of the conflict between natives and immigrants is in *Chia-hsing HC* (1906 ed.), 11, *passim.*

37. *Huang-ch'ao ching-shih-wen t'ung-pien*, "Geography," 10.15a–16a.

38. *Ibid.*, pp. 14a–14b.

39. *Ibid.*, pp. 15a–16a.

40. *Huang-ch'ao Tao Hsien T'ung Kuang tsou-i*, 29.

41. Lo, ed., *Chung-wang*, pp. 117–122.

42. *Hsiao-kan HC* (1883 ed.), 5.2b; *Hsien-feng HC* (1867 ed.), 7.3b–4a.

43. *Nan-k'ang FC* (1872 ed.), 11.19b.

44. *Nan-ch'ang FC* (1873 ed.), 18.68b–69a.

45. Lo, ed., *Chung-wang*, p. 134.
46. *Jui-chou FC* (1873 ed.), 6.12a–12b; *Kao-an HC* (1871 ed.), 9.9b.
47. Rockhill, "An Inquiry," p. 673.
48. Ch'in Han-ts'ai, *Tso-wen-hsiang-kung tsai hsi-pei* (Shanghai, 1946), p. 77.
49. *Baron Richthofen's Letters, 1870–1872* (Shanghai, 2nd ed., 1903), p. 143.
50. See Chapter VII.
51. *Chen-tsai hui-k'an* (1928), pp. 23–24.
52. Ch'in, *Tso-wen-hsiang-kung tsai hsi-pei*, pp. 135–136; also Tso's memorial in *Huang-ch'ao ching-shih-wen t'ung-pien*, "Geography," 10.13b.
53. Chang Hsiao-mei, *Ssu-ch'uan ching-chi ts'an-k'ao tzu-liao* (Shanghai, 1939), B.24–27.
54. *Chiang-hsi nien-chien* (1937), ch. 29, *passim.*
55. Ting Li, *Chung-Kung ti min-ping chih-tu* (Hong Kong, 1954), *passim.*
56. Hsi Hung, *Lao-hung-ch'ü hsing* (Hankow, 1953), p. 4.
57. Notably *Kuo-wen chou-pao*, vol. 10, no. 44, November 6, 1933.
58. Ku Kuan-chiao, *San-shih-nien-lai ti Chung-Kung* (Hong Kong, 1955), p. 92.
59. Hsi Hung, *Lao-hung-ch'ü hsing*; T'ang T'ieh-hai, *Chung-yang lao-ken-chü-ti yin-hsiang-chi* (Hankow, 1953); Ch'en Mu, *Nan-fang lao-ken-chü-ti fang-wen-chi* (Hankow, 1953); Yen Wei-ping, *Hui-tao Ching-kang-shan* (Hankow, 1950).
60. Ho Ying-ch'in, *Pa-nien k'ang-chan chih ching-kuo* (Nanking, 1946), appendix, table 1.
61. F. F. Liu, *A Military History of Modern China, 1924–1949* (Princeton, 1956), p. 137.
62. *K'ang-chan pa-nien-lai ti pa-lu-chün yü hsin-ssu-chün* (Political Department of the Communist 18th Army Group, 1945).
63. *Jen-min shou-ts'e* (1951); also *Kung-fei chung-yao tzu-liao hui-pien* (Taipei, 1952), no. 5.
64. *Ching-chou FC* (1880 ed.), 76, *passim.*
65. *Feng-yang FC* (1908 ed.), 4B, *passim.*
66. *Chen-tsai hui-k'an* (1928), pp. 10–12.

CHAPTER XI. Conclusion.

1. Van der Sprenkel, "Population Statistics of Ming China."
2. On the unfair fiscal burden of the lower Yangtze area, see Ku Yen-wu, *Jih-chih lu* (CPPY ed.), 10.7a–17b.
3. The best description of the general conditions in Yunnan in the Ming is Hsieh Chao-che, *Tien-lüeh.*
4. *HWHTK*, 13.
5. Chang Huang, *T'u-shu pien* (compiled between 1562 and 1577; T'ien-ch'i, 1621–27 ed.), 90.16b–17a.
6. *Wu-chin HC* (1605 ed.), 3.7a. The editor was T'ang Ho-cheng, son of T'ang Shun-chih.
7. *K'uai-chi HC* (1575 ed.), 5.2b; also *Shao-hsing FC* (1586 ed.), 15.12a.

8. *Lien-chiang HC* (1805 ed.), 2.2b–3b, cites a Ming edition.
9. *Ch'ang-chou FC* (1618 ed.), 4.4a–11b.
10. *Min-shu*, 39.2a.
11. *Fu-chou FC* (1613 ed.), 26.4a.
12. Tung Ch'i-ch'ang, ed., *Shen-miao liu-chung tsou-shu hui-yao* (Yenching University Press, 1937), vol. 6, p. 20b.
13. *Wu-tsa-tsu*, 4.33a–34b.
14. In the late Ming some Shantung people were keenly aware of the general contrast in land tenure between north and south China. See *THCKLPS*, 21.1a.
15. *Wu-tsa-tsu*, 4.36b–37a.
16. *Ibid.*, 3.23b–25b.
17. *THCKLPS*, 23.52a–53b; 24.111a–111b; 26.83a.
18. Ta Chen, *Population.*
19. *HCCSWP*, 30.2a–4a.
20. *CCWHTK*, 19.5025.
21. *Ibid.*, 3.4871 and 3.4874.
22. *HCCSWP*, 34.34a–36a.
23. Lo Erh-kang, "The Population Pressure in the Pre-Taiping Rebellion Years," *Chinese Social and Economic History Review*, vol. VIII, no. 1, p. 62.
24. Yang Hsi-fu, *Ssu-chih-t'ang wen-chi* (1806 ed.), 10.1a–9a.
25. Lo, "Population Pressure," pp. 61–63.
26. P'u Sung-ling, *Hsing-shih yin-yüan* (undated twentieth-century ed.), ch. 24.
27. P'u Sung-ling, *Liao-ch'ai ch'üan-chi* (Shih-chieh shu-chü 1936 ed.), Prose Works A, pp. 25–27.
28. *Hsing-shih yin-yüan*, ch. 90.
29. *Yüeh-chou FC* (1746 ed.), 16.3b.
30. *Kuei-yang CC* (Hunan, 1866 ed.), 20, *passim.*
31. *Ying-ch'eng HC* (1882 ed.), 1.3a–3b, cites the 1669 and 1815 editions.
32. *Yüeh-chou chih* (Shensi, 1762 ed.), 4.18b–19a.
33. *HCCSWP*, chapters on social customs.
34. On the interrelation of standards of living and population growth, the words of a modern economic historian are helpful: "A great deal depends too on the size of the gap between the minimum subsistence level and the customary standards; the lower the standard of living at the start of the population increase, the more likely was it that the increase would be halted by Malthus's positive checks." H. J. Habakkuk, "English Population in the Eighteenth Century," *Economic History Review*, December 1953.
35. Sir Alexander Carr-Saunders, *World Population, Past Growth and Present Trends* (Oxford, 1937), p. 21.
36. *Ibid.*, pp. 137–140, 330–331.
37. Indeed, not a few of the early nineteenth-century local histories still recorded continued economic prosperity. But in some hilly areas where the density of population per square mile of cultivated land was high, prosperity came to an end even before 1750. *Ch'üan-chou chih* (Kwangsi, 1799 ed.), 4.2b, for example, testifies that in the memory of local elders northeastern Kwangsi had been a paradise up to about the end of the Yung-cheng period; after the second quarter of the eighteenth century the economic situation steadily worsened.

38. Hung Liang-chi, *Chüan-shih-ko wen-chi* (SPTK ed.), Series A, pp. 8a–10b.

39. C. F. Lung, "A Note on Hung Liang-chi: the Chinese Malthus," *T'ien-hsia Monthly*, October 1935. The translation from Hung is my own.

40. Kung Tzu-chen, *Ting-an wen-chi* (SPPY ed.), Prose Works, B.6a.

41. *Yung-chou FC* (1825 ed.), A. 18b; *P'ing-chiang HC* (1877 ed.), 20.4b–5a; *Ch'eng-pu HC* (1867 ed.), 10.35a–44a.

42. *Kan-chou FC* (1871 ed.), 20.5a–5b, cites the 1848 edition.

43. *Ying-shan HC* (1871 ed.), 15.1a.

44. See Chapter X, Section III, and Appendix II.

45. Habbakuk, "English Population in the Eighteenth Century."

46. Wang Shih-to, *Wang Mei-ts'un hsien-sheng chi* (1881 ed.), 11.4b–5a; 12.1a–2a.

47. *I-ping jih-chi*, 3.26b–28a.

48. *Ibid.*, 3.28a–31a.

49. Hu Ch'uan, *Tai-wan chi-lu liang-chung* (Taipei, 1951), vol. II, pp. 53b–54a.

50. For facts and theories concerning the recent birth-control movement in China, see Irene B. Taeuber, "Population Policies in Communist China," *Population Index*, October 1956.

INDEX

INDEX

BIBLIOGRAPHY
PRIMARY CHINESE SOURCES

An-hsien chih 安縣志 . 1932.

An-hui-sheng Tang-t'u-hsien t'u-ti ch'en-pao kai-lüeh 安徽省當塗縣土地陳報概略. Ministry of Finance, 1935-1936.

An-lu hsien-chih pu-cheng 安陸縣志補正 . 1872.

Anhwei TC 安徽 . 1829 and 1877.

Anhwei t'ung-chih-kao 安徽通志稿 . Compiled in the 1930's; still uncompleted.

CCHWHTK: Ch'ing-ch'ao hsü-wen-hsien t'ung-k'ao 清朝續文獻通考 . CP ed.

CCWHTK: Ch'ing-ch'ao wen-hsien t'ung-k'ao 清朝文獻通考 . CP ed.

Chahar TC 察哈爾. 1935.

Chang-chou FC 漳州 . 1877.

Chang Huang 章潢 . T'u-shu pien 圖書編 . 1621-1627.

Chang-p'u HC 漳浦 . 1698.

Chang Shun-min 張舜民 . Hua-man lu 畫墁錄. Pai-hai 稗海 ed.

Chang Yü-shu 張玉書 . Chang-wen-chen-kung chi 張文貞公集 . 1772.

Ch'ang-chao ho-chih 常昭合志 . 1793.

Ch'ang-chou FC 常州 . 1618.

Ch'ang-hua HC 昌化 . 1822.

Ch'ang-ke HC 長葛 . 1931.

Ch'ang-lo HC 長樂 (Hupei). 1870.

Ch'ang-sha FC 長沙 . 1747.

Ch'ang-shan HC 常山 . 1585 ed., printed in 1660.

Ch'ang-shu HC 常熟. 1538.

Ch'ang-teh FC 常德. 1813.

Chao-ch'ing FC 肇慶. 1833.

Chao I 趙翼. Nien-erh-shih tsa-chi 廿二史劄記. Shih-chieh shu-chü 世界書局, 1947.

Chao Kuan 趙官. Hou-hu chih 後湖志. Rev. ed., 1621.

Chao Lien 昭槤. Hsiao-t'ing hsü-lu 嘯亭續錄. 1880.

Chao-yüan HC 招遠. 1660 ed., 1846 reprint.

Ch'ao-p'ing yüeh-fei fang-lüeh 剿平粵匪方略. 1872.

Chekiang TC 浙江. 1735 ed., CP reprint.

Chen-chiang FC 鎮江. 1685.

Chen-tsai hui-k'an 振災會刊. Honan-Shensi-Kansu Famine Relief Commission, 1928.

Ch'en-chou FC 陳州. 1746.

Ch'en-chou FC 陳州. 1765.

Ch'en Hung-mou 陳宏謀. P'ei-yüan-t'ang ou-ts'un-kao 培遠堂偶存稿. Late eighteenth century ed.

Ch'en K'uei-lung 陳夔龍, ed. Ch'ien-shih chi-lüeh hou-pien 黔詩紀略後編. 1911.

Ch'en Kuo-fu 陳果夫. Su-cheng hui-i 蘇政回憶. Taipeh, 1951.

Ch'en Mao-jen 陳懋仁. Ch'üan-nan tsa-chih 泉南雜志. TSCC ed.

Ch'en Mu 陳牧. Nan-fang lao-ken-chü-ti yin-hsiang-chi 南方老根據地印象記. Hankow, 1953.

Ch'eng-ch'eng HC 澄城. 1786.

Ch'eng-pu HC 城步. 1866.

Ch'eng-tu HC 成都. 1815.

Chi-fu TC 畿輔. 1732 ed., CP reprint.

Chi-hsi HC 績溪. 1581.

Chi-nan FC 濟南. 1839.

Chi-ning CC 濟寧. 1843.

Chi-yang HC 濟陽. 1763.

Ch'i-hsien chih 杞縣志. 1788.

Ch'i-min yao-shu 齊民要術. SPPY ed.

Ch'i-shui HC 蘄水. 1880.

Chia-hsiang HC 嘉祥. 1869.

Chia-hsing FC 嘉興. 1878.

Chia-hsing HC 嘉興. 1906.

Chia-shan HC 嘉善. 1623 and 1894.

Chia-ying CC 嘉應. 1898.

Chiang-chin HC 江津. 1875.

Chiang-hsi-sheng ti-cheng kai-k'uang 江西省地政概況. Kiangsi Provincial Government, 1941.

Chiang-p'u p'i-sheng 江浦埤乘. 1891.

Chiang-su-sheng Chiang-tu-hsien t'u-ti ch'en-pao kai-lüeh 江蘇省江都縣土地陳報概略. Ministry of Finance, 1935.

Chiang-su-sheng chien-fu ch'üan-an 江蘇省減賦全案. 1867.

Chiang-su-sheng Hsiao-hsien t'u-ti ch'en pao kai-lüeh 江蘇省蕭縣土地陳報概略. Ministry of Finance, 1935.

Chien-teh HC 建德 (Chekiang). 1919.

Chien-wei HC 犍為. 1937.

Ch'ien Ta-hsin 錢大昕. Shih-chia-chai yang-hsin lu 十駕齋養新錄. SPPY ed.

Ch'ien-t'ang HC 錢塘. 1609 ed., 1893 reprint.

Ch'ien-yang HC 黔陽. 1874.

Chin-t'an HC 金壇. 1888.

Chin-t'ang HC 金堂. 1811.

Ch'in-an HC 秦安 . 1535.

Ching-an HC 靖安 . 1870.

Ching-chou chih 靖州志 . 1837.

Ch'ing-ch'ao hsü-wen-hsien t'ung-k'ao, see **CCHWHTK.**

Ch'ing-ch'ao wen-hsien t'ung-k'ao, see **CCWHTK.**

Ch'ing-ho HC 清河 . 1928.

Ch'ing-shih-lu 清實錄 .

Ch'ing-yang HC 青陽 . 1891.

Ch'ing-yüan FC 慶遠 . 1828.

Chiu-chiang FC 九江 . 1873.

Ch'iu Chün 丘濬 . Ta-hsüeh yen-i pu 大學衍義補 . 1931 reprint.

Ch'iung-chou chih 邛州志 . 1818.

Chou-chih HC 盩厔 . 1785.

Chou-li 周禮 . Shih-san-ching chu-shu 十三經注疏 ed.

Chu-hsi HC 竹溪 . 1869.

Chu Kuo-chen 朱國禎 . Yung-ch'uang hsiao-p'in 湧幢小品 . 1622.

Chu-shan HC 竹山 . 1867.

Chu Yün-chin 朱雲錦 . Yü-sheng shih-hsiao lu 豫乘識小錄 . 1873.

Ch'u-chou chih 滁州志 . 1897.

Ch'u Yüan 褚淵 . Mu-mien p'u 木 譜 . Shang-hai chang-ku ts'ung-shu 上海掌故叢書 ed.

Chung-hua jen-min kung-ho-kuo ti-i-ke wu-nien-chi-hua ti ming-tz'u chieh-shih 中華人民共和國第一個五年計劃的名詞解釋 . Peking, 1955.

Chung-kuo ching-chi nien-chien 中國經濟年鑑 1935.

Ch'ung-ch'ing FC 重慶 . 1843.

Chü-yung HC 句容. 1904.

Ch'üan-chiao HC 全椒. 1920.

Ch'üan-chou chih 全州志. 1799.

En-p'ing HC 恩平. 1934.

Fang-hsien chih 房縣志. 1866.

Fang Kuan-ch'eng 方觀承. Fang-ch'üeh-min-kung tsou-i 方恪敏公奏議. 1851.

Fen-chou FC 汾州. 1609 and 1771.

Feng-huang-t'ing chih 鳳凰廳志. 1758.

Feng Kuei-fen 馮桂芬. Hsien-chih-t'ang kao 顯志堂稿. Late Ch'ing ed.

Feng-t'ai HC 鳳臺 (Anhwei). 1814 ed., 1936 reprint.

Feng-yang FC 鳳陽. 1908.

Fengtien TC 奉天. 1934.

Fu-chou FC 福州. 1613.

Fu-shun HC 富順. 1777 ed., 1882 reprint.

Fukien TC 福建. 1829 and 1942.

Hai-chou chih 海州志. 1811.

Hai-ning HC 海寧. 1557 ed., 1898 reprint.

Han-chung FC 漢中. 1813 ed., 1924 reprint.

Han-shu 漢書. SPPY ed.

Hang-chou FC 杭州. 1923 ed., compiled between 1879 and 1919.

HCCSWP: Ho Ch'ang-ling 賀長齡, ed. Huang-ch'ao ching-shih wen-pien 皇朝經世文編. 1886.

Heilungkiang t'ung-chih-kao 黑龍江通志稿. 1932.

Heng-chou FC 衡州. 1774 ed., 1875 reprint.

Ho Ch'ang-ling, see HCCSWP.

Ho-feng CC 鶴峰. 1822.

Ho-nan FC 河南. 1779.

Ho Ying-ch'in 何應欽. Pa-nien k'ang-chan chih-ching-kuo 八年抗戰之經過. Nanking, 1946.

Hsi Hung 西虹. Lao-hung-ch'ü hsing 老紅區行. Hankow, 1953.

Hsia-men chih 厦門志. 1839.

Hsia-p'u HC 霞浦. 1929.

Hsiang-t'an HC 湘潭. 1553.

Hsiang-yang FC 襄陽. 1760.

Hsiao-i-t'ing chih 孝義廳志. 1883.

Hsiao-kan HC 孝感. 1883.

Hsiao-shan HC 蕭山. 1751.

Hsieh Chao-che 謝肇淛. Tien-lüeh 滇略. Late Ming ed.

------- Wu-tsa-tsu 五雜組. 1895 Japanese ed.

Hsien-chü HC 僊居. 1608 ed., 1935 reprint.

Hsien-feng HC 咸豐. 1867.

Hsien-yu HC 僊遊. 1770 ed., 1873 reprint.

Hsin-fan HC 新繁. 1907.

Hsin-teng HC 新登. 1922.

Hsing-hua-ts'un chih 杏花村志. 1685.

Hsing-tzu HC 星子. 1871.

Hsiu-shui HC 秀水. 1684.

Hsü Hsi-lin 徐錫麟. Hsi-ch'ao hsin-yü 熙朝新語. Ch'ing-tai pi-chi ts'ung-k'an 清代筆記叢刊 ed.

Hsü Kuang-ch'i 徐光啟. Nung-cheng ch'üan-shu 農政全書.. 1843.

------- Hsü-wen-ting-kung chi 徐文定公集. Shanghai, 1933.

Hsü Shih-ch'ang 徐世昌. Tung-san-sheng cheng-lüeh 東三省政略. 1911.

Hsü Tung 徐棟, ed. Pao-chia shu 保甲書. 1848.

Hsü-wen-hsien t'ung-k'ao, see HWHTK.

Hsü Yu-ch'ü 徐有榘. Chung-shu p'u 種藷譜. 1834 Korean ed.

Hsün-chou FC 潯州. 1826.

Hsün-yang HC 洵陽. 1870.

Hu-chou FC 湖州. 1874.

Hu Ch'uan 胡傳. "Tun-fu nien-p'u" 鈍夫年譜. Manuscript.

------- T'ai-wan chi-lu liang-chung 台灣紀錄兩種. Taipeh, 1951.

Hu-nan min-cheng t'ung-chi 湖南民政統計. Department of Civil Affairs, Hunan Provincial Government, 1941.

Hunan TC 湖南. 1757 and 1885.

Hu-nan wen-cheng 湖南文徵. Late Ch'ing ed.

Hu-pei-sheng nien-chien 湖北省年鑑. 1935.

Hupei TC 湖北. 1921 ed., CP reprint.

Hu-pu tse-li 户部则例. 1776.

Hu Shih-ning 胡世寧. Hu-tuan-min tsou-i 胡端敏奏議. Chekiang shu-chü ed.

Hua-chou chih 華州志. 1572 and 1684 eds., both reprinted in 1882.

Hua-yang HC 華陽. 1934.

Hua-yang kuo-chih 華陽國志. SPPY ed.

Huai-an FC 淮安. 1748 and 1884.

Huang-ch'ao ching-shih wen-pien, see HCCSWP.

Huang-ch'ao ching shih-wen t'ung-pien 皇朝經世文統編. 1901.

Huang-ch'ao Tao Hsien T'ung Kuang tsou-i 皇朝道咸同光奏議. 1902.

Huang-Ch'ing tsou-i 皇清奏議. 1805 ed., 1936 reprint.

Huang Hsün 黃訓, ed. Huang-Ming ming-ch'en ching-chi lu 皇明

名臣經濟錄. 1549.

Huang Pen-chi 黃本驥. Hu-nan fang-wu chih 湖南方物志. 1846.

Huang Tso 黃佐. Nan-yung chih 南廱志. 1544.

Hui-chou FC 徽州. 1502 and 1827.

Hui-chou FC 惠州. 1881.

Hui-chou-fu fu-i ch'üan-shu 徽州府賦役全書. 1620.

Hun-yüan CC 渾源. 1661.

Hung Liang-chi 洪亮吉. Chüan-shih-ko wen-chi 卷施閣文集. SPTK ed.

HWHTK: Hsü-wen-hsien t'ung-k'ao 續文獻通考. CP ed.

I-ch'ang FC 宜昌. 1864.

I-chou FC 沂州. 1760.

I-hsing HC 宜興. 1590 and 1799.

I-wu HC 義烏. 1802.

I-yang HC 益陽. 1874.

Jen-min shou-ts'e 人民手册. Ta-kung-pao, 1951 and 1952.

Ju-ch'eng HC 汝城. 1907 and 1932.

Jui-chin HC 瑞金. 1603 and 1872.

Jui-chou FC 瑞州. 1628 and 1873.

K'ai-p'ing HC 開平. 1933.

Kan-chou FC 贛州. 1871.

K'ang-chan pa-nien-lai ti pa-lu-chün yü hsin-ssu-chün 抗戰八年來的八路軍與新四軍. Political Department of the Eighteenth Army Group, 1945.

Kansu TC 甘肅. 1909.

Kao-an HC 高安. 1871.

Kao-chou FC 高州. 1827.

Kao-ch'un HC 高淳. 1881.

Ke Shih-chün 葛士濬, ed. Huang-ch'ao ching-shih-wen hsü-pien 皇朝經世文續編. 1888.

Kiangsi fu-i ch'üan-shu 江西賦役全書. 1611.

Kiangsi nien-chien 江西年鑑 1937.

Kiangsi TC 江西 1881.

Kirin TC 吉林 1891.

Ku-su chih 姑蘇志. 1506.

Ku-t'ien HC 古田. 1606.

Ku Yen-wu 顧炎武, see THCKLPS.

------- Jih-chih lu 日知錄. SPPY ed.

Ku Ying-t'ai 谷應泰. Ming-shih chi-shih pen-mo 明史紀事本末. CP ed.

K'uai-chi HC 會稽. 1575.

Kuang-chou FC 廣州. 1879.

Kuang-hsi-sheng nung-ts'un tiao-ch'a 廣西省農村調查. Executive Yüan, 1935.

Kuang-hsi-sheng san-shih-i-nien-tu liang-shih tseng-ch'an shih-shih chi-hua kang-yao 廣西省三十一年度糧食增產實施計劃綱要. Kwangsi Provincial Government, 1942.

Kuang-hsin FC 廣信. 1873.

Kuang-hua HC 光化. 1882.

Kuang-p'ing FC 廣平. 1894.

Kuang-p'ing HC 廣平. 1608.

Kuang-shan HC 光山. 1936.

Kuang-teh CC 廣德. 1881.

Kuang-tung ching-chi nien-chien 廣東經濟年鑑. Kwangtung Provincial Bank, 1941.

Kuei-chou-sheng t'ung-chi tzu-liao hui-pien 貴州省統計資料彙編. Bureau of Statistics, Kweichow Provincial Government, 1942.

Kuei-hsien chih 貴縣志. 1893.

Kuei-yang CC 桂陽. 1866.

Kung-fei chung-yao tzu-liao hui-pien 共匪重要資料彙編. Taipeh, 1952.

Kung-hsien chih 鞏縣志. 1555 ed., 1935 reprint.

Kung Tzu-chen 龔自珍. Ting-an wen-chi 定盦文集. SPPY ed.

K'ung Shang-jen 孔尚任. Jen-jui lu 人瑞錄. Chao-tai ts'ung-shu 照代叢書 ed.

Kwangtung TC 廣東. 1840.

Lai-chou FC 萊州. 1740.

Lan Ting-yüan 藍鼎元. Lu-chou ch'u-chi 鹿洲初集. 1732.

------- Lu-chou tsou-shu 鹿洲奏疏. 1732.

Lei-chou FC 雷州. 1811.

Li Chao-lo 李兆洛. Yang-i-chai wen-chi 養一齋文集. 1936 reprint.

Li-ch'eng HC 歷城. 1640 and 1771.

Li-ch'üan HC 醴泉. 1935.

Li Fu 李紱. Mu-t'ang ch'u-kao 穆堂初稿. 1831.

Li-ling HC 醴陵. 1871.

Li-p'ing FC 黎平. 1845 and 1892.

Li-shui HC 溧水. 1883.

Li Yen-chang 李彥章. Chiang-nan ts'ui-keng k'e-tao pien 江南催耕課稻編. 1889.

Liang Chang-chü 梁章鉅. T'ui-an sui-pi 退庵隨筆. Ch'ing-tai pi-chi ts'ung-k'an ed.

Lien-chiang HC 連江. 1805.

Lien-chou FC 廉州. 1756.

Lin-ch'ing CC 臨清. 1749.

Lin-shui HC 鄰水. 1835.

Lin Tse-hsü 林則徐. Lin-wen-chung-kung cheng-shu 林文忠公政書. 1879.

Ling-hsien chih 酃縣志. 1873.

Liu-ho-hsien hsü-chih-kao 六和縣續志稿. 1920.

Liu Hsün 劉恂. Ling-piao lu-i 嶺表錄異. TSCC ed.

Liu Jui-t'u 劉瑞圖. Tsung-chih che-min wen-hsi 總制浙閩文檄. 1672.

Liu-yang HC 瀏陽. 1551.

Lo-ch'uan HC 洛川. 1806 and 1944.

Lo Yüan 羅願. Hsin-an chih 新安志. 1175 ed., 1888 reprint.

Lu Lung-ch'i 陸隴其. San-yü-t'ang wen-chi 三魚堂文集. 1868.

Lung-shan HC 龍山. 1818.

Lung-yen CC 龍巖. 1835.

Ma-ch'eng HC 麻城. 1877.

Ma Tuan-lin 馬端臨. Wen-hsien t'ung-k'ao 文獻通考. CP ed.

Mei Tseng-liang 梅曾亮. Po-hsien-shan-fang wen-chi 柏梘山房文集. 1856.

Meng-ch'eng HC 蒙城. 1915.

Mi-lo CC 彌勒. 1738.

Mien-yang CC 沔陽. 1531 ed., 1926 reprint.

Min-hsien hsiang-t'u chih 閩縣鄉土志. Ca. 1903.

Min-shu 閩書. 1629.

Ming-shih 明史. SPPY ed.

Ming-shih-lu 明實錄. Kiangsu Sinological Library photostat ed.

Nan-ch'ang FC 南昌. 1873.

Nan-ch'eng HC 南城. 1873.

Nan-k'ang FC 南康. 1872.

Nei-cheng nien-chien 內政年鑑. 1935.

Nei-wu fa-ling chi-lan 內務法令輯覽. Peking Government, 1918.

Ning-p'o FC 寧波. 1560.

Ning-shan-t'ing chih 寧陝廳志. 1829.

Pa-ling HC 巴陵. 1872.

Pa-min t'ung-chih 八閩通志. 1496.

Pao-ch'ing FC 寶慶. 1849.

Pao-ning FC 保寧. 1826.

Pao Shih-ch'en 包世臣. "Ch'i-min ssu-shu" 齊民四書, in An-wu ssu-chung 安吳四種. 1846.

Pao-ying HC 寶應. 1840.

Pi-hsien chih 郫縣志. 1813.

P'ing-chiang HC 平江. 1874.

P'ing-hsiang HC 萍鄉. 1784 and 1872.

Po-po HC 博白. 1832.

Po-yang HC 鄱陽. 1749.

P'u-chen chi-wen 濮鎮紀聞. Ch'ing manuscript, 1787.

P'u-ch'i HC 蒲圻. 1864.

P'u Sung-ling 蒲松齡. Hsing-shih yin-yüan 醒世姻緣. Kuang-i shu-chü ed.

------- Liao-chai ch'üan-chi 聊齋全集. Shih-chieh shu-chü ed.

P'u-t'ien HC 莆田. 1758 ed., 1879 reprint.

San-t'ai HC 三台. 1814.

Shan-hua HC 善化. 1877.

Shang-chiang liang-hsien chih 上江兩縣志. 1874.

Shang-hai HC 上海. 1588 and 1814.

Shang-yüan HC 上元. 1594.

Shansi TC 山西. 1892.

Shao-hsing FC 紹興. 1586.

Shao-yang-hsien hsiang-t'u-chih 邵陽縣鄉土志. 1907.

She-hsien chih 歙縣志. 1937.

She-hsien hui-kuan lu 歙縣會館錄. 1834.

Sheng-ching TC 盛京. 1684 ed., printed in 1711.

Sheng K'ang 盛康, ed. Huang-ch'ao ching-shih-wen hsü-pien 皇朝經世文續編. 1897.

Shensi TC 陝西. 1542 and 1735.

Shensi t'ung-chih-kao 陝西通志稿. 1934.

Shih-chi 史記. SPPY ed.

Shih-chu-t'ing chih 石柱廳志. 1843.

Shih-ch'üan HC 石泉 (Szechwan). 1833.

Shih-ch'üan HC 石泉 (Shensi). 1849.

Shih-liao hsün-k'an 史料旬刊, No. 21.

Shih-shou HC 石首. 1866.

Shou-chou chih 壽州志. 1890.

Shu-ch'eng HC 舒城. 1907.

Shu Wei 舒位. P'ing-shui-chai shih-chi 瓶水齋詩集. TSCC ed.

Shun-teh HC 順德. 1585.

Shun-t'ien FC 順天. 1593.

Singde 納蘭性德. Lu-shui-t'ing tsa-chih 淥水亭雜識. Ch'ing-tai pi-chi t'sung-k'an ed.

Ssu-ch'uan-sheng t'u-ti hsing-cheng kai-k'uang 四川省土地行政

概况. Bureau of Land Administration, Szechwan Provincial Government, 1940.

Ssu-hui HC 四會. 1896.

Ssu-hung ho-chih 泗虹合志. 1888.

Su-chou chih 宿州志. 1889.

Su-chou FC 蘇州. 1379, 1877, and 1883.

Sun Ch'eng-tse 孫承澤. Ch'un-ming meng-yü lu 春明夢餘錄. Ku-hsiang-chai 古香齋, 1913.

Sung-chiang FC 松江. 1819 and 1882.

Sung-hsien chih 嵩縣志. 1767.

Sung-shih 宋史. SPPY ed.

Sung Ying-hsing 宋應星. T'ien-kung-k'ai-wu 天工開物. 1637 ed., 1919 photostat reproduction.

Ta-Ch'ing fa-kuei ta-ch'üan 大清法規大全. 1901-1909.

Ta-Ch'ing hui-tien 大清會典. 1690, 1732, 1764, and 1818.

Ta-Ch'ing hui-tien shih-li 大清會典事例. 1818.

Ta-Ch'ing hui-tien tse-li 大清會典則例. 1764.

Ta-Ch'ing i-t'ung-chih 大清一統志. 1812 ed., CP reprint.

Ta-Ming hui-tien 大明會典. 1502 and 1587.

Ta-t'ung FC 大同. 1776.

Ta-yao HC 大姚. 1845.

Tai-chou chih 代州志. 1784.

T'ai-an HC 泰安. 1929.

T'ai-p'ing t'ien-kuo 太平天国. Documents edited by the Chinese Historical Society, Peking, 1952.

T'ai-ts'ang CC 太倉. 1629.

Tan Hsiang-liang 但湘良. Hu-nan Miao-fang t'un-cheng k'ao 湖南苗防屯政考. 1882.

Tan-tu HC 丹徒. 1879.

T'an Ts'ui 檀萃. Tien-hai yu-heng chih 滇海虞衡志.

TSCC ed.

Tang-yang HC 當陽. 1867.

T'ang Shun-chih 唐順之. Ching-ch'uan hsien-sheng wen-chi 荆川先生文集. SPTK ed.

T'ang T'ieh-hai 唐鐵海. Chung-yang lao-ken-chü-ti yin-hsiang chi 中央老根據地印象記. Hankow, 1953.

Tao-chou chih 道州志. 1877.

T'ao Chu 陶澍. T'ao-wen-i-kung ch'üan-chi 陶文毅公全集. Late Ch'ing ed.

THCKLPS: Ku Yen-wu 顧炎武. T'ien-hsia chün-kuo li-ping shu 天下郡國利病書. SPTK ed.

Ti-fang tzu-chih ch'üan-shu 地方自治全書. Shanghai, 1929.

"Tien-yeh hsü-chih lu" 典業須知錄. Undated Ch'ing manuscript.

Ting Yüeh-chien 丁曰健. Chih-T'ai pi-kao lu 治臺必告錄. 1867.

Tso Tsung-t'ang 左宗棠. Tso-wen-hsiang-kung tsou-kao 左文襄公奏稿. 1890.

Tsun-i FC 遵義. 1841.

Tsung Chi-ch'en 宗稷辰. Kung-ch'ih-chai wen-ch'ao 躬恥齋文鈔. 1851.

Tung Ch'i-ch'ang 董其昌, ed. Shen-miao liu-chung tsou-shu hui-yao 神廟留中奏疏彙要. Yenching University Press, 1937.

Tung-hua lu 東華錄. Hsien-feng period.

Tung-kuan HC 東莞. 1911.

Tung-yang HC 東陽. 1832.

T'ung-an HC 同安. 1929.

T'ung-ch'uan FC 潼川. 1786 and 1897.

Tzu-chung-hsien hsü-hsiu Tzu-chou chih 資中縣續修資州

志. 1928.

Tzu-yang HC 紫陽. 1843.

Wang Ch'ing yün 王慶雲. Hsi-ch'ao chi-cheng 熙朝紀政. 1898.

Wang Feng-sheng 王鳳生. Che-hsi shui-li pei-k'ao 浙西水利備考. 1823 ed., 1878 reprint.

------- Yüeh-chou ts'ung-cheng lu 越州從政錄. 1823.

------- Sung-chou ts'ung-cheng lu 宋州從政錄. 1826.

Wang Hsin 王訢. Ch'ing-yen lu 清煙錄. 1805.

Wang Hui-tsu 汪輝祖. Hsüeh-chih i-shuo 學治臆說. 1823.

Wang Shih-chün 王士俊. Li-chih hsüeh-ku pien 吏治學古編. Preface dated 1723-1724; printed later by Wang's son.

Wang Shih-mao 王世懋. Min-pu shu 閩部疏. TSCC ed.

Wang Shih-to 汪士鐸. Wang-mei-ts'un hsien-sheng chi 汪梅村先生集. 1881.

------- I-ping jih-chi 乙丙日記. Yenching University Press, 1935.

Wang Ying-chiao 汪應蛟. Wang-ch'ing-chien-kung tsou-shu 汪清簡公奏疏. Undated late Ming ed.

Wei Yüan 魏源. Ku-wei-t'ang wai-chi 古微堂外集. 1878.

Wei-yüan-ting chih 威遠廳志. 1839.

Wen-hsien chih 溫縣志. 1746.

Wen-shang HC 汶上. 1608.

Wu-ch'ang HC 武昌. 1885.

Wu-chin HC 武進. 1605.

Wu-hsi-chin-k'uei hsien-chih 無錫金匱縣志. 1883.

Wu-hsien chih 吳縣志. 1642.

Wu-hsing chih 吳興志. 1201 ed., 1914 reprint.

Wu-ling HC 武陵. 1863.

Wu-men pu-sheng 吳門補乘. 1829.

Wu Ting-ch'ang 吳鼎昌. Ch'ien-cheng wu-nien 黔政五年. Kweichow Provincial Government, 1943.

Wu-yang HC 舞陽. 1834.

Yang Ching-jen 楊景仁. Ch'ou-chi pien 籌濟編. 1879.

Yang-chou FC 揚州. 1685.

Yang-ch'ü HC 陽曲. 1843.

Yang Hsi-fu 楊錫紱. Ssu-chih-t'ang wen-chi 四知堂文集. 1806.

Yeh Meng-chu 葉夢珠. Yüeh-shih pien 閱世編. Shang-hai chang-ku ts'ung-shu ed.

Yen-chou FC 兗州. 1768.

Yen-ling wen-hsien chih 鄢陵文獻志. 1862.

Yen Wei-ping 嚴慰冰. Hui-tao ching-kang shan 回到井崗山. Hankow, 1950.

Yin-hsien t'ung-chih 鄞縣通志. Compiled in the 1930's, still uncompleted.

Ying-ch'eng HC 應城. 1882.

Ying-shan HC 應山. 1871.

Ying-shang HC 潁上. 1878.

Yu-hsien chih 攸縣志. 1871.

Yung-cheng chu-p'i yü-chih 雍正硃批諭旨. Undated late Ch'ing ed.

Yung-ch'eng HC 榮城. 1840.

Yung-ch'ing HC 永清. 1779 ed., printed in 1813.

Yung-chou FC 永州. 1393 and 1825.

Yung-p'ing FC 永平. 1711 and 1880.

Yü Cheng-hsieh 俞正燮. Kuei-ssu lei-kao 癸巳類稿. An-hui ts'ung shu 安徽叢書 ed.

Yü-chou chih 裕州志. 1546.

Yü-hang hsien-chih-kao 餘杭縣志稿. 1906.

Yü-i HC 盱眙. 1903.

Yü-kan HC 餘干. 1823.

Yü-shan HC 玉山. 1873.

Yü-tz'u HC 榆次. 1862.

Yüan-chou FC 袁州. 1760 and 1874.

Yüan-chou FC 沅州. 1790.

Yüan-ling HC 沅陵. 1871.

Yüan Mei 袁枚. Hsiao-ts'ang shan-fang shih-wen-chi 小倉山房詩文集. SPPY ed.

Yüan-wu HC 原武. 1595.

Yüeh-chou chih 耀州志. 1762.

Yüeh-chou FC 岳州. 1746.

Yüeh chüeh-shu 越絶書. SPPY ed.

Yün-nan-sheng nung-ts'un tiao-ch'a 雲南省農村調查. Executive Yüan, 1935.

SECONDARY CHINESE AND JAPANESE SOURCES

Chang Hsiao-mei 張肖梅. Kuei-chou ching-chi 貴州經濟. CP, 1939.

------- Ssu-ch'uan ching-chi ts'an-k'ao tzu-liao 四川經濟參考資料. CP, 1939.

Ch'en Han-sheng 陳翰笙. Mu-ti ch'a-i 畝的差異. CP, 1929.

Ch'en Kao-yung 陳高傭, et al. Chung-kuo li-tai t'ien-tsai jen-huo piao 中國歷代天災人禍表. Shanghai, 1939.

Cheng Ho-sheng 鄭鶴聲. Chung-kuo chin-shih-shih 中國近世史. Chungking, 1945.

Ch'eng Mao-hsing 程懋型. Hsien-hsing pao-an chih-tu 現行保安制度. Shanghai, 1936.

Chiang-hsi chih mi-mai wen-ti 江西之米麥問題. Nanch'ang, 1933.

Ch'in Han-ts'ai 秦翰才. Tso-wen-hsiang-kung tsai hsi-pei 左文襄公在西北. CP, 1946.

Ch'ing-nei-ko chiu-ts'ang han-wen huang-ts'e lien-ho mu-lu 清內閣舊藏漢文黃冊聯合目錄. Peiping: Palace Museum, 1947.

Chou Yin-t'ang 周蔭棠. T'ai-wan chün-hsien chien-chih shih 台灣郡縣建置史. Chungking, 1943.

Chung Hsin 鍾歆. Yang-tzu-chiang shui-li k'ao 揚子江水利考. CP, 1936.

Chung-hua jen-min kung-ho-kuo fen-sheng ti-t'u 中華人民共和國分省地圖. Ta-chung 大中 Book Company, 1953.

Chung-kuo jen-k'ou wen-t'i chih t'ung-chi fen-hsi 中國人口問題之統計分析 Directorate of Statistics, 1946.

Chung-kuo t'u-ti wen-t'i chih t'ung-chi fen-hsi 中國土地問題之統計分析. Directorate of Statistics, 1941.

Ch'üan Han-sheng 全漢昇. "Chung-ku tzu-jan ching-chi" 中國自然經濟; Chung-yang yen-chiu-yüan li-shih yü-yen yen-chiu-so chi-k'an 中央研究院歷史語言研究所集刊, Vol. 10.

------- "T'ang-Sung cheng-fu sui-ju yü huo-pi ching-chi ti kuan-hsi" 唐宋政府歲入與貨幣經濟的關係; Chung-yang yen-chiu-yüan li-shih yü ten yen-chiu-so chi-k'an, Vol. 20.

Fan Wen-lan 范文瀾. Han-chien k'uei-tzu-shou Tseng Kuo-fan ti i-sheng 漢奸劊子手曾國藩的一生. Hsin-hua Company, 1944.

Fang Hsien-t'ing 方顯廷, ed. Nan-k'ai ching-chi yen-chiu 南開經濟研究. CP, 1936.

Fu Chiao-chin 傅角今. Hu-nan ti-li chih 湖南地理志. Ch'ang-sha, 1933.

Fu I-ling 傅衣凌. Fu-chien tien-nung-ching-chi-shih ts'ung-k'ao 福建佃農經濟史叢考. Fukien Christian University, 1944.

------- Ming-Ch'ing shih-tai shang-jen chi shang-yeh tzu-pen 明清時代商人及商業資本. Peking, 1956.

Fujii Hiroshi 藤井宏. "Shinan shonin no kenkyu (1)" 新安商人研究(一); Tōyō gakuho 東洋學報 (June 1953).

Han Ch'i-t'ung 韓啟桐. Chung-kuo tui-jih chan-shih sun-shih chih ku-chi, 1937-1943 中國對日戰事損失之估計. CP, 1946.

Han Ch'i-t'ung and Nan Chung-wan 韓啟桐, 南鍾萬. Huang-fan-ch'ü ti sun-hai yü shan-hou chiu-chi 黃泛區的損害與善後救濟. CP, 1948.

Ho Wei 何畏. Wo-kuo tang-ch'ien ti liang-shih cheng-ts'e 我國當前的糧食政策. Peking, 1955.

Hsieh Hsing-yao 謝興堯. Ch'ing-ch'u liu-jen k'ai-fa tung-pei shih 清初流人開發東北史. Shanghai, 1948.

Hsin-hua pan-yüeh-k'an 新華半月刊. 1956, Nos. 2 and 15; 1958, No. 3.

Hua Shu 華恕. Wo-kuo ti-i-ke wu-nien-chi-hua chung ti nung-yeh tseng-ch'an wen-ti 我國第一個五年計劃中的農業增產問題. Shanghai, 1956.

Kato Shigeshi 加藤繁 Shina Keizai-shi Kōshō 支那經濟史考證. Tōyō Bunko 東洋文庫, 1953.

Ku Kuan-chiao 古貫郊. San-shih-nien-lai ti chung-kung 三十年來的中共. Hongkong, 1955.

Kuo-wen chou-pao 國聞週報, Vol. 10, No. 44 (1933).

Lang Ching-hsiao 郎擎霄. Pao-chia yün-tung chih li-lun yü shih-chi 保甲運動之理論與實際. Shanghai, 1930.

Li Chia-jui 李家瑞. Pei-p'ing feng-su lei-cheng 北平風俗類徵. CP, 1939.

Li Kuang-t'ao 李光濤. "Chang Hsien-chung shih-shih" 張獻忠史事; Chung-yang yen-chiu-yüan li-shih yü-yen yen-chiu-so chi-k'an, Vol. 25 (1953).

Liang Fang-chung 梁方仲, "Single-Whip System": "I-t'iao-pien fa" 一條鞭法; Chung-kuo she-hui ching-chi-shih chi-k'an 中國社會經濟史集刊, Vol. 4, No. 1.

------- "Ming-tai hu-k'ou t'ien-ti yü t'ien-fu t'ung-chi" 明代户口田地與田賦統計; Chung-kuo she-hui ching-chi-shih chi-k'an, Vol. 3, No. 1.

------- "Ming-tai kuo-chi-mao-i yü yin ti shu-ch'u-ju" 明代國際貿易與銀的輸出入; Chung-kuo she-hui ching-chi-shih chi-k'an, Vol. 6, No. 2.

------- "Shih i-t'iao-pien fa" 釋一條鞭法; Chung-kuo she-hui ching-chi-shih chi-k'an, Vol. 7, No. 1.

------- "Ming-tai shih-tuan-chin fa" 明代十段錦法; Chung-kuo she-hui ching-chi-shih chi-k'an, Vol. 7, No. 1.

------- "Ming Yellow Registers": "Ming-tai huang-ts'e k'ao" 明代黃冊考; Ling-nan hsüeh-pao 嶺南學報, Vol. 10, No. 2.

------- "Chronological Chart": "Ming-tai i-t'iao-pien fa nien-piao" 明代一條鞭法年表; Ling-nan hsüeh-pao, Vol. 12, No. 1.

Lien Heng 連橫. T'ai-wan t'ung-shih 臺灣通史. CP, 1947.

Liu Hsüan-min 劉選民. "Ch'ing-tai tung-san-sheng chih i-min yü k'ai-ken" 清代東三省之移民與開墾; Shih-hsüeh nien-pao 史學年報, Vol 2, No. 5 (1938).

Lo Erh-kang 羅尔綱. T'ai-p'ing t'ien-kuo shih-kang 太平天国史綱. CP, 1937.

------- "Ch'ing-chi ping-wei-chiang-yu ti ch'i-yüan" 清季兵為將有的起源; Chung-kuo she-hui ching-chi-shih chi-k'an, Vol. 5, No. 2 (1937).

------- Hsiang-chün hsin-chih 湘軍新志. CP, 1939.

------- "T'ai-p'ing t'ien-kuo ke-ming ch'ien ti jen-k'ou ya-p'o wen-ti" 太平天國革命前的人口壓迫問題; Chung-kuo she-hui ching-chi-shih chi-k'an, Vol. 8, No. 1 (1949).

-------, ed. Chung-wang Li Hsiu-ch'eng tzu-chuan yüan-kao chien-cheng 忠王李秀成自傳原稿箋證. Peking, 1951.

------- T'ai-p'ing t'ien-kuo shih-shih k'ao 太平天国史事考. Peking, 1955.

Lo Hsiang-lin 羅香林. K'e-chia yen-chiu tao-lun 客家研究導論. Kwangtung: Hsing-ning 興寧, 1933.

Lo Yü-tung 羅玉東. Chung-kuo li-chin shih 中國釐金史. CP, 1936.

Lou Yün-lin 樓雲林. Ssu-ch'uan 四川. Shanghai, 1941.

Lü Ssu-mien 呂思勉. Ch'in-Han shih 秦漢史. Shanghai, 1947.

------- Liang-Chin Nan-pei-ch'ao shih 兩晉南北朝史. Shanghai, 1948.

Meng Hsien-chang 孟憲章. Chung-kuo chin-tai ching-chi-shih chiao-ch'eng 中國近代經濟史教程 Shanghai, 1951.

Nei-ko-ta-k'u hsien-ts'un han-wen huang-ts'e mu lu 內閣大庫現存漢文黃冊目錄. Peiping: Palace Museum, 1936.

Nishijima Sadao 西嶋定生. "Mindai ni okeru momen no jukyu ni tsuite," 明代に於ける木棉の普及に就いて; Shigaku Zasshi 史學雜誌, Vol. 57, No. 4-5

Otake Fumio 小竹文夫. Kinsei Shina keizaishi kenkyu 近世支那經濟史研究. Tokyo, 1942.

P'an Kuang-tan 潘光旦. "Chin-tai su-chou ti jen-ts'ai" 近代蘇州的人才; She-hui k'e-hsüeh 社會科學, Vol. 1, No. 1 (1935).

Shen Nai-cheng 沈乃正. "Ch'ing-mo chih tu-fu chi-ch'üan, chung-yang chi ch'üan, yü t'ung-shu pan-kung" 清末之督撫集權中央集權與'同署辦公'; She-hui K'e-hsüeh 社會科學, Vol. 2, No. 2 (1937).

Shih-shih yüeh-pao 時事月報 (July 1930).

T'an Ch'i-hsiang 譚其驤. "Chung-kuo nei-ti i-min-shih: Hu-nan p'ien" 中國內地移民史:湖南篇; Shih-hsüeh nien-pao, Vol. 1, No. 4 (1932).

Teng Chih-ch'eng 鄧之誠. Ku-tung so-chi ch'üan-pien 骨董瑣記全編. Peking, 1955.

Ti-cheng yüeh-k'an 地政月刊, Vols. 1-4.

Ting Li 丁勵. Chung-kung ti min-ping chih-tu 中共的民兵制度. Hongkong, 1954.

Wan kuo-ting 萬國鼎. "Chung-kuo t'ien-fu niao-k'an chi ch'i kai-ke ch'ien-t'u" 中國田賦鳥瞰及其改革前途; Ti-cheng yüeh-k'an 地政月刊, Vol. 4, No. 2-3.

Wang Ch'ung-wu 王崇武. "Ming-tai hu-k'ou ti hsiao-chang"

明代户口的消長 ; Yen-ching hsüeh-pao 燕京學報, No. 20 (December 1936).

Wang Shih-ta 王世達. "Ming-cheng-pu hu-k'ou tiao-ch'a chi ke-chia ku-chi" 民政部户口調查及各家估計; She-hui k'e hsüeh tsa-chih 社會科學雜誌, Vol. 3, No. 3 (September 1932) and Vol. 4, No. 1 (March 1933).

Wang Ying 王瑛. "T'ai-p'ing t'ien-kuo ke-ming ch'ien-hsi ti t'u-ti wen-t'i" 太平天國革命前夕的土地問題; Chung-shan wen-hua chiao-yü-kuan chi-k'an 中山文化教育館季刊, Vol. 3, No. 1.

Wen Chün-t'ien 聞鈞天. Chung-kuo pao-chia chih-tu 中國保甲制度. CP, 1935.

Wu Ch'eng-hsi 吳承禧. "Hsia-men ti hua-ch'iao hui-k'uan yü chin-yung tsu-chih" 廈門的華僑滙款與金融組織; She-hui k'e-hsüeh tsa-chih 社會科學雜誌, Vol. 8, No. 2 (June 1937).

Wu Ch'i-chün 吳其濬. Chih-wu ming-shih t'u-k'ao 植物名實圖考. CP ed.

------- Chih-wu ming-shih t'u-k'ao ch'ang-pien 植物名實圖考長編. CP ed.

Wu Ch'uan-chün 吳傳鈞. Chung-kuo liang-shih ti-li 中國糧食地理. CP, 1948.

Wu Han 吳晗. Chu Yüan-chang chuan 朱元璋傳. Shanghai, 1949.

Yao Tseng-yin 姚曾廕. Kuang-tung-sheng ti hua-ch'iao hui-k'uan 廣東省的華僑滙款. CP, 1943.

Yen Chung-p'ing 嚴中平. Chung-kuo mien-fang-chih shih-kao 中國棉紡織史稿. Peking 1955.

------- et al., ed. Chung-kuo chin-tai ching-chi-shih t'ung-chi tzu-liao hsuan-chi 中國近代經濟史統計資料選輯. Peking, 1955.

WORKS IN WESTERN LANGUAGES

Annual Report of the Library of Congress (1940).

Ashton, T. S. The Industrial Revolution, 1760-1830. Oxford, 1948.

Buck, J. L. Land Utilization in China. 3 vols.; Chicago, 1937.

Cameron, M. E. The Reform Movement in China, 1898-1912. Stanford, 1931.

Carr-Saunders, Sir Alexander. World Population, Past Growth and Present Trends. Oxford, 1937.

Chen Ta. Chinese Migrations, with Special Reference to Labor Conditions. Washington, D. C., 1923.

------- Population in Modern China. Published as a special monograph in the American Journal of Sociology, Part II (July 1946).

Cheng, Y. K. Foreign Trade and International Development of China. Washington, D. C., 1956.

Chiang, S. T. The Nien Rebellion. Seattle, 1954.

China Handbook, 1950.

China International Famine Relief Commission, Annual Reports (1924-1937).

Chinese Maritime Customs. China's Foreign Trade (1929-1937). The Soya Beans of Manchuria, Special Series, No. 31, 1911.

Chou, S. H. "Financing the Economic Development of Manchuria, 1900-1945." A paper read at the Conference on the Chinese Economy in September 1956, sponsored by the Center for East Asian Studies, Harvard University.

Chu Co-ching. "Climatic Pulsations during Historic Time in China," Geographical Review, 16:274-282 (1926).

Clapham, Sir John H. The Economic Development of France and Germany, 1814-1914. Cambridge, England, 1946.

Davidson, James W. The Island of Formosa, Past and Present. New York, 1903.

Du Halde, P. J. B. A Description of the Chinese Empire and Chinese Tartary. 2 vols; London, 1738.

Eckstein, Alexander and Y. C. Yin. "Mainland China's Agricultural Product in 1952." Manuscript, Russian Research Center, Harvard University.

Grajdanzev, A. J. "Manchuria as a Region of Colonization," Pacific Affairs (March 1946).

Habakkuk, H. J. "English Population in the Eighteenth Century," Economic History Review (December 1953).

Ho, Franklin L. "Population Movement to the Northeastern Frontier of China," Chinese Social and Political Science Review, Vol. 15, No. 3 (October 1931).

Ho Ping-ti. "The Salt Merchants of Yangchou: A Study of Commercial Capitalism in Eighteenth-Century China," Harvard Journal of Asiatic Studies (June 1954).

------- "The Introduction of American Food Plants into China," American Anthropologist (April 1955).

------- "American Food Plants in China," Plant Science Bulletin (January 1956).

------- "Early-Ripening Rice in Chinese History," Economic History Review (December 1956).

Hosie, Sir Alexander. "Droughts in China, A. D. 620 to 1643," Journal of the North China Branch of the Royal Asiatic Society, 12:51-89 (1878).

------- Manchuria, Its People, Resources and Recent History. New York, 1904.

Hsiao Chien. How the Tillers Win Back Their Land. Peking, 1951.

Hsiao, K. C. "Rural Control in Nineteenth-Century China," Far Eastern Quarterly (February 1953).

Hu Shih. The Chinese Renaissance. Chicago, 1935.

Huc, Father M. A Journey through the Chinese Empire. 2 vols.; New York, 1855.

Hummel, Arthur W., ed. Eminent Chinese of the Ch'ing Period. 2 vols.; Washington, D. C., 1943-1944.

Jones, F. C. Manchuria since 1931. London, 1949.

Krotevich, S. "Vsekitayskaya perepis' naseleniya 1953 g.," Vestnik Statistiki, No. 5, pp. 31-50 (September-October 1955).

Liu, F. F. A Military History of Modern China, 1924-1949. Princeton, 1956.

Lung, C. F. "A Note on Hung Liang-chi: The Chinese Malthus," T'ien-hsia Monthly (October 1953).

MacNair, H. F., ed. China. Berkeley, 1946.

------- The Chinese Abroad. New York, 1933.

Mallory, Walter H. China: Land of Famine. New York, 1928.

Manchoukuo Yearbook, 1942.

Morse, Hosea B. The Trade and Administration of China. New York, 1920.

Norman, H. E. Japan's Emergence as a Modern State. New York, 1940.

Parker, E. H. "Notes on Some Statistics Regarding China," Journal of the Royal Statistical Society (1899).

People's China. 1956, No. 10.

Pioneer Settlement Cooperative Studies. New York, 1932.

Purcell, Victor. The Chinese in South-East Asia. London, 1951.

Remer, F. C. Foreign Investments in China. New York, 1933.

Ricci, Matthew. China in the Sixteenth Century: The Journal of Matthew Ricci. New York, 1953.

Richthofen, Baron Ferdinand. Reports on the Provinces of Hunan, Hupeh, Honan, and Shansi. Shanghai, 1870.

------- Letters on the Provinces of Chekiang and Nganhwei. Shanghai, 1871.

------- Baron Richthofen's Letters, 1870-1872. 2nd ed.; Shanghai, 1903.

1903.

Rockhill, W. W. "An Inquiry into the Population of China," Annual Report of the Smithsonian Institution (1905).

Sarton, George. Introduction to the History of Science, Vol. 3, Part 2. Baltimore, 1948.

Shabad, Theodore. China's Changing Map, A Political and Economic Geography of the Chinese People's Republic. New York, 1956.

------- "Counting 600 Million Chinese," Far Eastern Survey (April 1956).

Shen, T. H. Agricultural Resources of China. Ithaca, 1951.

Sprenkel, Otto van der. "Population Statistics of Ming China," Bulletin of the School of Oriental and African Studies, University of London, Vol. 15, Part 2 (1953).

Statistics of China's Foreign Trade during the Past Sixty-Five Years. CP, 1931.

Taeuber, Irene B. "Population Policies in Communist China," Population Index (October 1956).

------- "A Note on the Population Statistics of Communist China," Population Index (October 1956).

Tawney, R. H. Land and Labour in China. London, 1937.

Trevelyan, G. M. English Social History. New York, 1946.

Vavilov, N. I. Selected Writings of N. I. Vavilov; Chronica Botanica. Vol. 13, No. 1-6.

Wang, Y. C. "The Rise of Land Tax and the Fall of Dynasties in Chinese History," Pacific Affairs, 1938, pp. 201-220.

Wilcox, W. F. "A Westerner's Effort to Estimate the Population of China and Its Increase since 1650," Journal of the American Statistical Association (1930).

------- International Migrations. New York, 1931.

Williams, S. Wells. The Middle Kingdom. 2 vols.; New York, 1907.

Wong, K. C. and L. T. Wu. A History of Chinese Medicine.

Tientsin, 1932.

Wu, Y. L. An Economic Survey of Communist China. New York, 1956.

Yang, L. S. "Notes on the Economic History of the Chin Dynasty," Harvard Journal of Asiatic Studies (June 1946).

------- "Notes on Dr. Swann's Food and Money in Ancient China," Harvard Journal of Asiatic Studies (December 1950).

------- Money and Credit in China. Cambridge, Mass., 1952.

GLOSSARY OF CHINESE CHARACTERS NOT INCLUDED IN THE BIBLIOGRAPHY

Ah Ts'ao 阿操
Ah Yin 阿尹
Champa 占城
Chang Hua 張華
Chang Teh-chien 張德堅
Ch'ao-chin hsü-chih ts'e 朝覲須知册
chen 鎮
ch'eng-ting nai-tzu 成丁男子

ch'i-ling 畸零
chia 甲
chiang-chi 匠籍
ch'ing-hsiang-chü 清鄉局
Chou Chen 周忱
Chung-shu-sheng 中書省
ch'ü 區
chün-chi 軍籍
chün-yao 均徭
fang 坊
Fo-pao 佛保
hao-hsien 耗羨
Ho Ch'iao-yüan 何喬遠
Ho-shen 和珅
Ho T'ao 霍韜
hsiang 廂
hsiang 鄉
Hsü Shou-heng 徐壽衡
hsün-huan-ts'e 循環册
hu-k'ou-jen-ting 户口人丁
hu-t'ieh 户帖
Huang-ts'e 黃册
huo-hao 火耗
I, Prince 怡賢親王
i-li ta-liang ti 一例大糧地
k'e-tzu 客子
keng 秔
Ku Tsung 顧琮
kuan-ting 官丁
Kuan-yin-pao 觀音保
Kuei-ho 貴和
Kuei-mao 貴懋
kung 弓
Lang Li-ch'ing 郎禮卿
li 里
li-chai 力差
li-chia 里甲
Li Shih-chen 李時珍
Li Wei 李衛
Li Yüan-tse 李元則

liang-ju i-wei-ch'u 量入以為出
lien-pao 聯保
lin 鄰
Liu Kuo-fu 劉國黻
Liu Ming-ch'uan 劉銘傳
liu-yü 流寓
lü 閭

Mai Chu 邁柱
Min-cheng-pu 民政部
min-chi 民籍

Nei-cheng-pu 內政部
Nien 捻

P'an Ku 盤古
pao 保
pao-an-tui 保安隊
pao-chia 保甲
pao-wei-t'uan 保衛團
Pen-ts'ao kang-mu 本草綱目
peng-min 棚民
Po-wu chih 博物志
pu 步
pu-ch'eng-ting 不成丁

san-teng chiu-tse 三等九則
Shen Pao-chen 沈葆楨
shih-mu 市畝
Shih Yün-chang 施閏章
shu-shu 蜀黍
ssu-ting 私丁
Su Lin-po 蘇霖渤
Sun Chia-kan 孫嘉淦

T'ang Ho-cheng 唐鶴徵
T'ang Pin 湯斌
T'ien Wen-ching 田文鏡
ting 丁
Tse-ch'ing hui-ts'uan 賊情彙纂
ts'un 村
Tuan Yü-ts'ai 段玉裁
t'uan 團
t'uan-lien 團練
t'un-hu 屯戶

Wang K'ai-yün 王闓運
Wang Lun 王倫

yang-lien 養廉
Yang Yung-t'ai 楊永泰
Yeh P'ei-sun 葉佩蓀
Yen Jo-chü 閻若璩
yin-chai 銀差
Yü Ch'eng-lung 于成龍
Yü-lin t'u-ts'e 魚鱗圖冊
Yü T'eng-chiao 俞騰蛟
Yüan Shu-ku 袁樹穀